# Globalisation and Livelihood Transformations in the Indonesian Seaweed Industry

I0042037

This book explores the rapidly changing seaweed industry in Indonesia, the largest global producer of carrageenan-bearing seaweeds.

Seaweed production in Indonesia has grown exponentially over the last twenty years, and rural communities across the country have embraced this new livelihood activity. This book begins with an examination of the global carrageenan seaweed industry, from the global market for carrageenan in processed foods, to the national and regional contexts in Indonesia across which it is farmed, processed, and traded. It then explores the ways that rural communities have reshaped their lives around seaweed production, with chapters on agrarian transformations, negotiations over access to sea space, farmer decision-making in presence of environmental, social, and economic constraints, the role of women and casual labourers in the industry, and the marketing of seaweed through social networks. Based on a multi-disciplinary research initiative, this book demonstrates the interrelatedness of environmental, social, and economic dynamics on seaweed production, processing, and trade, and argues for key policy interventions to support the sustainable development of the industry in the context of climate change. It also provides a lens for understanding and improving the broader processes of sustainable rural development in a rapidly globalising and commercialising world.

This book will be of great interest to students and scholars of aquaculture, food systems, agricultural economics, rural studies and sustainable development.

**Zannie Langford** is Research Fellow at the Griffith University Asia Institute and an Honorary Research Fellow in the School of Agriculture and Food Sustainability at the University of Queensland, where she undertook the research for this book. Her current research explores shifts in development financing in Indonesia and the Pacific. She has also undertaken a range of applied research projects focusing on land tenure, global value chains, smallholder agribusiness and rural development financing in Northern Australia, Indonesia and the Pacific.

# Earthscan Food and Agriculture

**Climate Neutral and Resilient Farming Systems**
Practical Solutions for Climate Mitigation and Adaptation
*Edited by Sekhar Udaya Nagothu*

**Sustainable Apple Breeding and Cultivation in Germany**
Commons-Based Agriculture and Social-Ecological Resilience
*Hendrik Wolter*

**Food Policy in the United Kingdom**
An Introduction
*Martin Caraher, Sinéad Furey and Rebecca Wells*

**Peasants, Capitalism, and the Work of Eric R. Wolf**
Reviving Critical Agrarian Studies
*Mark Tilzey, Fraser Sugden, and David Seddon*

**Peasants, Capitalism, and Imperialism in an Age of Politico-Ecological Crisis**
*Mark Tilzey and Fraser Sugden*

**Genome Editing and Global Food Security**
Molecular Engineering Technologies for Sustainable Agriculture
*Zeba Khan, Durre Shahwar, and Yasmin Heikal*

**University Engagement with Farming Communities in Africa**
Community Action Research Platforms
*Edited by Anthony Egeru, Megan Lindow, and Kay Muir Leresche*

For more information about this series, please visit: www.routledge.com/
books/series/ECEFA/

# Globalisation and Livelihood Transformations in the Indonesian Seaweed Industry

**Edited by**
**Zannie Langford**

Routledge
Taylor & Francis Group
LONDON AND NEW YORK

earthscan
from Routledge

First published 2024
by Routledge
4 Park Square, Milton Park, Abingdon, Oxon OX14 4RN

and by Routledge
605 Third Avenue, New York, NY 10158

*Routledge is an imprint of the Taylor & Francis Group, an informa business*

*British Library Cataloguing-in-Publication Data*
A catalogue record for this book is available from the British Library

*Library of Congress Cataloging-in-Publication Data*
Names: Langford, Zannie, editor.
Title: Globalisation and livelihood transformations in the Indonesian seaweed industry / edited by Zannie Langford.
Description: New York : Routledge, 2024. | Includes bibliographical references and index.
Identifiers: LCCN 2023035357 (print) | LCCN 2023035358 (ebook) | ISBN 9781032025469 (hardback) | ISBN 9781032025490 (paperback) | ISBN 9781003183860 (ebook)
Subjects: LCSH: Marine algae industry—Indonesia. | Marine algae industry—Environmental aspects—Indonesia. | Marine algae industry—Social aspects—Indonesia. | Marine algae industry—Economic aspects—Indonesia.
Classification: LCC SH390.5.I5 G46 2024 (print) | LCC SH390.5.I5 (ebook) | DDC 641.6/9809598—dc23/eng/20231107
LC record available at https://lccn.loc.gov/2023035357
LC ebook record available at https://lccn.loc.gov/2023035358

ISBN: 978-1-032-02546-9 (hbk)
ISBN: 978-1-032-02549-0 (pbk)
ISBN: 978-1-003-18386-0 (ebk)

DOI: 10.4324/9781003183860

Typeset in Times New Roman
by codeMantra

# Contents

# Figures

# Tables

# Contributors

**Risya Arsyi Armis** worked as a Junior Scientific Officer on the Partnership for Australia-Indonesia Research Commodities group project on which this research is based from August 2021–December 2022. She holds a Bachelor of Fisheries from Hasanuddin University, Indonesia, and previously worked on a women's empowerment project for coastal communities in Ujung Tanah, Makassar, in PT Pertamina Region VII.

**Eko Ruddy Cahyadi** is an Associate Professor in Production and Operation Management at IPB University, Indonesia. He obtained his PhD in Economics from Leibniz University in Hannover, Germany. His research interests include operation and supply chain management, impact evaluation and sustainability. He also works for the United Nations Industrial Development Organization (UNIDO) as a national monitoring and evaluation expert in a project focusing on strengthening selected aquaculture and seaweed value chains.

**Boedi Julianto** is Director of PT Jaringan Sumber Daya. He has seventeen years of expertise in seaweed research and development, having worked as Development Officer with the International Finance Corporation and Consultant for World Fish, Aqua Spark, Hatch Blue and the University of Queensland. He also serves as Indonesia's National Seaweed Value Chain Expert on the Global Quality Standard Program.

**Adam M. Komarek** works as a Teaching and Research Academic at the University of Queensland. His research focuses on: (1) the economics of sustainable agriculture at the field and farm scale especially for issues of risk and resilience and rural livelihoods, (2) agrifood supply chain design and performance, and (3) foresight analysis within agrifood systems. He teaches on the university's Agribusiness degrees.

**Zannie Langford** is Research Fellow at the Griffith University Asia Institute and an Honorary Research Fellow in the School of Agriculture and Food Sustainability at the University of Queensland, where she undertook the research for this book. Her current research explores shifts in development financing in Indonesia and the Pacific. She has also undertaken a range of applied and academic research projects focusing on land tenure, global value chains, smallholder agribusiness and rural development financing in Northern Australia, Indonesia and the Pacific.

**Imran Lapong** worked as a Junior Scientific Officer on the Partnership for Australia-Indonesia Research Commodities group project on which this research is based from August 2021–December 2022. He holds a Bachelor of Marine Science from Hasanuddin University, Indonesia, and a Master's degree from James Cook University, Australia.

**Yanti N. Muflikh** is a Faculty Member of the Department of Agribusiness at IPM University (*Institut Pertanian Bogor*), Indonesia. She holds an undergraduate degree from Bogor Agricultural University (1999), a Master's degree (2008) and a PhD (2021) from the University of Queensland. Her research focuses on analysing value chains in agribusiness products, with a particular interest in systems thinking and dynamics. She has published papers in *Agricultural Systems* and the *Journal of Agribusiness in Developing and Emerging Economies*.

**Nunung Nuryartono** is outgoing Dean of the Faculty of Economics and Management at IPB University (*Institut Pertanian Bogor*), Indonesia, and currently Deputy for Coordination of Social Welfare Improvement at the Coordinating Ministry for Human Development and Cultural Affairs, Indonesia. His areas of expertise include development economics, and public policy. Nunung is a Senior Fellow in the Partnership for Australia-Indonesia Research where he co-leads the Commodities Domain.

**Syamsul Pasaribu** is a Faculty Member of the Department of Economics, at IPB University, Indonesia, where he is also an Executive Secretary at the International Center for Applied Finance and Economics (InterCAFE). His main areas of research interest are development economics, labour economics and financial economics.

**Risti Permani** is a Senior Lecturer in Agribusiness at the School of Agriculture and Food Sustainability at the University of Queensland with extensive research experience in Indonesian and Australian agricultural trade, value chains and policies. She is a co-founder of AgLive Indonesia and currently serves as a member of the Board of Directors at the Centre for Indonesian Policy Studies (CIPS).

**Radhiyah Ruhon** worked as a Junior Scientific Officer on the Partnership for Australia-Indonesia Research Commodities group project on which this research is based from August 2021–December 2022. She holds a Bachelor of Biological Science from Universitas Hasanuddin, Indonesia. Her Master's degree, majoring in Marine Biology, was undertaken at the University of Western Australia, in which she worked on coastal carbon study.

**Irsyadi Siradjuddin** is a Lecturer in Urban and Regional Planning at Universitas Islam Negeri Alauddin, Indonesia, and part of the JaSuDa team. He earned his Bachelor's degree from IPB University (*Institut Pertanian Bogor*), Indonesia, and his Master's degree at Universitas Hasanuddin, Indonesia. He is currently pursuing a PhD in Earth and Environmental Technology at Universitas Hasanuddin. His interests include rural economics and agropolitan development. He also has experience in seaweed research and development.

**Fikri Firmansyah Sjahruddin** is a PhD candidate at the University of Queensland, focusing on sustainable coral reef fisheries management. Prior to starting his current studies, Fikri has been extensively involved in marine conservation in Indonesia. His career in this field began when he worked for the Fauna & Flora International Aceh Program and the World Wide Fund for NatureIndonesia.

**Scott Waldron** is an Associate Professor in Agricultural Economics at the School of Agriculture and Food Sustainability at the University of Queensland. Scott teaches on agricultural development, policy and trade and has conducted twenty-five agricultural development projects in China, Southeast Asia and the Pacific. Scott is a Senior Fellow in the Partnership for Australia-Indonesia Research where he co-leads the Commodities Domain.

**Zulung Zach Walyandra** worked as a Junior Scientific Officer on the Partnership for Australia-Indonesia Research Commodities group project on which this research is based from August 2021–December 2022. He holds a Bachelor of Fisheries from Hasanuddin University, Indonesia, and previously worked as Field Researcher and Program Facilitator for university and non-government organisations.

**Jing Zhang** is an Agricultural Economist with a diverse range of expertise encompassing agricultural policy, natural resource management, farming systems and the international trade. She has a Master's degree in financial management from Northwest A&F University, China, and a doctorate in agricultural economics from the University of Queensland. She has developed advanced skills in statistical, market and policy analysis, coupled with extensive experience and valuable connections within the Chinese agricultural sector.

# Preface

## The seaweed expansion

On coastlines across the tropics, a change is taking place. With growing global demand for processed foods, a market has grown for a type of seaweed little known fifty years ago: the eucheumatoid, or 'carrageenan' seaweeds. Early each morning, the residents of Pitu Sunggu village in South Sulawesi rise and begin their work of planting and harvesting seaweed. Marine seaweed farming came to Pitu Sunggu less than twenty years ago and has rapidly changed the way that people make their living – it is now the main livelihood in the village, and a source of income for more than 62,000 people across Indonesia. Seaweed farming has driven farmers to claim large areas of previously communal sea space for private use, has pushed incomes higher and has led to the employment of large numbers of casual wage labourers from surrounding villages. These rapid changes have transformed coastal livelihoods dramatically over the past two decades as increasing numbers of Indonesian villages have been incorporated into global seaweed supply chains. However, to understand how this came to be the case, it is necessary to look to where this story begins: 12,000 kilometres away, in the chilly waters off the coast of Ireland.

A seaweed grows there, known as carrageen, or Irish Moss.[1] Clinging to rocky outcrops, with purplish or reddish-green fan-shaped fronds, it grows wild across the icy coastlines (Davidson 2014, p. 146). For centuries Irish people have visited the shoreline at low tide, and, using a small, sharp knife, taken cuttings of the seaweed for use at home (McMonagle and Morrison 2020, p. 1289). In the Irish town of Donegal, seaweed collectors wander along 'a tiered ledge of rock that stretch[es] right out into the bay … ankle-deep in swathes of tangle weed' picking out 'dark curly tuft[s]' of carrageen (Connell 2015, p. 47). They take these home and spread them outside to dry, leaving it 'for several days, day and night, in the rain and dew as well as the sun … [because] carrageen ought to be bland, the flavours of the sea washed away with the rain' (Connell 2015, p. 49). Irish Moss is used as a folk remedy for the common cold (McMonagle and Morrison 2020, p. 1295, Davidson 2014, p. 146) and is an important ingredient in traditional Irish cooking, where it is used to make jellied milk puddings (O'Connor 2017, p. 118; and see Allen 1977). When boiled in water, it releases a gel known as carrageenan which thickens liquids to a smooth and consistent texture. It is this gel which has become a common ingredient in many processed foods, produced in huge volumes and traded around the world, finding its way into the households of a vast number of consumers globally.

Carrageenan derives its name from the Irish name for Irish Moss – *carraigı́n* ('little rock') or *carraigeen* ('moss of the rock') (Mitchell and Guiry 1983). Carrageenan is a hydrocolloid – a compound used to thicken, gel and stabilise processed foods, as well as in other products such as cosmetics and pharmaceuticals. Carrageenan improves the texture and shelf life of processed foods – it stops ice cream from becoming grainy, makes meat products juicier and is used in a range of specialist products such as gluten-free breads, vegan meat analogues and non-dairy milks and cheeses. Carrageenan is now used in a huge range of products, from Ben and Jerry's ice cream to Campbell's soups.[2] It is prized for the smooth texture it adds to processed foods, and as a vegan alternative to animal derived gelatin. It is most widely used in the meat and dairy industries (Bixler and Porse 2011; Grand View Research 2023) where it is valued for its ability to mimic a 'fatty' feeling that consumers enjoy.

The uses of carrageenan have been known since the 1800s, but initially it was not widely used because agar, a gel with similar properties extracted from different types of seaweed, could be obtained more cheaply from Japan (Craigie, Cornish and Deveau 2019). Agar became increasingly popular when its use in bacteriology was discovered in 1882, and through the early 1900s it was imported to Europe for scientific purposes, as well as for use in a growing range of prepared foods. However, the supply of agar from Japan was interrupted during World War II. Because of its importance to bacteriology, agar was quickly designated a 'critical war material' and it was no longer permitted to be used for purposes other than bacteriological culture (Humm 1947). Industry and government agencies began searching for new sources of agar for bacteriology, and for agar alternatives to replace its use in foods (Humm 1947, p. 317). Carrageenan was a possible alternative, and it was discovered that although agar and carrageenan work in similar ways, the gel produced by carrageenan was superior for many applications.

The discovery of many uses of carrageenan in food processing drove a harvesting frenzy across European and Canadian coastlines during the twentieth century. Irish Moss was harvested in increasingly large quantities from the 1940s (Pringle and Mathieson 1987), and by the mid-1960s it had become clear that wild stocks harvested from temperate waters could not support the growing global demand for carrageenan much longer (Craigie, Cornish and Deveau 2019, p. 4). Researchers turned their attention to searching for another type of carrageenan seaweed – one that could be found or farmed in quantities large enough to support the growing global industry.

There were several early candidates – *Iridaea*, for example, grew wild on the coasts of Chile, and *Gigartina* could be sourced from Spain (Qin 2018). However, wild harvesting of these temperate-water species could not supply large volumes of the product. Eventually, in the 1960s, researchers in the Philippines turned their focus to *Kappaphycus alvarezzi* (known colloquially as 'cottonii') and *Eucheuma denticulatum* (known colloquially as 'spinosum'). Thanks to the warm waters, abundant sunlight and nutrient rich oceans, these species grew quickly in the tropics and could be found growing wild on reefs and shallow lagoons across Indonesia and the Philippines (Imeson 2009 p. 165). In an interview with the author in 2023, well-known Indonesian seaweed industry professional Iain Neish described his work on an early mission searching for seaweeds containing carrageenan in Eastern Indonesia:

> My first time in Indonesia was June 1974 when I spent three months in Maluku Utara on a seaweed survey ... At that time, commercial farming of the [carrageenan seaweeds] had not happened yet ... So I spent three months on [a] boat ... did about 400 dives ... We were focused on searching for the 'motherload' of wild seaweed.

As the mission progressed, the team learned that these tropical seaweeds did not grow in large beds but in small patches. They did not find large areas where the carrageenan seaweeds could be harvested at scale, instead they noted that the seaweeds were labour intensive to find and gather. It became clear that sailing around searching for wild seaweeds was not a process that would be able to meet the demands of the growing carrageenan industry – what was needed was a way of farming seaweed on a large scale. Efforts to farm seaweed had been underway in the Philippines since 1965 with limited success. Cultivation sites were affected by frequent typhoons, excessive grazing by fish, and management issues, and several sites were abandoned (Neish et al. 2017, p. 5). Poor performance of trial farms saw development programmes underperforming, and many programmes were in danger of being abandoned (Neish et al. 2017, p. 6). Eventually, in the early 1970s, an ideal site was found in the south Philippines, where there were vast areas of shallow coral reefs, clear, flowing waters and large numbers of coastal residents open to new livelihood activities. Farming of the tropical carrageenan seaweeds in significant quantities was achieved for the first time in this location in 1974. The success of the farm trial reverberated through the industry and the Indonesian wild seaweed survey was abandoned. As Neish described:

> Suddenly we could buy farmed seaweed. The Philippines [had] produced three or four times the annual amount of [wild] seaweed that we were buying from Indonesia and the Philippines [in total]. So basically, it swamped the market ... they produced this glut of seaweed [and] the market collapsed because nobody knew what to do with all of [it]. That's when it was obvious that we should not bother to keep looking for wild seaweed beds ... it was basically a futile effort and we should focus on farming.

The price of the carrageenan seaweeds crashed as supply suddenly far exceeded demand, and existing supplies were worth next to nothing. However, the price crash was temporary and over the following decades the carrageenan processing industry expanded, supporting a growing seaweed farming industry in the Philippines.

It took another twelve years for seaweed farming to succeed in Indonesia. As Neish relates:

> The Danish guy Hans Porse, a stubborn guy … I think his boss kind of told him, 'Hans, lay off all this seaweed farming in Indonesia'. But Hans kept it going … [and] by 1986, finally, after 12 years of persistent effort, Hans managed to get the seaweed growing in [Bali] … So … he sent out a message to the others … He said, 'Come to Bali! Finally, it's working. We have to talk about where-to from here.' So we all went down there and well, what can I say? It was marvellous to see.

Having developed the ability to farm carrageenan seaweed in Indonesia, the Indonesian researchers working on the project sought to establish trial plots across the country. As Neish explained, the Philippines had enjoyed a 'total monopoly' on carrageenan seaweed production for the past decade and 'had basically a cartel of seaweed suppliers'. They 'knew the middlemen were taking a big cut', so through the late 1980s worked to establish trial plots across Indonesia in order to open up new sources. By 1990 they managed to successfully farm the seaweed in a number of locations across the country. Through the early 1990s this was supported by carrageenan companies who helped new seaweed farmers to establish plots by providing them with letters of credit, consequently seaweed farmers and processors had close supply relationships.

However, in the mid-1990s, the industry underwent a shift. A new, low-cost method for processing the seaweed into a 'semi-refined' product was developed, and it suddenly became much cheaper to process seaweed. Carrageenan processors began to increase in number, boosting demand for raw materials (Zhang et al. 2023). With the viability of farming established, and a new, cheaper method for producing carrageenan become available, the 1990s saw an explosion in the number of carrageenan factories. As Neish put it, the industry 'became a zoo' and many of the companies who had developed and maintained farm plots left the industry as they were no longer guaranteed their supply. The quantities produced were still low, but with proof-of-concept farms established across Indonesia, and a large, competitive market for the products, by the year 2000 the stage was set for an explosion of seaweed farming. Production took off, and by 2015 Indonesia was producing over ten million tonnes of seaweed per year, making it the largest carrageenan seaweed producer in the world. Within two decades, seaweed went from being a heavily supported and uncertain commodity in Indonesia to being one of its largest aquacultural industries, grown across its diverse coastlines and supporting the livelihoods of over 62,000 coastal households (BPS 2022). Carrageenan seaweed farming has now expanded into tropical coastal regions around the world – to Zanzibar, Tanzania, Madagascar, Papua New Guinea, the Solomon Islands, China, Malaysia, Cambodia and Venezuela – although at least 92 per cent of the global supply still comes from Indonesia and the Philippines.[3]

With such a rapid expansion, there has been little time to explore this new phenomenon, and for much of the development of the industry 'speculation flowed freely while the scientific testing of theories and hypotheses attracted little financial support' (Neish et al. 2017, p. 1). How did the seaweed grow best? Why would it suddenly turn white and die off in huge areas? What was the effect on the coastal environment? How did coastal households adapt and how did it change the work that women and men did, the foods they ate and the way they lived? Who are the entrepreneurs who sprang up overnight to negotiate between farmers and seaweed buyers? What drove the rapid development of seaweed factories across Indonesia and what is their future? And how does the global food market today, driven by consumer preferences, hold the future of tropical coastlines and the thousands of farmers who depend on them for their survival in the balance? This book seeks to understand how this product moves from seaweed farming sites in the shallow shores off a small village on the island of Sulawesi, Indonesia, through the seaweed value chain, through layers of middlemen and processors, exporters, refiners, blenders, food processors and retailers until it finally reaches the kitchens of consumers of perfectly textured food products. We explore how this global value chain is grounded in local places, how it has driven the transformation of coastal spaces and livelihoods and how a shift in consumer preferences could again transform these local places and the livelihoods of the people who depend on them.

## Notes

1 Irish Moss refers to the species of both *Chondus crispus* and *Mastocarpus stellatus*, and 'Irish harvesters collect both seaweeds indiscriminately as carrageen' (McMonagle and Morrison 2020, p. 1289).
2 As a Ben and Jerry's representative explains, 'We use carrageenan as a stabilizer in our product. The purpose is to bond with water molecules and thereby inhibit the grown of ice crystals as the ice cream freezes. This helps to offer some protection from iciness due to temperature fluctuations during distribution' (Lindsay Bumps, cited in DiSalvo 2016, n.p.). Campbell's (2023) explain of their use of carrageenan, 'We use it to keep our chicken meat juicy'.
3 FAO data (FAO 2023) suggests that 97 per cent of global marine carrageenan seaweed production occurs in Indonesia and the Philippines. However, there are known reporting issues for Indonesia, the Philippines, Malaysia and China hydrocolloid seaweed production data reported to the FAO (Hatch 2023a). If FAO data on global carrageenan seaweed production is adjusted for Indonesia, the Philippines and Malaysia to match Hatch industry estimates (Hatch 2023b), the proportion of global marine carrageenan seaweed production derived from Indonesia and the Philippines is 92 per cent. See Appendix 1 for further information on statistical issues.

## References

Allen, Myrtle. 1977. *The Ballymaloe Cookbook*. London: Agri-books.
Bixler, Harris J. and Hans Porse. 2011. A decade of change in the seaweed hydrocolloids industry. *Journal of Applied Phycology* 23, 321–335. https://doi.org/10.1007/s10811-010-9529-3
BPS (*Badan Pusat Statistik*). 2022. Hasil Survei Komoditas Perikanan Potensi Rumput Laut 2021 Seri 2. *Badan Pusat Statistic*. https://www.bps.go.id/publication/2022/08/29/269de33babc6e3d52bbae5b6/hasil-survei-komoditas-perikanan-potensi-rumput-laut-2021-seri-2.html

Campbell's. 2023. About our ingredients. https://www.campbellsoup.ca/about-us/whats-in-our-food/our-ingredients Accessed 6 January 2023.

Connell, Monica. 2015. *Gathering Carrageen: A Return to Donegal.* Sheffield: Sandstone Press Ltd.

Craigie, James S., M. Lynn Cornish and Louis E. Deveau. 2019. Commercialization of Irish Moss aquaculture: the Canadian experience. *Botanica Marina* 62(5), 411–432. https://doi-org.ezproxy.library.uq.edu.au/10.1515/bot-2019-0017

Davidson, A. (2014 [1999]). *The Oxford Companion to Food,* 3rd edition. Edited by Tom Jaine. New York: Oxford University Press.

DiSalvo, David. 2016. Dear Ben & Jerry's, Why is there seaweed in my ice cream? *Forbes,* 31 August 2016. https://www.forbes.com/sites/daviddisalvo/2016/08/31/dear-ben-jerrys-why-is-there-seaweed-in-my-ice-cream/?sh=143576871d6f Accessed 6 January 2023.

FAO (Food and Agriculture Organisation of the United Nations). 2023. FishStatJ (softward for FAO'S Fisheries and Aquaculture statistics). https://www.fao.org/fishery/en/statistics/software/fishstatj

Grand View Research. 2023. Carrageenan: market estimates and trend analysis. Purchased from *Grand View Research.*

Hatch (2023a). Global production overview. https://seaweedinsights.com/global-production Accessed 6 June 2023.

Hatch (2023b). Eucheumatoids. https://seaweedinsights.com/global-production-eucheumatoids Accessed 6 June 2023.

Humm, H. (1947). Agar – a pre-war Japanese monopoly. *Economic Botany* 1(3), 317–329. http://www.jstor.com/stable/4251862

Imeson, A. O. 2009. Carrageenan and furcellaran. In *Handbook of Hydrocolloids,* 2nd edition. Edited by G. O. Phillips and P. A. Williams. 164–185. Cambridge: Woodhead Publishing.

McMonagle, Micheal and Morrison, Liam. (2020). The seaweed resources of Ireland: a twenty-first century perspective. *Journal of Applied Phycology* 32, 1287–1300. https://doi.org/10.1007/s10811-020-02067-7

Mitchell, M. E. and Michael D. Guiry. 1983. Carrageen: a local habitation or a name? *Journal of Ethnopharmacology* 9(2), 347–351. https://doi.org/10.1016/0378-8741(83)90043-0

Neish, Iain C., Miguel Sepulveda, Anicia Q. Hurtado and Alan T. Critchley. 2017. Reflections on the commercial development of eucheumatoid seaweed farming. In *Tropical Seaweed Farming Trends, Problems and Opportunities*: *Focus on Eucheuma and Kappaphycus of Commerce.* Edited by Anicia Q. Hurtado, Alan T. Critchley and Iain C. Neish. 1–28. Cham, Switzerland: Springer International Publishing.

O'Connor, Kaori. 2017. *Seaweed: A Global History.* London: Reaktion Books.

Pringle, J. D. and A. C. Mathieson. 1987. *Chondrus crispus* Stackhouse. In *Case Study of Seven Seaweed Resources.* Edited by M. S. Doty, J. F. Cady and B. Santelices. Rome: FAO Fisheries Technical Report 281, pp. 49–122.

Qin, Yimin. 2018. Seaweed bioresources. In *Bioactive Seaweeds for Food Applications.* Edited by Yimin Qin. London: Academic Press. https://doi.org/10.1016/C2016-0-04566-7

Zhang, Jing, Scott Waldron, Zannie Langford, Boedi Julianto, and Adam Martin Komarek. 2023. China's growing influence in the global carrageenan industry and implications for Indonesia. *Journal of Applied Phycology* June, 1–22. https://doi.org/10.1007/s10811-023-03004-0

# Acknowledgements

This book draws on research undertaken by a team of researchers from the Commodities Domain of the Partnership for Australia Indonesia Research (PAIR). PAIR is an initiative of the Australia-Indonesia Centre (AIC), supported by the Australian government and run in partnership with the Indonesian Ministry of Research and Technology, the Indonesian Ministry of Transport, the South Sulawesi provincial government and many organisations and individuals from communities and industry. The AIC is a bilateral research consortium established in 2014 to advance the people-to-people and institutional links between the two nations in the fields of science, technology, education, innovation and culture (www.ausindcentre.org). AIC and PAIR are comprised of a consortium of eleven universities in Australia and Indonesia, including the University of Queensland (UQ). The authors gratefully acknowledge the support and funding of PAIR, AIC and UQ for the research. We would also like to thank the reviewers of our published work for their observations and insights. Finally, we would like to thank the people who participated in our research, including those from government and industry bodies, but particularly the people of Pitu Sunggu and Laikang villages who generously hosted four of the authors and shared their knowledge and experience.

# Abbreviations

| | |
|---|---|
| AIC | Australia-Indonesia Centre |
| ARLI | *Asosiasi Rumput Laut Indonesia* (Indonesian Seaweed Association) |
| ASTRULI | *Asosiasi Industri Rumput Laut Indonesia* (Indonesian Seaweed Industry Association) |
| ATCC | Alkali Treated Cottonii Chips |
| BLG | Biota Ganggang Laut |
| BPS | *Badan Pusat Statistik* (Central Bureau of Statistics) |
| DFAT | Department of Foreign Affairs and Trade of the Australian Government |
| FAO | Food and Agriculture Organisation of the United Nations |
| FDI | Foreign Direct Investment |
| Kemenperin | *Kementerian Perindustrian Republic Indonesia* (Ministry of Industry of the Republic of Indonesia) |
| KKP | *Kementerian Kelautan dan Perikanan Republik Indonesia* (Ministry of Marine Affairs and Fisheries of the Republic of Indonesia) |
| MKPRI | *Menteri Kelautan dan Perikanan Republik Indonesia* |
| MLA | Meat and Livestock Australia |
| Mt | Megatonnes |
| PAIR | Partnership for Australia Indonesia Research |
| Pangkep | *Kabupaten Pangkajene dan Kepulauan* |
| PPP | Purchasing Power Parity |
| RC | Refined Carrageenan |
| SRC | Semi-Refined Carrageenan |

# Introduction

*Zannie Langford*

Seaweed seems to be the new 'super crop'. With land beset with so many competing uses, the idea of cultivating the ocean holds a lot of appeal. What if pressure
could be taken off the land by producing crops in the sea, and these crops could
be used for such varied uses as biofuels, carbon capture and food? Seaweed is increasingly promoted for a vast range of environmental benefits: to capture carbon,
for methane-reducing cattle feed, as a sustainable food source, as a fertiliser and to
replace non-renewable sources in the making of bioplastics and biofuels. It is a key
component of many blue economy development programmes, and the development of oceans has been described by some as the 'blue revolution'. Yet seaweed
farming is a relatively new endeavour globally. Despite the current enthusiasm for
the seaweed industry, there is relatively little public understanding of it globally,
how it looks today and how it is likely to develop in the future. This chapter introduces some 'seaweed fundamentals', providing a snapshot of the global industry.
It situates the Indonesian seaweed industry within this broader global market and
describes the approach and structure of the book.

## A snapshot of the global seaweed industry

Seaweed farming is a relatively new phenomenon. Although domestication of land
plants has been undertaken for thousands of years, there are only a few examples
of successful domestication of seaweeds prior to the nineteenth century[1] – namely
the cultivation of *Porphyra* in Japan (best known today for its use in making *nori*
for sushi rolls). In 1950, the global seaweed industry was still very small, and
mostly consisted of harvesting wild seaweeds (Figure I.1). Over the seventy years
that followed, new cultivation methods and growing demand for seaweed products
saw seaweed cultivation expand rapidly. By 2021, marine seaweed farming was a
US$15 billion industry, with the Food and Agriculture Organisation of the United
Nations (FAO) estimating that over 33 million tonnes (megatones (Mt)) are produced annually (FAO 2023).

Of the world's seaweed, 99 per cent is produced in Asia in just six countries:
China, Indonesia, South Korea, the Philippines, North Korea and Japan. The industry is strikingly concentrated, with just six species accounting for 95 per cent of
global production (FAO 2023). Three of these are food seaweeds – the well-known

DOI: 10.4324/9781003183860-1

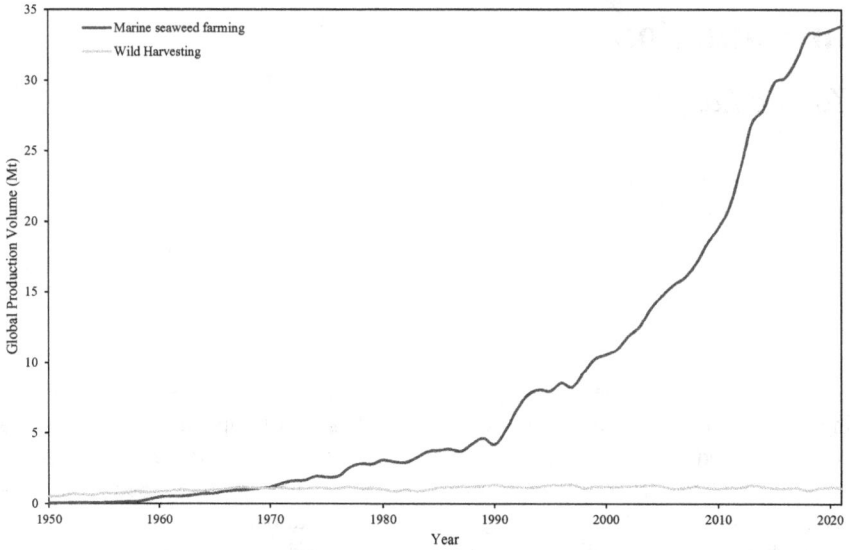

*Figure I.1* Global seaweed production from farming and wild harvesting
Source: Data from FAO (2023).

varieties used to produce *nori* (used in sushi roles), *wakami* (used in seaweed salads) and *kombu*[2] (to flavour soups) (Table I.1).[3] These three food species represent the majority of global seaweed production and are produced almost exclusively in the temperate waters of China, Japan, South Korea and North Korea. The other three main commercial species are tropical seaweeds used to extract the hydrocolloids carrageenan and agar, and are produced mainly in Indonesia, the Philippines and China.

Perhaps surprisingly, many more widely known species of seaweeds are only produced in very small quantities. *Spirulina*, for example, represents only 0.2 per cent of the global industry (FAO 2023), and *Asparagopsis*, known for its applications in cattle feed to reduce methane emissions, is still exceedingly difficult to grow and process, and had its first commercial sale in 2022 (MLA 2022). China dominates production of all major seaweeds except the carrageenan seaweeds, of which the majority are produced in Indonesia and the Philippines. These carrageenan seaweeds have contributed an increasing proportion of global seaweed production over the last twenty years, and now represent 26 per cent of the world's seaweed industry by volume (Figure I.2).

Ninety-two per cent of the global production of the carrageenan seaweed species *Eucheuma* and *Kappaphycus* occurs in Indonesia and the Philippines, which contribute 81 per cent and 15 per cent of global supply respectively (Figure I.3).[4] Other carrageenan seaweed producers include Malaysia (2 per cent of global production), Tanzania (including Zanzibar) (1 per cent of global production), the Solomon Islands, Madagascar, China, Venezuela, Papua New Guinea, Brazil, Cambodia, Kenya, Timor-Leste, Viet Nam and a few other countries producing less than 200

*Table I.1* Intensively cultivated commercial seaweeds

| Use | Species | Main products | Proportion of global market | Main producers and proportion of global production (2021) |
|---|---|---|---|---|
| Food | *Saccharina japonica* | Kombu | 39% | China (90%) South Korea (5%) North Korea (5%) |
| | *Undaria pinnatifida* | Wakame | 8% | China (77%) South Korea (21%) Japan (2%) |
| | *Porphyra* spp. | Nori | 8% | China (71%) South Korea (20%) Japan (9%) |
| Hydrocolloid | *Eucheuma* spp. *Kappaphycus* spp. | Carrageenan | 26% | Indonesia (81%) Philippines (15%) |
| | *Gracilaria* spp. | Agar | 14% | China (85%) Indonesia (15%) |
| – | All other species | – | 5% | – |

Source: Data from FAO (2023).

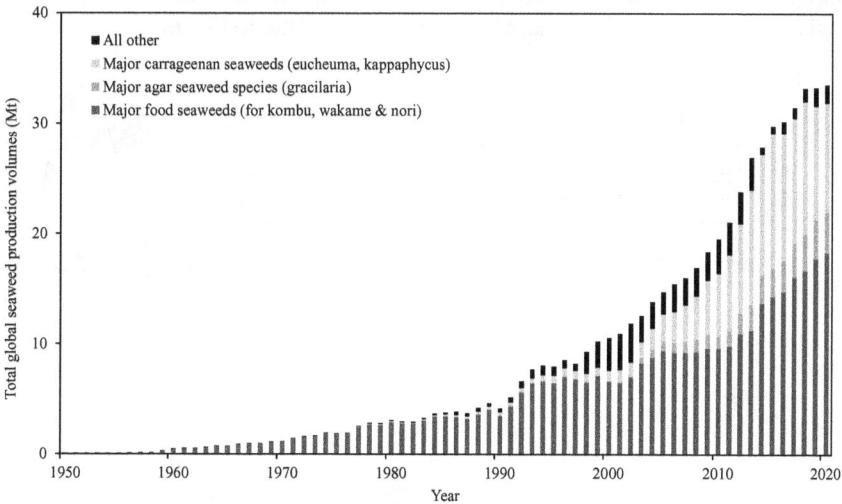

*Figure I.2* Global marine seaweed production by end use
Source: Data from FAO (2023).

tonnes per year. The dominance of Indonesia, the Philippines, Malaysia and Tanzania is linked in part to their ability to compete on price as a result of exchange rate dynamics. Each of these countries have purchasing power parity (PPP) significantly lower than their exchange rates, meaning that farmers in these countries can buy more with income earned on global markets. In exchange for a kilogram of seaweed purchased for a world market price of US$2, a farmer in Indonesia can buy goods worth US$6.03, a farmer in the Philippines goods worth US$5.14, a farmer in Malaysia goods worth US$5.20 and a farmer in Tanzania goods worth US$5.16. This is in contrast with other countries – for example, farmers in Papua New Guinea can purchase goods worth only US$3.02 with income earned from the same quantity of seaweed, and farmers in the Solomon Islands goods worth only US$2.30 (World Bank 2023). This means that seaweed prices may not be high enough to incentivise widespread uptake of the product in areas with lower exchange rate to PPP ratios, although this may change if prices rise over the long term.

In Indonesia, seaweed was reported to be a US$2 billion industry in 2021 (FAO 2023). This income is derived mostly from carrageenan seaweeds (79 per cent of national production value), as well as the agar producing species *Gracilaria* (19 per cent of national production value), small amounts of *Sargassum* and small volumes of *Caulerpa* (sea grapes), which are sold for consumption in salads in local markets. Between the years 2000 and 2015, Indonesian carrageenan seaweed production grew from 0.2 to 10.1Mt, before declining again to 7.1Mt in 2021. Despite the recent decline in production, it supports the livelihoods of around 62,000 coastal households (BPS 2022 and see Appendix 1 for a detailed discussion of statistical issues in the seaweed industry). As a result of the importance of seaweed farming to coastal livelihoods, the Government of Indonesia has outlined ambitious plans to further increase seaweed production (Presidential Decree 33–2019), particularly in the Eastern Indonesian provinces of West Papua, Maluku and North Maluku (Figure I.4).[5]

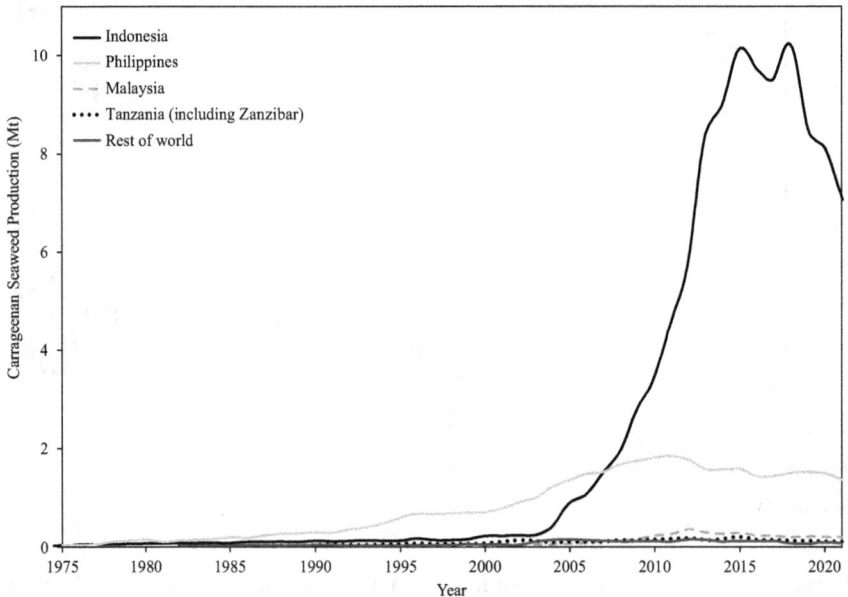

*Figure I.3* Global carrageenan seaweed production

Source: Data from FAO (2023).

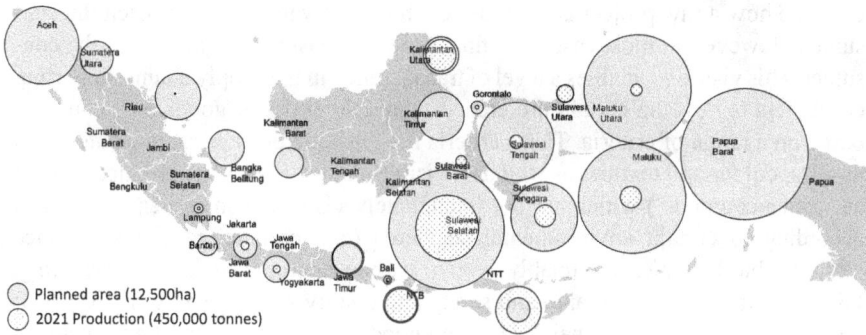

Labels visible on map: Aceh, Sumatera Utara, Riau, Sumatera Barat, Jambi, Bengkulu, Sumatera Selatan, Bangka Belitung, Lampung, Banten, Jakarta, Jawa Barat, Jawa Tengah, Yogyakarta, Jawa Timur, Kalimantan Barat, Kalimantan Tengah, Kalimantan Selatan, Kalimantan Timur, Kalimantan Utara, Gorontalo, Sulawesi Utara, Sulawesi Tengah, Sulawesi Barat, Sulawesi Selatan, Sulawesi Tenggara, Bali, NTB, NTT, Maluku, Maluku Utara, Papua Barat, Papua

○ Planned area (12,500ha)
○ 2021 Production (450,000 tonnes)

*Figure I.4* Current and planned seaweed production in Indonesia
Source: Data from Presidential Decree 33–2019 and FAO (2023).

With such rapid expansion, coastal communities across the country have experienced dramatic changes in the last two decades. Carrageenan seaweeds, although relatively low value, are incredibly shelf-stable – after being harvested, they are sun dried, and when dried properly can be stored for many months before being sold. This makes these seaweeds suitable for cultivation in remote areas, including by farmers who may only have sporadic access to markets. As a result, carrageenan seaweed production has been widely taken up in Indonesia, even in remote areas. Indonesia produces these seaweeds relatively cheaply, and the cost-competitiveness of Indonesian farmers also affects its market share.

Many coastal fishermen and farmers have transitioned their livelihood strategy partially or completely into seaweed farming, and as a result have been able to build more elaborate houses, send their children to school and purchase cars and motorbikes (Langford, Turupadang, Oedjoe et al. 2022; Langford, Waldron, Nuryartono et al. 2023). In other areas, communities experienced a seaweed farming 'boom' followed by a 'bust' resulting from environmental collapse (Steenbergen et al. 2017). The farming of carrageenan seaweed is accompanied by a wide range of social, economic and environmental changes, and systems of livelihoods and community governance have had to change to accommodate this new form of cultivation. However, these livelihoods do not have a certain future: the carrageenan seaweed industry is based on consumer acceptance of the food additive carrageenan, and demand for foods with the properties that carrageenan can impart. Understanding the future of the carrageenan seaweed industry, and how the livelihoods of the farmers who depend on it may be supported, relies on an understanding of the full value chain: from global industry to local activity.

## Carrageenan value chains: from global to local

Descriptive studies of how global value chains are grounded in local places can reveal important insights into how social lives are reorganised around new products (see, e.g., Dixon 2002; Tsing 2015; Weiss et al. 2016; West 2012). Such studies all examine how a raw project is transformed through a value chain to reach the consumer. However, in most cases, the final project is easily recognisable to the consumer. This visibility enables a level of transparency in the supply chain, supporting consumers to develop certain preferences about the types of goods they consume based on a range of criteria. These criteria may include perceptible attributes such as taste, colour and texture, as well as invisible characteristics (of products known as 'credence goods') which are not immediately visible, such as being produced according to certain environmental standards (e.g. organic, rainforest alliance certified, biodynamic, sustainably grown, carbon neutral), social standards (e.g. fair-trade, free-range, humane) and safety and quality standards (e.g. BEIC 2023; FSSC 2023). The supply chains for some goods are relatively short and methods for tracking these criteria have been developed – for example, it is possible to buy free-range chicken, grass-fed beef and fair-trade coffee. In each of these cases, it is fairly clear what the product is, and there is some understanding of what production criteria are being met. Carrageenan seaweed is quite different from these products as most people do not realise that they are consuming it. Carrageenan appears on ingredient listings as E407 and E407a in Europe and is found in a wide range of products, but consumers are often not aware they are consuming these products, and as such there is very little transmission of consumer preferences.

Carrageenan is a particular type of food additive known as a hydrocolloid: *hydro-* meaning water, and *colloid* meaning a dispersion of one substance in another substance – such as a gel or emulsion. A *hydrocolloid* is therefore a substance which, when combined with water, acts as a thickener, gelling agent or stabiliser. Hydrocolloids have long been used in processed foods and are not limited to carrageenan – other common hydrocolloids include gelatin, pectin, guar gum, cellulose gum, xanthum gum, arabic gum, agar, alginate and locust bean gum (Table I.2). These additives each have different properties which mean they can give foods different textures.

The most common hydrocolloids are guar gum (made from guar beans), gelatin (produced mainly from cows and pigs), xanthum gum (produced by fermenting sugars from crops such as wheat, corn and soy), cellulose gum (often produced from wood pulp or cotton seeds) and arabic gum (produced from acacia trees). These five hydrocolloids together make up 89 per cent of the global hydrocolloid market. Carrageenan makes up approximately 3 per cent of the hydrocolloid market by volume and 8 per cent by value, and in 2022 the market was worth an estimated US$872 million (Grand View Research 2023a). Carrageenan is particularly widely used in the meat and dairy industries, and as a substitute for gelatin in vegetarian and vegan foods. Because of its specific gelling properties it has been growing in popularity, projected to become a US$1.3 billion dollar market by 2030 (Grand View Research 2023b). Carrageenan is used mostly in the meat and dairy industries, but also in water gels (such as jellies, confectionary and shelf-stable desserts), toothpaste, beer and petfood (Campbell and Hotchkiss 2017). It is often used in blends with other gelling agents to achieve precise textures in processed foods (Blakemore and Harpell 2009; Thomas 1997).

*Table I.2* Common hydrocolloids

|  | Common sources | ~Market share (by volume) |
|---|---|---|
| Guar gum | Guar beans | 21% |
| Gelatin | Animal collagen, mainly from cows and pigs | 20% |
| Xanthum gum | Fermentation of sugars (e.g. from wheat, corn, soy) | 19% |
| Cellulose gum | Wood pulp, cotton seeds | 17% |
| Arabic gum | Acacia trees | 12% |
| Pectin | Citrus fruit peels, apples | 6% |
| Carrageenan | Various seaweeds, primarily *Eucheuma* and *Kappaphycus* species | 3% |
| Alginates | Various brown seaweeds | 2% |
| Locust bean | Carob tree seeds | 1% |
| Agar | Various seaweeds, primarily *Gracilaria* species | 1% |

Source: Data from Grand View Research (2023a)

The demand for carrageenan is linked to demand for hydrocolloids in general (e.g., with long-term trends such as increasing global consumption of processed foods) as well as relative demand for carrageenan over competing hydrocolloids. The demand for certain types of hydrocolloids over others is linked to both the properties of specific hydrocolloids and the demand for the products in which they are used. For example, part of the growth of the carrageenan industry is attributable to the growing demand for vegetarian and vegan foods, for which animal-derived gelatin is not suitable. It is also linked to consumer taste and texture preferences, since different hydrocolloids give foods different textures– for example, carrageenan is able to mimic a 'fatty' texture which many consumers enjoy and as a result is widely used in meat and dairy products, and, as such, growing demand for meat and dairy products could be expected to bolster demand for carrageenan.

The demand for carrageenan is also linked to consumer preferences against certain products. In 2016, the US National Organics Standards Board voted to recommend that carrageenan be removed from the United States Department of Agriculture (USDA) list of organic food additives, as a result of public concern over potential health impacts (NOSB 2016). In the months that followed, Indonesian seaweed farmer and industry groups advocated against this recommendation, on the grounds that it could significantly affect the industry (Mudassir 2018; Dwijayanto 2018). This recommendation was ultimately not adopted because there are limited other options to carrageenan to provide necessary functions in processed foods, and there is a dearth of scientific evidence supporting claims of negative health impacts (USDA 2018). However, if carrageenan were to be rejected by consumers on a large scale, this could have a reverberating effect on the carrageenan seaweed value chains and the villages that grow it. Notably, the scale of this effect would depend on how widespread consumer preferences are: a rejection from consumers in the United States, for example, would have impacts that would be contained if these preferences did not extend to consumers in Asia. Chapter 1 explores these dynamics further.

The development of the carrageenan industry has involved significant amounts of 'work' at all levels of the value chain – to establish the physical possibility of seaweed farming, to develop methods for processing it cheaply in large quantities, developing products which use it and maintaining its social acceptability in processed foods. Of particular interest to this book are the thousands of Indonesian farmers who produce it – who have reorganised their social and economic lives around this new commodity, and who depend on the industry for their livelihoods. This book is organised in two parts. Part I traces the carrageenan value chain from the global to the local level. Part II examines the village-level transformations which have taken place to enable the large-scale production of this commodity. The next section describes the methodological approach taken.

## Background and methods

This research was undertaken as part of a research programme known as the 'Partnership for Australia Indonesia Research' (PAIR), funded by the Australian Government Department of Foreign Affairs and Trade (DFAT) via the Australia-Indonesia Centre (AIC). The programme ran from 2019 to 2023 and brought together researchers from four Australian and seven Indonesian universities, as well as industry and government stakeholders including the Indonesian Ministry of Research and Technology (RISTEK-BRIN) and the South Sulawesi Provincial Government (see PAIR 2023). The research was divided into four streams: this book draws on research conducted by the 'commodities' research team through a series of interrelated packages of work drawing on a range of different types of data (for published reports on these projects see Abdul Aziz et al. 2023; Cozzolini et al. 2023; Hovey et al. 2023; Komarek et al. 2023; Langford et al. 2021; Langford, Turupadang et al. 2022; Langford Zhang et al. 2022; Langford , Waldron et al. 2023; Langford, Turupadang and Waldron 2023b; Langford et al. 2024; Permani et al. 2023; Stone et al. 2023; Waldron et al. 2022; Zhang et al. 2023). This book is structured as an edited monograph to facilitate contributions from a large cohort of contributors to the research programme.

The PAIR programme was established with the support of the South Sulawesi governor and takes this province as its primary location of research. South Sulawesi is the largest seaweed producing province in Indonesia. It is home to the major port of Makassar, which has recently been redeveloped to support much greater volumes of trade and direct export to international locations. Within South Sulawesi, the regencies of *Maros*, *Baru* and *Pangkajene dan Kepulauan* (hereafter 'Pangkep') were identified as priority areas for the programme of research. The commodities group was tasked with investigating seaweed production in this region and focused research on the regency of Pangkep due to the large number of seaweed farmers in this area. In this regency, the village of Pitu Sunggu (Figure I.6) was selected as the case study location following a survey of the seaweed production characteristics of villages in mainland Pangkep (see Langford, Waldron, Nuryartono et al. 2023 for full details). A second site was identified for comparison – the village of Laikang in Takalar Regency, an established seaweed growing region which produces the

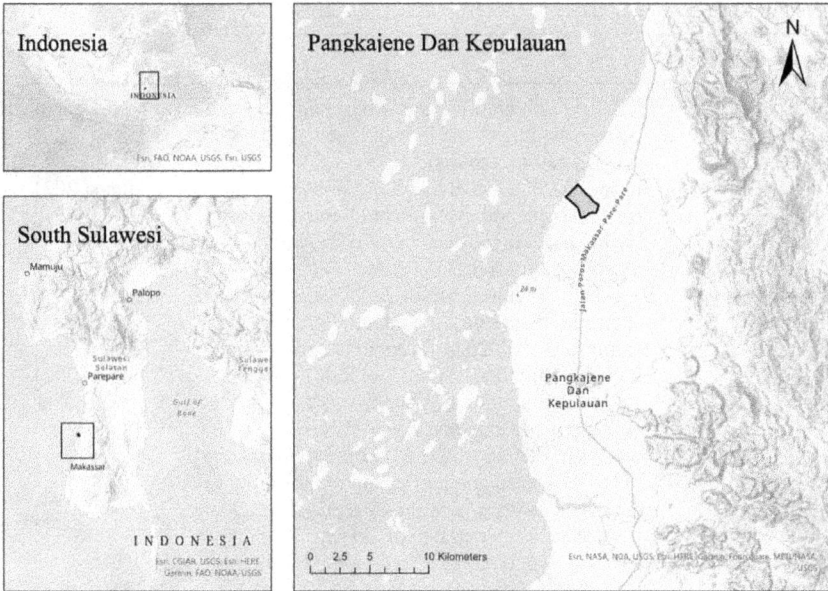

*Figure I.5* Location of Pitu Sunggu village within Indonesia

Source: Map created by Alexandra Langford using ARCGis Pro.

largest quantities of seaweed in South Sulawesi. This site was chosen to facilitate comparisons with the less substantial seaweed cultivation site of Pitu Sunggu.

This book draws primarily on research collected through the main project associated with this programme of research (see Langford, Waldron, Nuryartono et al. 2023), with methods including a structured household survey, 215 semi-structured interviews and 16 months each of ethnographic research by four field researchers. The book also draws on research undertaken through three shorter projects on policy (see Permani et al. 2023), value chain margins (see Komarek et al. 2023) and on farmer resilience in NTT (see Langford, Waldron et al. 2022; Langford, Turupadang, and Waldron 2023). Full details of the methodological approach and findings of these projects can be found in the published reports from these projects, and are summarised in Table I.3.

This approach to long-term, in-depth, qualitative social research offers a few key methodological advantages:

1. Long-term social research produces a holistic understanding of seaweed farmer livelihoods

    Much social research with seaweed farmers is based on short periods of fieldwork, and as such relies on farmers' reports of their motivations, perceptions and goals. These reports are snapshots in time and vary considerably as the circumstances change. Our research has the benefit of providing a more long-term view of the livelihoods of seaweed farmers over the course of more

*Table I.3* Research methods

| Project component | Details | Timeline |
| --- | --- | --- |
| Pilot project | Desktop research providing a baseline understanding of the industry (see Nuryartono et al. 2020 for results). | July–December 2020 |
| Village survey | Survey of village characteristics in coastal villages of Pangkep, and assessment of community willingness to participate in the research (internal reports on each region produced). | August 2021 |
| Structured household survey | Extended structured survey of 273 seaweed farming households in Pitu Sunggu and Laikang (see Langford et al. 2024 for method and results). | October–December 2021 |
| Ethnographic research | 16 months of ethnographic research by four of the contributors to this book, two based in each village (R. Ruhon and Z. Z. Wulyandra in Pitu Sunggu, R.A. Armis and I. Lapong in Laikang). They observed and participated in the daily life of seaweed farming communities and recorded their observations in detailed fieldnotes. | August 2021–December 2022 |
| Semi-structured interviews | 215 semi-structured interviews (in Pitu Sunggu (n = 89), Laikang (n = 82) and NTT (n = 44)), with seaweed farmers, village residents and local government workers. Interviews were undertaken in either Indonesian or local language according to interviewee preference, were transcribed in Indonesian and analysed thematically. | January–December 2022 |
| Satellite imagery analysis | Satellite imagery for Pitu Sunggu in 2022 was analysed manually (see Langford et al. 2021 for method). | January–December 2022 |
| Industry personal communications | Extended personal communications with industry stakeholders including collection of information on seaweed sourcing and processing. | January 2021–June 2023 |
| Policy document analysis and interviews with government officials | A comprehensive investigation into 67 policy documents sourced from both desktop research and interviews with key informants (see Permani et al. 2023 for full methods). | April–October 2022 |
| Value chain analysis | 34 face-to-face interviews with actors in the value chain, including 5 seedling suppliers, 15 farmers, 12 traders (including 10 village traders, 1 Takalar-level trader and 1 Makassar-based exporter) and 2 processors (see Komarek et al. 2023 for full method). Data from these interviews was assessed using descriptive statistics. | June–August 2022 |

than a year, which allowed observation of changing seasons, as well as farmers' responses to weather and price events, and variations through different cultural and religious periods. This allowed us to observe the variations in livelihoods that occurs over time – through the wet season of high waves and frequent occurrence of diseases, the transition season in which one by one, farmers stopped cultivating one species and started cultivating another, through the dry season in which high water temperatures led to outbreaks of epiphytes and poor growth of seaweed and then back to the wet season, when farmers again moved their plots to adjust to the changing weather. This long-term observation meant that field researchers gained a detailed understanding of seaweed farming livelihoods and the strategies that seaweed farmers employ to produce seaweed year-round, despite drastically changing oceanic conditions. It also means that we gained a greater understanding of how seaweed production techniques changed throughout the year, and the interconnected ways that biophysical, social and economic phenomena affect farmer decision-making.

2. Grounded in the technicalities of seaweed production

All four of the field researchers hold qualifications in marine sciences, and as such are attentive to the technical details of seaweed production, including the characteristics of the species being grown, the diseases and epiphytes farmers experience, the differences in productive strategies they employ (such as farm plot locations, planting spacings, seedling sizes, yields). This has allowed them to critically engage with farmers' choices in order to understand how social factors – such as the management of sea space – lead to different production choices (such as rope spacing) and generate different results (such as more intensive use of labour and lower yields by newer entrants to the industry). These insights are invaluable in understanding the factors which contribute to different experiences of seaweed farming livelihoods. Our multi-disciplinary approach allows a more holistic analysis of the interactions between social, economic and environmental dynamics in the villages. The focus on seaweed farmers, rather than more generally on livelihoods, allows a targeted analysis of how this crop is experienced by the people who grow it.

3. Triangulation of multiple perspectives

By undertaking research with not only farmers, but also the people they sell to (local traders), the people they employ (seaweed binders, who tie seaweed propagules to ropes) and professionals involved in seaweed industry governance (such as village and provincial government professionals), this study is able to contribute a balanced understanding of the roles of different village actors in the Indonesian seaweed industry. Personal communications were undertaken with a range of industry actors to gain greater insights into the structure of the industry, and focus groups in conjunction with government provided insights into challenges and priorities in the governance of the industry. This provides an in-depth understanding of the various actors involved in negotiating the structure of the Indonesian seaweed industry.

### Translations and use of gendered language

All quotes are translations by the authors from either Indonesian or the local language of the respondent (Makasarese or Buginese). We have endeavoured to reflect the style of speech, emphasis and meaning of the speaker in these translations. In the quotes and discussion of the book, where gendered language reflects the gendered nature of work involved, this language is used in the translations – for example, all crab netters in Pitu Sunggu are men, so when respondents describe the activities of certain men in crab netting, gendered language is used in the translations. In the chapter exploring the work of local traders, which includes male and female participants, gendered language is avoided to protect the anonymity of the traders involved. Where names of farmers are used, these are pseudonyms.

## Notes on statistics

Indonesian seaweed data is reported by a range of agencies (including the Ministry of Marine Affairs and Fisheries, the Central Bureau of Statistics and the Ministry of Industry) for several production indicators (production volume, export volume, processing volumes, area under production, number of households engaged in production, and number of farmers engaged in production). Attempts to reconcile data from these different sources demonstrate that they rely on very different assumptions and therefore generate vastly different estimates of the size of the seaweed industry. Appendix 1 provides a full outline of data produced by different sources that was consulted in the process of researching this book and demonstrates the inconsistencies between them.

This chapter has outlined the features of the global seaweed industry using FAO data. For Indonesia, we have suggested that these are probably overestimated by around 4.8x (see Appendix 1). Indonesia is not the only country to inaccurately report seaweed production statistics to the FAO. Hatch (2023a; 2023b) recently compared industry and production estimates for major seaweed reporting countries and found that data reported by China, Indonesia, the Philippines, and Malaysia were inconsistent.[6] They estimated that in 2021, Indonesian carrageenan seaweed production statistics were overestimated by 5.1x, Philippines carrageenan seaweed production statistics by 2.6x and Malaysian seaweed production statistics by 6 times. Unfortunately, they did not estimate the overestimation of Chinese seaweed production. This makes it difficult to reconcile inaccuracies across countries. As such, FAO statistics are used in this chapter despite known inaccuracies, because they highlight the dominance of Indonesia and the Philippines in the carrageenan seaweed industry. Their dominance is so great that even if they were revised down using the overestimation factors provided by Hatch, they would still represent 92 per cent of the global carrageenan industry. As such, the figures provided in this chapter provide a realistic insight into the concentration of carrageenan seaweed production in these two countries. For the remainder of the book, data is used selectively as follows.

Chapter 2 relies only on *Badan Pusat Statistik* (BPS) (Central Bureau of Statistics, Indonesia) (2022) survey data, which as Appendix 1 describes, is realistic. Chapter 3 examines the South Sulawesi seaweed industry and therefore uses data from the South Sulawesi Ministry of Marine Affairs and Fisheries (*Kementerian Kelautan dan Perikanan* (KKP)), reminding the reader that of this data, household participation data is likely to be realistic, while production volumes and cultivation areas are not (but still demonstrate the geographic distribution of production around the province). Part II of the book focuses mainly on village-level livelihoods and uses BPS (2022) survey data to contextualise these where appropriate.

## Structure of the book

This book is structured in two parts. Part I provides an overview of the development of the global value chains that have emerged to drive the global carrageenan seaweed industry, while Part II explores the negotiation of these changes at the village level. Readers primarily interested in the global dynamics of carrageenan use and production are advised to start with Part I. Readers who would like to focus more particularly on the village-level changes driven by the industry are advised to read Part II.

## Part I: The global carrageenan seaweed value chain

The first part of the book telescopes down from the global to the provincial level to explore how global carrageenan seaweed value chains are organised at different scales.

### Chapter 1: The global carrageenan market

Chapter 1 explores the long-term drivers of carrageenan demand, production and trade at the international level, demonstrating how the industry has been transformed by decades of sustained growth and development, and the dominance of Indonesia, China and the Philippines within it.

### Chapter 2: The Indonesian seaweed industry

Given the global context and trade patterns outlined in Chapter 1, this chapter examines developments in the Indonesian seaweed industry at the national level. It provides industry-wide context, examines the sectors of production, marketing and processing and then describes some of the cross-cutting issues in zoning, investment, product development and food safety.

### Chapter 3: The South Sulawesi seaweed industry

Chapter 2 explored how Indonesia has worked to support the development of the seaweed industry through investments in production, marketing, processing and research nationally. This chapter looks in more detail at the provincial level, focusing on South Sulawesi Province, outlining key features of the South Sulawesi seaweed industry, including production, trade, processing and export.

## Part II: Livelihood transformations

The second part of the book focuses on one seaweed farming village: Pitu Sunggu, in South Sulawesi, Indonesia, and explores in detail the environmental, social and economic transformations which have taken place to enable the production of this commodity, beginning in the sea, where the seaweed is grown, and moving up to examine changes in social and economic organisation resulting from the industry and systems for marketing the product to the traders and processors who use it.

*Chapter 4: Export commodity frontiers and the transformation of village life*

This chapter explores the local transformations that have occurred in Pitu Sunggu village to accommodate the introduction of new commodities over the last century. It describes a series of livelihood transformations which have taken place – from field rice to wet rice, from wet rice to fish and shrimp farming, and from marine fishing to seaweed farming, and how this history of change has shaped the structure of the seaweed industry in Pitu Sunggu today.

*Chapter 5: From communal access to private ownership:*
*Negotiating access to the sea*

Rights to the sea in South Sulawesi are widely communal and non-exclusive: anyone may make their living from the sea, and in Pitu Sunggu this has traditionally involved fishing and catching crabs. Yet to farm seaweed, a person must be able to claim exclusive rights to an area of the sea, prevent others from using it and ensure the exclusive right to harvest the seaweed when it is grown. This chapter explores how village members have negotiated these transitions, in just two decades transforming the sea from a communal resource to individual ownership of plots which may be bought, sold and rented.

*Chapter 6: Environmental and socio-economic constraints to seaweed farming*
*and Chapter 7: Farmer decision-making in the Indonesian seaweed industry*

Having established the right to sea space in which to farm seaweed, the next challenge is to grow the seaweed. These chapters examine the environmental and bio-physical challenges of seaweed farming that farmers have had to solve to make seaweed driven livelihood transformations happen. Growing at often unpredictable rates throughout the year as the ocean salinity changes with the tropical monsoon, and frequently dying and turning white and clear like ice, this is a crop that has often defied attempts to control and promote its growth. These chapters explore how rapidly changing ocean conditions affect seaweed growth, and how farmers make decisions within environmental, economic and social constraints.

*Chapter 8: Gendered work and casual labour in the Indonesian seaweed*
*industry*

Growing seaweed is not just about production being physically possible: to be produced in large quantities over the long term, seaweed households must also reorganise their labour and relations in order to enable the work required to take place. This chapter explores how a range of people, including seaweed farmers, but also women and children in and outside of seaweed farming households, and casual labourers from neighbouring villages, have come to be incorporated in the seaweed industry, and how their work is integral to the industry.

*Chapter 9: Seaweed marketing: village-based traders as financial and market*
*intermediaries*

The production of seaweed depends on the ability to sell it in the marketplace. For carrageenan seaweeds the final buyers in Indonesia are processors and exporters.

However, farmers do not sell to these entities directly, but through several layers of middlemen. This chapter explores the work of traders who have established operations in the village to connect farmers with global markets. It explores their work buying, packaging and transporting seaweed, and the tactics they employ to capture supply from farmers, including providing a range of financial and retail services. These key intermediaries finally provide the last connection of the village to the wider industry, bringing the seaweed into the global value chains described in Part I.

### Chapter 10: Conclusion

This chapter reflects on the industry as presented through the chapters of this book. This is a study of a product and a crop not well known to consumers – an ingredient which reaches households through many layers of processing and blending to appear towards the end of ingredient listings as an unassuming code, with no clear connection to the coastal households across tropical Southeast Asia who produced it. This book shows how global trends, such as in food manufacturing, are transmitted through value chains to meso and local levels, and how changes in local dynamics form part of feedback loops which in turn create changes through the value chain. It has important practical implications in understanding Indonesian efforts to 'upgrade' these value chains. It also highlights that paths of translation are not frictionless, but negotiated by a range of actors operating at different scales and with different goals. This chapter analyses the key findings from the book, provides policy recommendations and suggests areas for further research.

Through this approach, Part I of this book first traces the contours of the global carrageenan seaweed industry, from global trade and consumption of carrageenan to the development of the Indonesian carrageenan seaweed production and processing industries, to the provincial dynamics of the South Sulawesi seaweed industry and the position of the region of Pangkep, and the village of Pitu Sunggu, within it. Part II explores the day-to-day work of seaweed production in Pitu Sunggu, how it has reshaped village livelihoods and use of the sea, how farmers have developed strategies to minimise risk and increase production, how casual wage labourers, women, children and people with disabilities have been incorporated into the industry and how all these activities respond to price signals transmitted by local traders via regional warehouses. This book seeks to demonstrate the interrelatedness of environmental, social and economic dynamics on seaweed production, and also argues for key policy interventions to support the sustainable development of the industry in the face of climate change.

## Notes

1 Domestication involves control of the reproductive cycle, as opposed to merely propagating from cuttings (Wikfors and Ohno 2001).
2 *Saccharina japonica* is used to produce *kombu*, but also to produce a type of hydrocolloid known as alginate. While estimates of the scale of production of *Saccharina japonica* derived alginates are difficult to reach, it appears that the amounts used for alginate production are not large (Peteiro 2017).
3 *Saccharina japonica* also known as Japanese kelp, and formerly known as *Laminaria japonica*. *Porphyra* spp. here includes FAO categorisations of 'laver (nori)' and 'nori

nei'. *Eucheuma* spp here includes FAO classifications of 'eucheuma spp nei' and 'spiny eucheuma'. The statistics for *Eucheuma* and *Kappaphycus* species are combined here to avoid issues arising from reporting inconsistencies between countries.

4  Note that until the year 2000, Indonesia did not report carrageenan and agar containing seaweeds separately, but reported both as 'Red seaweeds'. Data prior to 2000 is for 'Red seaweeds' and from 2000 onwards is for '*Eucheuma* seaweeds nei'.

5  Planned area data from Presidential Decree 33–2019. Existing production data from BPS 2023. The planned area is the total maximum planned area for each province. The area and quantity estimates were made equivalent at 1 ha = 36 tonnes of seaweed using the 2016 reported quantity from *Badan Pusat Statistik* yearbook (9,773,055 tonnes) at the 2016 estimate of current area under production noted in Presidential Degree 33–2019 (271,336 ha). It is notable that Nusa Tenggara Timor appears to have already exceeded the planned production area. This is probably due to overestimation in provincial production data (see Appendix 1A).

6  The data for China for *Undaria*, *Saccharina* and *Gracilaria* were reported to be 'inconsistent', with data for the carrageenan seaweeds ('eucheumatoids') were of 'unknown' accuracy. The data for Indonesia, the Philippines and Malaysia for the agar containing seaweed *Gracilaria* and the eucheumatoids were reported to be 'inconsistent'.

# References

Abdul Aziz, Ammar, Pramaditya Wicaksono, Sanjiwana Arjasakusuma, Fathing Ayuni Azizan, Scott Waldron, Scott Chapman and Alexandra Langford. 2023. *Capacity Building and Knowledge Transfer in Seaweed Mapping in Indonesia*. Melbourne, Australia: Australia-Indonesia Centre.

BEIC (British Egg Industry Council). 2023. Gold Standard for British Lion. Available at https://britisheggindustrycouncil.com/blog/gold-standard-for-british-lion

Blakemore, William R. and Alan R. Harpell. 2009. Carrageenan. In *Food Stabilisers, Thickeners and Gelling Agents,* edited by Alan Imeson. Sussex: Blackwell Publishing Ltd.

BPS (*Badan Pusat Statistik*). 2022. Hasil Survei Komoditas Perikanan Potensi Rumput Laut 2021 Seri 2. *Badan Pusat Statistik*. Available at https://www.bps.go.id/publication/202 2/08/29/269de33babc6e3d52bbae5b6/hasil-survei-komoditas-perikanan-potensi-rumput-laut-2021-seri-2.html

BPS (*Badan Pusat Statistik*). 2023. Production of Aquaculture by Province and Type of Cultivation (Tons). Accessed 6 June 2023. Available at https://www.bps.go.id/indicator/56/1509/1/production-of-aquaculture-by-province-and-type-of-cultivation.html

Campbell, Ross and Sarah Hotchkiss. 2017. Carrageenan Industry Market Overview. In *Tropical Seaweed Farming Trends, Problems and Opportunities: Focus on Eucheuma and Kappaphycus of Commerce,* edited by Anicia Q. Hurtado, Alan T. Critchley and Iain C. Neish. Cham, Switzerland: Springer International Publishing.

Cozzolino, Daniel, Mohamad Rafi, Wahyu Ramadhan, and Rudi Heryanto. 2023. *Technology to Improve Seaweed Marketing, Prices and Smallholder Incomes*. Melbourne, Australia: Australia-Indonesia Centre.

Dixon, Jane. 2002. *The Changing Chicken: Chooks, Cooks and Culinary Culture*. Sydney: UNSW Press.

Dwijayanto, A. 2018. Delisting dicabut, ekspor rumput laut ri ke Amerika berlanjut. Available at https://industri.kontan.co.id/news/delisting-dicabut-ekspor-rumput-laut-ri-ke-amerika-berlanjut

FAO (Food and Agriculture Organisation (of the United Nations)). 2023. FishStatJ (software for FAO's Fisheries and Aquaculture statistics). Available at https://www.fao.org/fishery/en/statistics/software/fishstatj

FSSC (Food Safety System Certification). 2023. About us. Available at https://www.fssc.com/about

Grand View Research. 2023a. *Hydrocolloids Market Size, Share & Trends Analysis Report by Product (Gelatin, Xanthan Gum, Carrageenan, Alginates, Pectin, Guar Gum, Carboxy Methyl Cellulose), by Function, by Application, and Segment Forecasts, 2020–2025*. Accessed 6 January 2023. Available at https://www.grandviewresearch.com/industry-analysis/hydrocolloids-market

Grand View Research. 2023b. *Carrageenan Market Size, Share & Trends Analysis Report by Processing Technology (Semi-refined, Gel Press, Alcohol Precipitation), by Function, by Product Type, by Application, by Region, and Segment Forecasts, 2020–2028*. Accessed 6 January 2023. Available at https://www.grandviewresearch.com/industry-analysis/carrageenan-market/toc

Hatch. 2023a. *Global Production Overview*. Accessed 6 June 2023. Available at https://seaweedinsights.com/global-production

Hatch. 2023b. *Eucheumatoids*. Accessed 6 June 2023. Available at https://seaweedinsights.com/global-production-eucheumatoids

Hovey, R., S. Werorilangi, W. Umar, H. Hasyim, N. M. R. Harusi, K. F. Dexter, R. Ruhon, Z. Zach, and A. Langford. 2023. *End of Life of Plastics Used in Seaweed Aquaculture in South Sulawesi*. Melbourne, Australia: Australia-Indonesia Centre.

Komarek, A., E. R. Cahyadi, J. Zhang, A. Fariyanti, B. Julianto, R. Arsyi, I. Lapong, A. Langford, S. Waldron, and M. Grist. 2023. *Increasing Incomes in Carrageenan Seaweed Value Chains in Takalar, South Sulawesi*. Melbourne, Australia: Australia-Indonesia Centre. Available at https://pair.australiaindonesiacentre.org/research/increasing-incomes-in-carrageenan-seaweed-value-chains-in-takalar-south-sulawesi

Langford, A., W. Turupadang, M. D. R. Oedjoe, C. Liufeto, B. Bire, M. Yohanis, Y. Suni, and S.Waldron, S. 2022. *Seaweed Farmer Resilience in Eastern Indonesia after COVID-19*. Canberra, Australia: Australian National University Indonesia Project.

Langford, A., W. Turupadang, and S. Waldron, 2023. Interventionist Industry Policy to Support Local Value Adding: A Case from the Eastern Indonesia Seaweed Industry: Implications for Coastal Communities. *Marine Policy* 151(105561): 1–11.

Langford, A., S. Waldron, N. Nuryartono, S. Pasaribu, B. Julianto, I. Siradjuddin, R. Ruhon, Z. Zach, I. Lapong, and R. Armis. 2023. *Sustainable Upgrading of the South Sulawesi Seaweed Industry*. Melbourne, Australia: Australia-Indonesia Centre. Available at https://pair.australiaindonesiacentre.org/research/sustainable-upgrading-of-the-south-sulawesi-seaweed-industry-2

Langford, A, S. Waldron, Sulfahri, and H. Saleh. 2021. Monitoring the COVID-19-Affected Indonesian Seaweed Industry Using Remote Sensing Data. *Marine Policy* 127 (104431): 1–10. doi.org/10.1016/j.marpol.2021.104431

Langford, Alexandra, Scott Waldron, Jing Zhang, Radhiyah Ruhon, Zulung Zach Walyandra, Risya Arsyi Armis, Imran Lapong, Boedi Julianto, Irsyadi Siradjuddin, Syamsul Pasaribu, and Nunung Nuryartono. (2024). "Diverse Seaweed Farming Livelihoods in Two Indonesian Villages." In *Tropical Seaweed Cultivation – Phyconomy: Proceedings of the Tropical Phyconomy Coalition Development, TPCD 1, Held at UNHAS University, Makassar, Indonesia. July 7th and 8th* edited by A. Hurtado, A. Critchley, and I. Neish. 00-00. New York: Cham, Switzerland: Springer Nature.

Langford, A., J. Zhang, S. Waldron, B. Julianto, I. Siradjuddin, I. Neish, and N. Nuryartono. 2022. Price Analysis of the Indonesian Carrageenan Seaweed Industry. *Aquaculture* 550(737828). doi.org/10.1016/j.aquaculture.2021.737828

MLA (Meat and Livestock Australia). 2022. *Asparagopsis Now Commercially Available to Beef Producers.* Accessed 11 January 2022. https://www.mla.com.au/news-and-events/industry-news/asparagopsis-now-commercially-available-to-beef-producers

Mudassir, R. 2018. Ekspor rumput laut ke as: Kembali mas daftar produk organik, simak paparan kemendag. 8 April. *Ekonomi.* Available at https://ekonomi.bisnis.com/read/20180408/12/781748/ekspor-rumput-laut-ke-as-kembali-masuk-daftar-produk-organik-simak-paparan-kemendag

NOSB (National Organic Standards Board). 2016. *Sunset 2018 Review Summary: NOSB Final Review Handling Substances §205.605(a), §205.605(b), §205.606 November 2016.* Available at https://www.ams.usda.gov/sites/default/files/media/HS2018SunsetReviews.pdf

Nuryartono, N., S. Waldron, A. Langford, Sulfahri, K. Tarman, S. H. Pasaribu, U. J. Siregar, and M. F. D. Lusno. 2020. *A Diagnostic Analysis of the South Sulawesi Seaweed Industry.* Melbourne, Australia: Australia-Indonesia Centre.

Peterio, César. 2017. Alginate Production from Marine Macroalgae, with Emphasis on Kelp Farming. In *Alginates and Their Biomedical Applications,* edited by Bernd H. A. Rehm and M. Fata Moradali, pp. 27–66. Available at https://doi.org/10.1007/978-981-10-6910-9_2

PAIR (Partnership for Australia-Indonesia Research). 2023. About the Partnership for Australia-Indonesia Research (PAIR). Available at https://pair.australiaindonesiacentre.org/about

Permani, Risti, Yanti Nuraeni Muflikh, Fikri Sjahruddin, Nunung Nuryartono, Scott Waldron, Alexandra Langford, and Syamsul Pasaribu. 2023. *The Policy Landscape and Supply Cahin Governance of the Indonesian Seaweed Industry: A Focus on South Sulawesi.* Melbourne, Australia: Australia-Indonesia Centre. Available at https://pair.australiaindonesiacentre.org/research/the-policy-landscape-and-supply-chain-governance-of-the-indonesian-seaweed-industry-a-focus-on-south-sulawesi

Presidential Decree 33–2019. *Peraturan Presiden Republik Indonesia Nomor 33 Tahun 2019 tentang Peta Panduan Pengembangan Industri Rumput Laut Nasional Tahun 2018–2021.*

Steenbergen, D. J., C. Marlessy, and E. Holle. 2017. Effects of Rapid Livelihood Transitions: Examining Local Co-Developed Change Following a Seaweed Farming Boom. *Marine Policy,* 82, 216–223. https://doi.org/10.1016/j.marpol.2017.03.026

Stone, Serafina, Zannie Langford, Imran Lapong, Risya Arsyi Armis, Zulung Zach Walyandra, Radhiyah Ruhon, Annie Wong, Boedi Julianto, Irsyadi Siradjudding, and Scott Waldron. (2023) Technology Uptake by Smallholder Farmers: The Case of the Indonesian Seaweed Industry. *Journal of Agribusiness in Developing and Emerging Economies.* Available at https://doi.org/10.1108/JADEE-01-2023-0011

Tsing, Anna Lowenhaupt. 2015. *The Mushroom at the End of the World : On the Possibility of Life in Capitalist Ruins.* Princeton, NJ: Princeton University Press.

USDA (United States Department of Agriculture). 2018. *Federal Register, vol. 83, no. 65, Wednesday, 4 April 2018/Rules and Regulations.* Agricultural Marketing Service, USDA, Department of Agriculture. Available at https://www.govinfo.gov/content/pkg/FR-2018-04-04/pdf/2018-06867.pdf

Waldron, Scott, Nunung Nuryartono, Alexandra Langford, Syamsul Pasaribu, Kustiariyah Tarman, Ulfah J. Siregar, Muhammad Farid Dimjati Lusno, Sulfahri, Boedi Sarjana Julianto, Irsyadi Siradjuddin, Radhiyah Ruhon, Zulung Zach Walyandra, Muhammad Imran Lapong, Risya Arsyi Armis, Eugene Sebastian, Helen Brown, Fadhilah Trya Wulandari,

Hasnawati Saleh, and Steve Wright. 2022. Policy Brief: Sustainable Upgrading of the South Sulawesi Seaweed Industry. Melbourne, Australia: Australia-Indonesia Centre. Available at https://pair.australiaindonesiacentre.org/wp-content/uploads/2022/11/SIP-1-EN-ONLINE.pdf

Weiss, Brad, Eliza Maclean, Jennifer Curtis, John O'Sullivan, Kevin Callaghan, Ross Flynn, Sam Suchoff, Sarah Blacklin, Vimala Rajendran, and Will Cramer. 2016. *Real Pigs: Shifting Values in the Field of Local Pork*. Durham, NC: Duke University Press.

West, Paige. 2012. *From Modern Production to Imagined Primitive*. Amsterdam: Duke University Press. Available at https://doi.org/10.1515/9780822394846

Wikfors, Gary H. and Ohno Masao. 2001. Impact of Algal Research in Aquaculture. *Journal of Phycology* 37: 968–974. Available at https://doi.org/10.1046/j.1529-8817.2001.01136.x

World Bank. 2023. Price Level Ratio of PP Conversion Factor (GDP) to Market Exchange Rate. Available at https://data.worldbank.org/indicator/PA.NUS.PPPC.RF

Zhang, Jing, Scott Waldron, Alexandra Langford, Boedi Julianto, and Adam Martin Komarek. 2023. "China's Growing Influence in the Global Carrageenan Industry and Implications for Indonesia." *Journal of Applied Phyconomy*. Available at https://doi.org/10.1007/s10811-023-03004-0

# Part I

# Globalisation and the Indonesian seaweed industry

# 1    The global carrageenan industry

*Jing Zhang, Zannie Langford, and Scott Waldron*

## Introduction

Local-level activity is impacted directly and indirectly by developments at national and global levels. This chapter explores the long-term drivers of carrageenan demand, including the properties of the product, market dynamics, competition and substitution with other hydrocolloids, evolving processing methods, shifting consumption trends, and trade patterns. As a highly globalised industry, the chapter focuses on industry developments at an international level by triangulating detailed trade, industry, and policy data through forensic open-source research in multiple languages, and then telescopes down to national and provincial levels in subsequent chapters.

## Market demand

Market demand for carrageenan and other hydrocolloids is difficult to estimate. This section describes the global hydrocolloid market and draws on market research reports by Grand View Research[1] (2023a; 2023b).

### Major hydrocolloids

Seaweeds cultivated for the manufacturing of hydrocolloids such as carrageenan and agar make up a significant proportion of overall global seaweed production, as described in the Introduction. Seaweed inputs, alongside a range of other plant, animal, and microbial sources, are the basis of the broader hydrocolloids market. Different types of hydrocolloids have different properties and prices patterns that make them partial substitutes and competitors in the global hydrochlorides market. Data showing the price and market share of selected hydrochlorides are presented in Figure 1.1. For more detailed analysis of carrageenan prices paid to different value chain actors in Indonesia, see Langford et al. 2022 and Komarek et al. 2023. Market share and specific gelling, thickening, and stabilising applications for the hydrochlorides are shown in Table 1.1.

The five most widely used hydrocolloids – guar gum, gelatin, xanthan gum, cellulose gum, and arabic gum account for 89 per cent of the global market volume

DOI: 10.4324/9781003183860-3

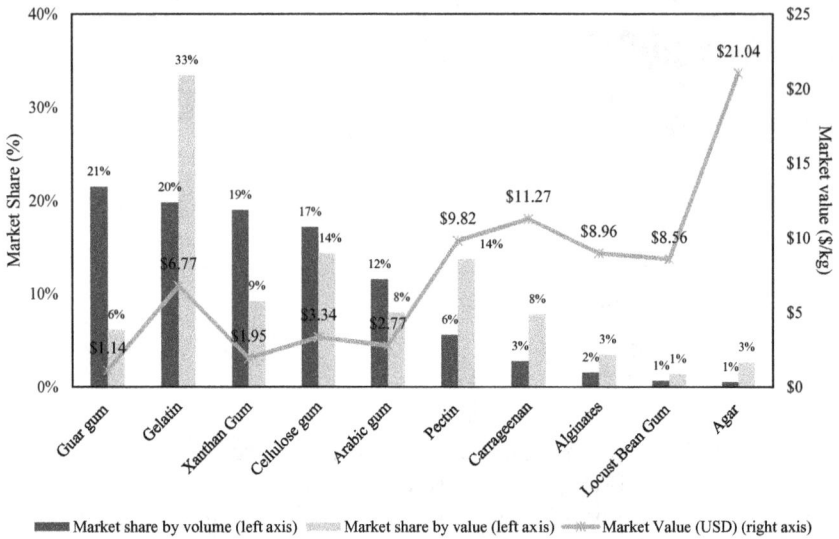

*Figure 1.1* Major hydrocolloids by market share and market value

Source: Data from Grand View Research (2023a).

and on average sell for $3.20/kg (Figure 1.1). The other five major hydrocolloids represent only 11 per cent of the global market share by volume, but 29 per cent by value, selling for an average price of $10.50/kg. Carrageenan falls into the category of a higher-price, lower-volume hydrocolloid, with a market share by volume of just 2.8 per cent, but a much higher market share by value of 8 per cent. The hydrocolloids extracted from seaweeds, carrageenan, alginates, and agar are some of the most expensive commercially important major hydrocolloids, selling for an average price per kilogram of $11.27, $8.96, and $21.04 respectively.

In some applications, these hydrocolloids are used as substitutes for each other, while others are used in blends as complementary goods. As substitutes, guar gum and locust bean gum exhibit similar thickening behaviours (highly shear thinning, high to low shear viscosity) but guar gum is lower-priced and therefore can replace locust bean gum in many applications. Some hydrocolloids are partial substitutes for each other. For example, xanthum gum, locust bean gum, and guar gum are all highly shear thinning, however xanthum gum maintains viscosity at high temperatures and at wide pH ranges, where guar gum and locust bean gum do not. Similarly, agar and carrageenan both form thermoreversible gels on cooling, which means they may behave similarly in some (but not all) applications. Cellulose gum and carrageenan are both used as stabilisers in shampoos, and a range of hydrocolloids are used in lotions. Xanthum gum, cellulose gum, and guar gum are all used in drilling fluids in oil and gas applications, fuelling demand for these products.

Many hydrocolloids are complimentary goods in certain applications and are combined in tailored blends to achieve specific thickening, gelling, and stabilising

*Table 1.1* Common hydrocolloids by source, main uses, market share, average price, and projected growth

| | Common sources | Main uses | Market share by value | Market share by volume | Average price (US$) | Projected volume growth |
|---|---|---|---|---|---|---|
| Gelatin | Animal collagen, mainly from cows and pigs | Gelling agent in food and beverages, nutraceuticals, healthcare, personal care, photography. Pharmaceuticals such as wound dressings, blood volume substitutes, homeostatic sponges. | 33% | 20% | $6.77 | 4.8% |
| Cellulose gum | Wood pulp, cotton seeds | Food and beverages, especially low-fat and frozen foods (e.g. salad dressings, gravies, dairy products, puddings, ice creams, creams, peanut butter, chocolate, frozen desserts, margarine, ketchup). Pharmaceuticals (tablet coatings) Oil drilling Cosmetics and personal care (e.g. toothpaste, shampoo, hair gels, body lotions, shower gels, face creams, ointments). | 14% | 17% | $3.34 | 3.8% |
| Pectin | Citrus fruit peels, apples | Food and beverages – thickener in fruit-based products (e.g. jams, fruit fillings, jellies), glazes, milk-based desserts, and stabiliser in fruit juices, acidic protein beverages, dairy products, confectionary. Personal care and cosmetics (e.g. lotions, aftershave creams, and gels). | 14% | 6% | $9.82 | 8% |
| Xanthan gum | Fermentation of sugars (e.g. from wheat, corn, soy) | Food and beverages (e.g. toppings, sauces, non-fat milk, dairy, ice cream, soups, gravies, instant beverages, ketchup), often used alongside locust bean gum. Oil and gas industries (e.g. drilling fluids, fracturing fluids, displacement agents). Personal care products (creams, toothpaste, lotions, shampoos, sunscreen, mascara, body washes). Other applications (paints, adhesives). | 9% | 19% | $1.95 | 6.1% |

Table 1.1 (Continued)

| | Common sources | Main uses | Market share by value | Market share by volume | Average price (US$) | Projected volume growth |
|---|---|---|---|---|---|---|
| Arabic gum | Acacia trees | Food and beverages (e.g. candy coatings, bakery toppings to prevent sugar crystallisation, soft drinks, foaming alcoholic beverages). Pharmaceuticals (e.g. drug suspensions, cough syrups, tablets). | 8% | 12% | $2.77 | 5.4% |
| Carrageenan | Various seaweeds, primarily *Eucheuma* and *Kappaphycus* species | Food and beverages (especially with dairy) (e.g. ice cream, dairy desserts, dairy beverages, cheeses, dips, sauces, puddings) but also meats, jams, bread, beverages, powdered foods and beverages, tofu, and pet food). Personal care and cosmetics (e.g. toothpaste, lotions, shaving gels, shampoos). Pharmaceuticals to control drug release in microspheres and microcapsules. | 8% | 3% | $11.27 | 3.8% |
| Guar gum | Guar beans | Food and beverages (e.g. dairy products, ketchup, fruit juices, cake batters, and pudding powders). Shale and gas industries for hydraulic fracturing. Pharmaceuticals. | 6% | 21% | $1.14 | 4% |
| Alginates | Various brown seaweeds | Food and beverages (e.g. instant noodles, ice creams, acid milk drinks, jellies, dressings, beers). Pharmaceuticals (e.g. wound dressings, drug tablets). Cosmetics (e.g. lipstick). Other applications (e.g. animal feed, textile printing, wastewater treatment). | 3% | 2% | $8.96 | 3.0% |
| Agar | Various seaweeds, primarily *Gracilaria* and *Gelidium* species | Bacteriology Food and beverages (e.g. ice creams, baked desserts, pie fillings, meringues) | 3% | 1% | $21.04 | 3.7% |
| Locust bean gum | Carob tree seeds | Food and beverages (e.g. ice cream (prevents formation of ice crystals), baked goods, milk products, frozen dairy desserts, soft cheeses, fruits and juices, alcoholic beverages), often used alongside xanthum gum, also used alongside agar and carrageenan. | 1% | 1% | $8.56 | 3.9% |

Source: Data from Grand View Research (2023a).

properties. For example, locust bean and xanthum gum are often combined in gels, while carrageenan can be blended with a wide range of other hydrocolloids for use in different applications. For example, it can be used with gelatin to improve food texture and stability (Wang et al. 2015), with agar in food packaging and wound dressings (Rhim 2013; Rhim and Wang 2013), with xanthum gum in jellies (Brenner et al. 2015), and with pectin to form biodegradable composite films (Alves et al. 2010). Xanthum gum can also be used in combination with starch, particularly as a fat replacer and in dairy products (Huc et al. 2014; Matignon, Barey et al. 2014; Matignon, Moulin et al. 2014).

The global hydrocolloids market is large and continues to grow. In 2022, Grand View Research (2023a) valued the industry at US$ 11.2 billion and have forecast growth of 4.9 per cent in volume and 6 per cent in value from 2023 to 2030 (Figure 1.2). This growth is driven by a range of factors, including growing demand for processed foods and oil and gas applications. The fastest projected annual growth rate is for pectin (8 per cent CAGR) and xanthan gum (6.1 per cent CAGR). Carrageenan is expected to grow by 3.8 per cent in volume and 5.4 per cent in value, suggesting possible improvements in the grade or quality of carrageenan being produced. Most hydrocolloids (73.5 per cent) are used in the food and beverage industries, with 12.5 per cent used in pharmaceuticals, 7.8 per cent in personal care and cosmetics, and 6.3 per cent in other applications such as oil and gas, textile printing, and construction coatings. This composition is not expected to change dramatically before 2030, with similar growth rates of 4.7–5.7 per cent projected for each of these applications. Although this appears to be a simple and linear market forecast (Figure 1.2).

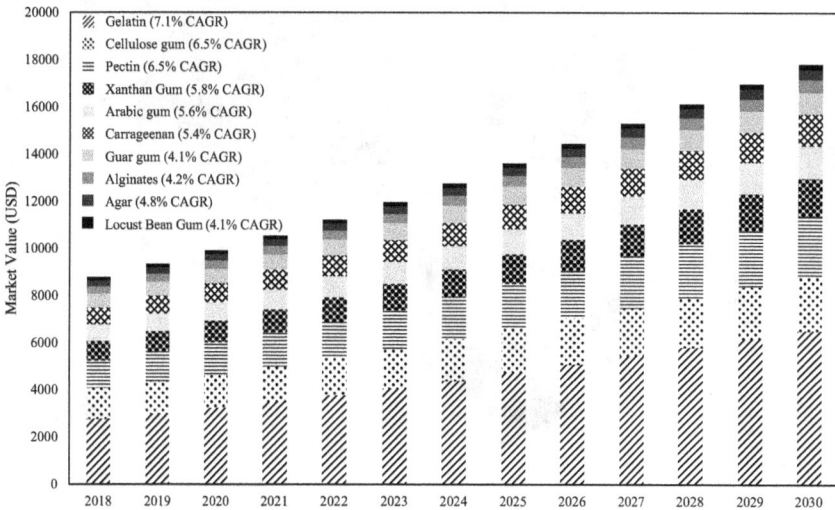

*Figure 1.2* Projected growth in hydrocolloid market value

Source: Data from Grand View Research (2023a).

## Applications of carrageenan

Carrageenan is the sixth largest hydrocolloid by volume and an industry worth an estimated $872 million in value. The main uses of carrageenan are in the food and beverage industry, which absorbs 77 per cent of global supply (Figure 1.3). Of this, 29 per cent of carrageenan is used in meat products, where it is added as a binding agent to increase water retention and improve texture. Another 18 per cent is used in the dairy industry to bind milk proteins and form a gel at low concentrations. In meat and dairy applications, carrageenan produces a 'mouthfeel' which simulates a 'fatty' feel (Weenen et al. 2005). As a result, it is often a substitute for fats and is used to thicken low fat dairy products and dairy replacement products. Carrageenan is also used in confectionary production to improve the rheological properties and stability of these products and represents 7 per cent of market use. In addition to uses in foods and beverages, demand for carrageenan is bolstered by the growing demand for health products such as confectionary using alternative sweeteners, which often require special formulations to maintain texture and mouth feel. A further 10 per cent of global supply enters the pharmaceutical industry where carrageenan is used for a range of functions including the control of drug release rate. A further 10 per cent is used in personal care and cosmetics, such as toothpaste, lotions, shaving gels, and shampoos, including many 'natural' products as a replacement for synthetic substances.

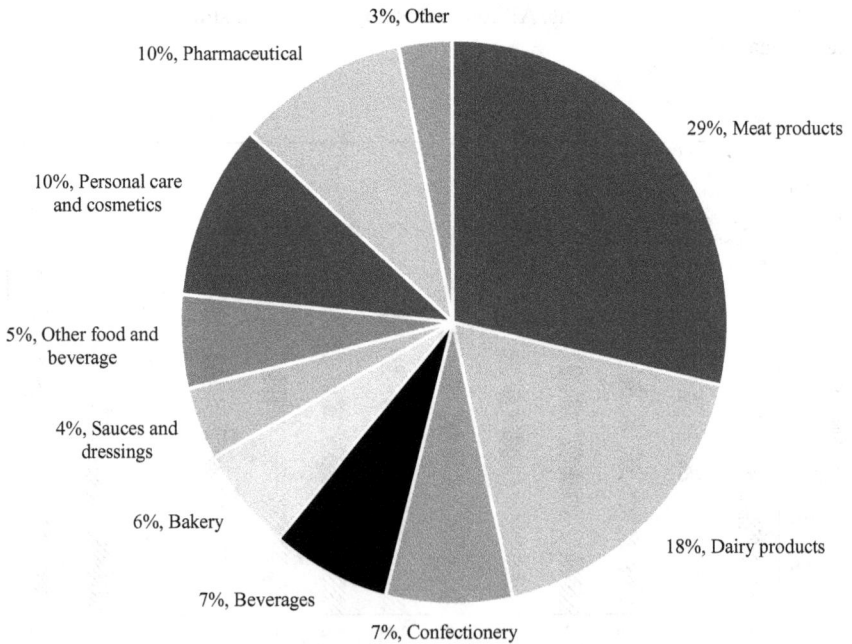

*Figure 1.3* Applications of carrageenan (market share by volume)
Source: Data from Grand View Research (2023b).

This snapshot of carrageenan usages may be subject to significant change (Figure 1.4). The proportion of carrageenan used in the food and beverage industry increased from 88per cent in 1999 to 91 per cent in 2009 and declined to 74 per cent in 2019. For example, the use of carrageenan in processed meats increased from 34 per cent in 1999 to 41 per cent in 2009 but declined to 28 per cent in 2019. The absolute volume increased by 85 per cent between 1999 and 2009 and plateaued in 2019 at 18,706 tonnes. The use of carrageenan in dairy products declined from 40 per cent to 31 per cent and then to 17 per cent of total carrageenan use in 2019, while absolute volumes plateaued between 2009 and 2019. The use of carrageenan in food and beverages, including bakery, confectionery, sauces and dressings, and beverages, increased from 14 per cent in 1999 to 19 per cent in 2009 and then 29 per cent of the total carrageenan application in 2019, with absolute volume usage almost doubling every decade. This is attributed to advancements in technology and research on the use of carrageenan in food and beverage applications as emulsions, gels, and stabilisers. The growth in the use of carrageenan for personal care and cosmetics and pharmaceuticals (i.e., pill coatings and drug capsules) has been particularly pronounced in the last decade and accounted for 12 per cent and 11 per cent respectively of the total carrageenan use in 2019.

## Types and grades of carrageenan

Three types of carrageenan are widely used in commercial applications: kappa-carrageenan (κ-carrageenan), iota-carrageenan (ι-carrageenan), and lambda-carrageenan (λ-carrageenan). These various types of carrageenan have quite different properties and are used in diverse applications (Table 1.2). The most

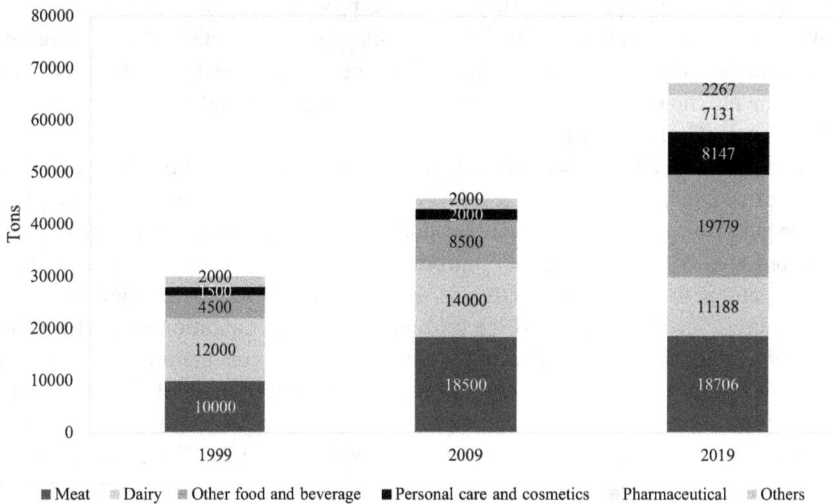

*Figure 1.4* Changing trends in global carrageenan application by sector, 1999–2019
Source: Data from ICF (2011) and Grand View Research (2023b).

*Table 1.2* Main types of carrageenan

| Type | Properties | Applications | Main seaweed sources | Global market share by volume |
|---|---|---|---|---|
| κ-carrageenan | Forms a brittle, stiff gel with calcium salts and a rigid, elastic gel with potassium salts | • Dairy products (chocolate milk, yoghurt, ice cream)<br>• Processed meat (sausages, poultry rolls, turkey breast, cooked cured ham)<br>• Dessert jellies<br>• Shampoo, toothpaste, lotions<br>• Pet food | *Kappaphycus alvarezii* (known colloquially as 'cottonii')<br>*Kappaphycus striatus* (known colloquially as 'sacol') | 67% |
| ι-carrageenan | Forms a soft, elastic gel with calcium salts | • Dairy desserts, dips, caramel sauces, beverages<br>• Nasal sprays | *Eucheuma denticulatum* (known colloquially as 'spinosum') | 24% |
| λ-carrageenan | Forms a highly viscous solution rather than a gel; Requires high temperatures for hydration | • Instant powder mix products | *Sarcothalia cripsata*<br>Gigartina skottsbergii | 9% |

widely produced and used is κ-carrageenan, which produces the most rigid gel. κ-carrageenan is produced mostly from the *Kappaphycus* species, of which Indonesia is the largest producer. Indonesia also produces ι-carrageenan containing seaweed *Eucheuma denticulatum,* but in smaller quantities due to its lower price. κ-carrageenan is used primarily in the food and beverage industry (78 per cent), but also in the pharmaceutical industry (9 per cent), and in personal care and cosmetics (10 per cent) (Grand View Research 2023b).

Carrageenan is also processed using different technologies to produce different grades of product. Refined carrageenan (RC) is produced using alcohol precipitation or gel press technology which are both higher cost processes that produce a purer product. Alcohol precipitation produces the highest purity and most refined product, and is used in pharmaceuticals, toothpaste, and high-value food products such as jams, jellies, soups, and ice creams. Alcohol precipitation is used for the extraction of all three types of carrageenan but is a high-cost method which limits its suitability for many applications. Around 19 per cent of the global carrageenan market is produced using alcohol precipitation, and it is expected to grow at 2.9 per cent annually in volume until 2030 (Grand View Research 2023b). Refined carrageenan can also be produced using the gel press method, which can be performed at lower cost as it does not use alcohol in the process. κ-carrageenan extracted using the gel press method creates a product with high gel strength and purity which is particularly prized in dessert jellies in the Asian market and is also used in meat and dairy products. Currently 29 per cent of carrageenan is produced using this method and that figure is expected to grow at around 4.1 per cent annually until 2030 (Grand View Research 2023b).

Most carrageenan seaweeds are processed into semi-refined carrageenan (SRC). SRC uses a lower-cost process to produce a lower purity product. The development of this technology led to significant reorganisation of the carrageenan seaweed industry in Indonesia in the 1990s. This technology led to the proliferation of small, low-technology processing companies across Indonesia and the Philippines (as described in the Preface). The main use of SRC is in pet food, although a food-grade SRC can be used in a range of products and appears on labels as E407a. It is also known as Processed Eucheuma Seaweed (PES) or Philippine Natural Grade (PNG) carrageenan. SRC currently accounts for 52 per cent of the global carrageenan market by volume and is expected to grow at a rate of 3.9 per cent annually (Grand View Research 2023b).

## Geographic distribution of carrageenan use

According to data from the United Nations (UN) Comtrade database (Figure 1.5), Germany and Japan were the two largest importers of carrageenan in 1990. Their combined import value was equivalent to the total combined imports of the third to tenth largest importing countries. These import patterns have since shifted, with increased imports between 1990 and 1995 (especially by the United States,

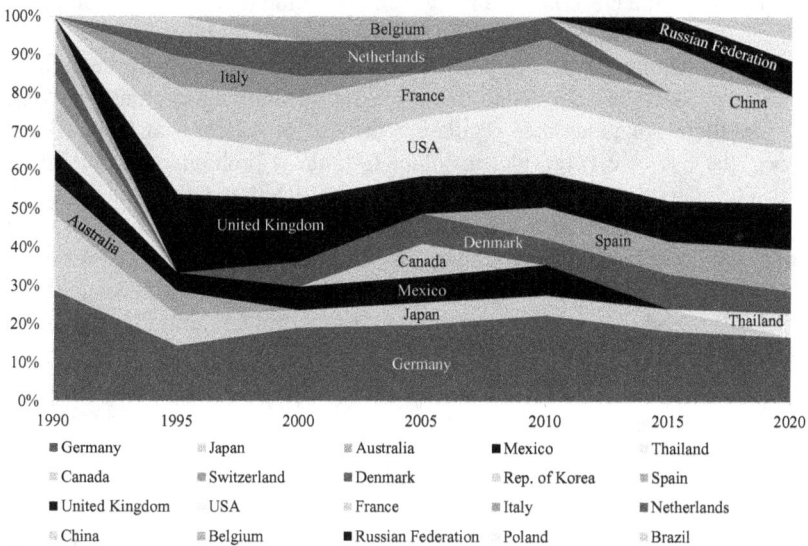

*Figure 1.5* Top 10 carrageenan importers by volume in 1990, 1995, 2000, 2005, 2010, 2015, and 2020

Source: Data from UN Comtrade (2023).

the United Kingdom, and France), 1995–2000 (Denmark, Italy, the Netherlands, and Belgium), 2000–2005 (Mexico and Canada), 2005–2010 (Spain in 2010), 2010–2015 (Russian Federation and China), and 2015–2020 (Poland, Brazil, and Thailand).

These shifts in carrageenan import patterns are forged by factors that include changes in demand, supply, trade policies, and technological advancements. Two broad observations arise from the trade data, one is the continued dominance of Europe and the United States, and the other is increasing demand in emerging markets. Germany remained the largest importer of carrageenan, as of 2021, in terms of volume with 14,884 tonnes imported, valued at approximately US$114 million. China and the United States followed, with import volumes of 12,028 and 11,352 tonnes, valued at around US$80 million and US$115 million, respectively. In contrast, Japan's carrageenan imports in 2020 were 3,226 tonnes, valued at approximately US$30 million, which has not changed significantly in the last 20 years. There has been a growing demand for carrageenan in countries such as China, Thailand, Brazil, and the Russian Federation, which were amongst the largest importers of carrageenan to the end of 2022. Economic growth and changing food consumption patterns will forge global trends in the future.

## Future trends

As the largest users of carrageenan, future demand is closely linked to developments in the food and beverage sector. Carrageenan was approved for use as a food additive by the US Food and Drug Administration (FDA) in 1961 and by the European Commission (EC) Scientific Committee in 1977. It was listed in Codex Alimentarius Commission CAC/GL 32 in 1999 and has been permitted for use in organic products in many countries since 2005 (Marinalg International 2023). However, there are guidelines regulating the use of carrageenan in foods. For instance, the use of carrageenan in infant formula is prohibited in Europe (European Parliament and Council 1995) and is restricted at a maximum usage of 0.3g/L in China's National Food Safety Standards (GB 1886.169–2016). These national standards diverge from findings of the UN expert committee on food additives from WHO/FAO, which reviewed scientific studies on carrageenan in 2014 (JECFA 2015) and found that its use in infant formula at concentrations up to 1g/L was 'not of concern'.

### Characteristics in demand

As consumers are demanding larger quantities of processed foods, demand for hydrocolloids is growing as they are needed to create suitable texture, mouthfeel, and product stability in processed foods. Demand for carrageenan has been bolstered by demand for low-fat products and dairy replacement products, as it is able to mimic a 'fatty' mouth feel in these foods. In addition, growing demand for 'natural' and 'organic' products has led to an increase in demand for carrageenan to replace synthetic alternatives, while growing demand for vegan and vegetarian products has seen carrageenan used as an alternative to animal-derived gelatin (Grand View Research 2023b).

### Safety concerns

The use of carrageenan in foods has recently received negative publicity. A degraded form of carrageenan, known as poligeenan, has the potential to cause inflammation and gastrointestinal issues in some individuals and concerns have been raised that carrageenan may degrade to poligeenan in the stomach (Cohen and Nobuyuki 2002; Weiner 2014; Bixler 2017; Martino et al. 2017; David et al. 2018). As a result of these health concerns, in 2016 the US National Organics Standards Board voted to recommend that carrageenan be removed from the United States Department of Agriculture (USDA) list of organic food additives (NOSB 2016). Following these recommendations, Indonesian seaweed farmer and industry groups mounted advocacy campaigns against the recommendation, which they felt was not based on sufficient evidence and would have significant negative effects on the industry (Dwijayanto 2018). The recommendation for the removal of carrageenan from the organic food additives list was ultimately not adopted. The reasons given were a lack of scientific evidence supporting the claims of negative health impacts, and a lack of suitable alternatives to carrageenan to provide necessary functions in processed foods (USDA 2018). Nonetheless, these safety concerns could have a significant effect on the carrageenan seaweed industry in the future, and some manufacturers have sought to reduce or eliminate the use of carrageenan, particularly for health food products where consumer awareness may be higher.

### Blends

Food and beverage manufacturers increasingly need tailored hydrocolloid products to achieve specific textures. As a result, there is increasing demand for hydrocolloid blends, which combine two or more hydrocolloids to achieve high performing and closely tailored products (CBI 2019). An increasing amount of carrageenan is used in blends for use by food and beverage manufacturers with specific gelling, texturising, stabilising, and thickening behaviours.

### Substitutes

Carrageenan is a relatively high per-unit cost hydrocolloid. However, hydrocolloids are used and blended in different quantities to achieve cost efficiencies. As hydrocolloids are only used in small quantities in processed foods, demand may not be highly price-elastic for many food and beverage applications, especially given the high cost of reformulating products and the risk of consumer acceptance. This is particularly the case for the makers of specialty products that incorporate carrageenan in specific blends to achieve narrowly defined properties in their products. Nonetheless, carrageenan is just one of an increasing range of hydrocolloids that can be incorporated and competes with these products on performance, price, and health and food safety characteristics or perceptions.

## Pet food

A move away from canned pet food towards the increasing use of dry pet food has also affected demand for carrageenan, especially SRC. Depending on the growth in demand in non-food and food applications, there may be a shift in demand for carrageenan of differing product grades, leading to changes in the proportion of RC and SRC being produced.

In conclusion, the global market for carrageenan is therefore expected to achieve continued growth based on solid demand from a range of sectors. While carrageenan can be substituted in some applications, it has specific gelling, thickening, and stabilising behaviours which differ from other hydrocolloids and will probably secure its continued use in a range of applications. Further, it is used in a wide range of products and has a diversified market. Substituting it for other hydrocolloids would require product reformulation which is likely to be expensive. Demand for carrageenan products in Asian, African and South American markets may also not be subject to the same consumer preferences as those in the European and North American markets. These factors suggest that demand for carrageenan is likely to remain diversified and will continue to grow in the future. However, its trajectory will be influenced by several factors, including price fluctuations, the availability of substitutes, geographic variations in demand, evolving consumer preferences, and its incorporation in blended products. Discussion will now turn to the global patterns of carrageenan production, shedding light on its intricate dynamics to contribute to a comprehensive understanding of the commodity's market landscape.

## Carrageenan production

The analysis suggests a buoyant future demand for carrageen seaweeds. However, this demand will be mediated by the carrageenan processing sector.

## Manufacturer numbers and capacity

The three main carrageenan-producing countries in the world are the Philippines, China, and Indonesia. These countries play a central role in meeting the global demand for carrageenan. Processing plants in these countries produce a full range of carrageenan products including alkali treated cottonii (ATC), food grade and non-food grade SRC, as well as gel-pressed and alcohol-precipitated RC. The specific composition and distribution of carrageenan types varies between countries, reflecting the preferences of local markets, production capabilities, and technological advancements unique to each nation.

## China

China has established a strong presence in the carrageenan market in relation to processing companies, throughput, and exports (Zhang et al. 2023). By mid-2021, China had approximately 150 manufacturers engaged in carrageenan processing, a significant increase from the 50–60 in operation in 2007 (IFC 2007). Additionally, approximately 250 Chinese companies are involved in the distribution or wholesale of carrageenan. The majority of Chinese carrageenan manufacturers are relatively small, with only around one-quarter having a registered capital base exceeding US$1.5 million.

Chinese carrageenan manufacturers are predominantly concentrated in southeast and central China, with easy access to ports to facilitate the import of seaweed and the export of carrageenan. Production is focused on SRC and RC for various industries and markets. State-owned enterprises have exited the industry, resulting in an ownership structure dominated by limited liability companies, either privately owned by individuals or mainland or Hong Kong companies.

Other details on the nine largest Chinese carrageenan processors are provided in Table 1.3, mainly gathered from company websites except for publicly listed companies like Green Fresh which discloses annual financial reports. While data on actual production is difficult to quantify, the capacity of production is stated, which ranges from 1,000–23,000 tonnes per year of SRC or RC. It is worth noting that very few Chinese processers specialise exclusively in carrageenan, with the vast majority licensed to produce a range of other gelling products, including agar, alginates, and konjac gum.

## The Philippines

The Philippines has 18 carrageenan processors, including 5 multinational companies and 13 domestic companies (BFAR 2022) with processing capacities that range from 1,500 to 7,200 tonnes (Table 2.4). The majority of the processors focus on the production of SRC, with 6 companies specialising in non-food grade carrageenan and 15 companies producing food-grade carrageenan. Refined carrageenan production is limited to Shemberg, Marcel, and W Hydrocolloids, with Shemberg manufacturing alcohol precipitated RC and Marcel and W Hydrocolloids supplying gel-pressed RC (BFAR 2022). According to the Philippine Seaweed Industry Roadmap (2022–2026), processing plants in the Philippines, which currently operate at an average capacity utilisation of 65 per cent, are expected to increase production and maximise plant capacities, and are to be supported by an adequate supply of raw seaweed materials.

## Indonesia

The majority of seaweed cultivated in Indonesia is exported for processing, with only 35 per cent of carrageenan seaweed being processed domestically (Zhang et al. 2023). The Indonesian government sought to build a processing sector in

*Table 1.3* Major carrageenan manufactures in China

| Company name | Location | Year established | Registered capital | Company type | Main products | Annual production capacity of carrageenan |
|---|---|---|---|---|---|---|
| BLG (Brilliant) | Shanghai | 1996 | 20,000,000 CNY | Ltd. (Private Chinese) | Carrageenan, konjac gum, agar and blend products for various applications | 23,000 tonnes SRC/RC |
| Green Fresh/Green Future | Fujian | 2007 | US$24,490,000 | Public limited company (Invested by Hong Kong company) | Carrageenan, agar, konjac gum, and their compound products | 10,355 tonnes SRC/RC (2022 actual output) |
| Longrun-Newstar | Guangxi | 2014 | 100,000,000 CNY | Joint-stock Ltd. (unlisted, private Chinese) | Konjac powder, konjac gum, and carrageenan | 4,000 tonnes RC 6,000 tonnes SRC |
| Gather Great Ocean | Shandong | 2000 | 52,940,000 CNY | Ltd. (Private Chinese) | Sodium alginate, carrageenan, agar, etc. | 4,000 tonnes SRC/RC |
| Zhenpai | Fujian | 1985 | 30,000,000 CNY | Ltd. (Private Chinese) | Carrageenan and agar | 3,800 tonnes SRC/RC |
| Global Ocean | Fujian | 2010 | 68,880,000 CNY | Joint venture (Mainland China-Hongkong) | Agar and carrageenan | 1,000 tonnes SRC/RC |
| Xieli/Sheli Hydrocolloids | Shandong (Yantai) | 2005 | 40,420,000 CNY | Ltd. (Invested by Hong Kong Sheli Ltd.) | Agar, carrageenan, konjac gum, alginate | 3,000 tonnes SRC/RC |
| Lvli Biotechnology (Green One) | Fujian | 2007 | 36,000,000 CNY | Ltd. (Invested in by Hong Kong company) | Carrageenan | 5,000 tonnes RC 1,500 tonnes SRC |
| Huixiang Haizao | Guangdong | 1991 | 119,344,479 CNY | Ltd. (private Chinese) | Carrageenan, agar | 720 tonnes RC 1,000 tonnes SRC |

Source: Data from the websites of the headquarters and branches of each company, documented by the authors.

*Table 1.4* Major carrageenan manufacturers in the Philippines

| Processor | Products | Production capacity (Tonnes/year) |
|---|---|---|
| Shemberg | SRC/RC | 3,600/2,600 |
| Marcel Food Sciences Inc. | SRC/RC | 5,400/1,800 |
| W Hydrocolloids (PBI) | SRC/RC | 2,400/1,500 |
| Ceamsia Asia, Inc. | SRC | 1,800 |
| Accel Carrageenan Corporation | SRC | 1,500 |
| MCPI Corporation | SRC | 1800 |
| Mioka Biosystems Corporation (Marcel) | SRC | 1,800 |
| TBK Manufacturing Corporation | SRC | 2500 |
| Mega Pollygums Corporation | SRC | 3,600 |
| LM Zamboanga Carrageenan Manufacturing Corporation | ATC/SRC | 600/1800 |
| Froilan Trading Corporation | ATC/SRC | 1,200/1,800 |
| Cebu Carrageenan Corporation | ATC/SRC | 1,600/800 |

Source: Data from BFAR (2022).

Indonesia, through subsidised support for domestic state-owned companies, and more recently, domestic trade restrictions (Langford, Turupadang et al. 2023). However, this has been largely unsuccessful. Indonesia has, however, attracted foreign investment into the sector through six Chinese companies (Zhang et al. 2023). At present, carrageenan processing in Indonesia involves 26 domestic companies and 7 foreign-invested companies as of 2022 (JaSuDa 2022). The foreign-invested companies have larger production capacities (up to 8,000 tonnes/year) compared to local manufacturers (no more than 1,500 tonnes/year). The average capacity utilisation of the companies currently ranges from 50– 60 per cent (Table 1.4). While four domestically owned plants process RC, their production capacity is smaller than foreign investors. Other local manufacturers primarily produce industrial grades of carrageenan for pet food and ATC. In comparison to China and the Philippines, Indonesia has a larger number of companies (ten processors) focused on ATC production. Indonesian carrageenan manufacturing is discussed further in Chapter 2 and a list of Indonesian carrageenan processors is provided in Appendix 4.

As is common in many sectors, carrageenan processors in all three countries tend not to operate at maximum capacity. This can be attributed to factors such as market demand, seasonal variations, production capabilities, business strategies, and investment considerations. In both the Philippines and Indonesia, multinational companies represent a substantial portion of their carrageenan processing industry. The foreign investment in the Philippines primarily originates from the United States and Europe. Within Indonesia, investment is predominantly China based. Carrageenan production in China is mainly owned by Chinese investors from mainland China and Hong Kong. In addition, the main focus of the carrageenan processing industry in the Philippines and Indonesia is on the production of ATC and SRC. However, there is a notable difference between the two countries in terms of their product specialisation with the Philippines having a larger number of companies engaged in the processing of SRC compared to Indonesia. In contrast, China has a higher concentration of companies involved in the processing of RC compared to the Philippines and Indonesia.

## Competitive dynamics

While there is considerable differentiation and segmentation, companies in China, Indonesia, and the Philippines compete for global market share on the basis of product quality, production capabilities, cost efficiency, and market access. There is considerable pressure to develop technologies, achieve efficiency, and meet quality and environmental standards. Global competitive trends are now discussed.

### *Rising prominence of Chinese processers*

The carrageenan processing sector has historically been dominated by multinational corporations from western countries, particularly the United States and Europe, with significant direct foreign investments in the Philippines (Richards-Rajadurai 1990; Blanchetti-Revelli 1997; Neish et al. 2017; Palanca-Tan 2018). Processors from China have emerged from a modest starting point and are now playing an increasingly influential role in the sector as processors, traders, market participants, and sources of outward investment (Zhang et al. 2023; Bixler and Porse 2011; Campbell and Hotchkiss 2017; Hurtado et al. 2019).

The increasing prominence of Chinese processors in the downstream sectors of the global carrageenan industry has significant implications. Firstly, China's emergence as the world's largest carrageenan processor, coupled with the resulting structural changes in the industry and patterns of outward investment, has the potential to significantly impact the demand for raw seaweed and the prices received by small coastal seaweed farmers (Blanchetti-Revelli 1997; Porse and Rudolph 2017). Secondly, Chinese outward investment presents both opportunities and challenges for policy makers in countries seeking industry development, employment generation, and tax revenue, while also facing pressure from domestic industry interests. Thirdly, the rise of Chinese carrageenan processors has attracted the attention of large corporate interests in western countries, who view them as either competitors or potential partners. Finally, China's increasing influence in global trade governance, food safety, and environmental protection has significant and wide-ranging implications for the entire carrageenan industry (Chan et al. 2011; Liu et al.2019; Langford, Waldron et al. 2023; Coenen et al. 2021; Dong and Li 2021; Waldron et al. 2023).

### *Trend towards processing in countries of origin*

The processing of seaweed extracts in countries of origin has increased in recent years, which has the potential to alter the structure and dynamics of the supply chain. In particular, Indonesia and the Philippines have traditionally been major seaweed producers and exporters of raw dried seaweed. However, these countries are building capacity in the carrageenan processing sector in a bid to capture more value domestically. If this can be achieved at scale, it may impose limits on the supply of raw dried seaweed for countries that previously sourced raw materials from Indonesia and the Philippines. Chinese investment in Indonesia can be seen as a means of securing supply. At the same time, Chinese carrageenan processors are under pressure from increasing labour costs and environmental standards, which act as drivers for outward-bound investment in Indonesia and other countries.

This movement within the supply chain requires major investment and upgrade capacity. This includes significant investment in technology and research and development from public investment and technology transfer from foreign-invested companies. It would require adherence to standards (e.g. International Organization 22000:2018 and 9001 certification, Food Safety Systems Certification FSSC, 22000) and the utilisation of mechanisms to facilitate access to markets such as the European Union (EU) (CBI 2019).

### *Shift towards RC*

Another form of chain advancement is a shift towards the production of higher-value RC, a stated goal of all three major carrageenan-producing countries. This shift is probably driven by a range of factors, including the rising demand for RC in food and beverage, pharmaceuticals, and personal-care products due to its higher purity, enhanced functional properties, and a broader range of applications. This contrasts with stagnant demand for SRC from the processed meat and dairy primary sectors (Figure 1.4), where visual clarity is not a crucial factor (Hotchkiss et al. 2016). Recognising these market dynamics and the changing demand landscape, major carrageenan processors are investing to upgrade their processing capabilities and infrastructure to enable the production of RC. Of course, this poses entry barriers and additional costs. If the additional revenues from the production of higher-priced RC outweigh the costs, companies may have an incentive to remain in the lower-value SRC market.

The shift towards RC signifies a strategic move by major processors to increase the value-added and competitiveness of their carrageenan industries. They all utilise alkaline mixtures in the processing of SRC. However, the production of RC through alcohol precipitation, is predominantly undertaken by Chinese processors and Shemberg Biotech in the Philippines. Compared to Indonesia and the Philippines, China has taken the lead in this specific process.

### *Effluent management*

Carrageenan extraction requires the input of alkali, acids, salts, water, and energy for heating and subsequent purification of carrageenan from seaweed biomass (Olatunji 2020). The process generates significant amounts of toxic waste, including wastewater, exhaust gas, solid waste, and noise. Zhang et al. (2023) highlights that wastewater is primarily produced during the washing, dehydration, and heating stages. Exhaust gas is emitted during the crushing and grinding of raw materials, as well as during the alkali and wastewater treatment, leading to odorous emissions. Solid waste consists of sediment washed from seaweed, filter residue from the filter press, recycled dust, raw material packaging (barrels and bags), and wastewater treatment sludge, which can potentially be repurposed as agricultural crop substrates or soil amendments. Moreover, the mechanical operation of production equipment, such as pulverisers, colloid mills, and centrifuges, generates noise levels measured at approximately 80 dB when assessed at a one-metre distance from noise level devices.

The management of waste disposal poses persistent challenges and represents a significant cost for carrageenan processors if they comply with environmental regulations. Compliance may include the construction of wastewater treatment facilities, the installation of dust collectors, the implementation of anti-seepage and hardening measures in factory areas and workshop grounds, and the establishment of general and hazardous waste temporary storage facilities.

Competition in the carrageenan production sector will force manufacturers to increase competitiveness by improving product quality, product diversification, and exploring new applications for carrageenan. This will encourage investment in research and development, infrastructure, and technology, leading to advancements within the industry, processing facilities, and improved industry capability. Moreover, companies could strive to capture larger market share by exploring new markets and expanding distribution networks, making carrageenan products more accessible globally. These could benefit stakeholders throughout the supply chain and the global carrageenan market by driving efficiency, fostering collaboration, and ensuring a wide range of carrageenan products to meet consumer demands. Changing environmental regulations may affect the profitability of carrageenan processing differently in different countries (Zhang et al. 2023).

The sector will also face opportunities and challenges arising from changing regulations, sustainability concerns, and evolving customer expectations. There are opportunities to demonstrate compliance with regulatory standards, which can enhance reputation and secure a competitive edge by meeting regulatory requirements. Furthermore, changing customer expectations require product reformulation, innovation, and responsiveness to emerging trends. However, manufacturers must strategically navigate these opportunities and challenges to balance economic viability and ensure long-term success.

**Trade**

Global production of carrageenan seaweed is mainly concentrated in Indonesia and the Philippines which is reflected in their exports. Available data from the UN Comtrade database reveals that the international trade volumes, values, and prices of carrageenan seaweed and carrageenan have fluctuated substantially over the last three decades. This is due to factors including market dynamics, environmental conditions, and trade policies. Understanding these factors aids in foreseeing and managing future trends in the industry.

### Carrageenan seaweeds

Over the course of three decades, both the volume and value of carrageenan sea-weed exports from Indonesia have undergone substantial increases (Figure 1.6), rising from around 12,085 tonnes and US$5.9 million in 1989 to 187,662 tonnes and US$219 million in 2021. This represents a compound annual growth rate of around 44 per cent in volume and 110 per cent in value over the past 33 years. The impact of the COVID-19 pandemic on exports was short-lived (declining 7.3 per cent in 2020) and was followed by a rebound the following year (see Langford, Waldron et al. 2021; Langford et al. 2022).

The volume and value of carrageenan seaweed exports from the Philippines are much smaller than Indonesian exports. The export of seaweed from the Philippines has increased slowly for more than two decades, with considerable fluctuations (Figure 1.7). According to the available data, the first peak in seaweed exports was recorded in 2000, with a total volume of 49,080 tonnes and a value of US$46.5 mil-lion. Export volumes then decreased, hitting a low of 10,823 tonnes in 2009, before experiencing a resurgence in the early 2010s. This was then followed by another decline to approximately 3,100 tonnes in 2017. Export volumes remained below 15,000 tonnes from 2017 through to the end of 2022.

According to UN Comtrade statistics, the average trade price of seaweed prod-ucts has been highly unstable in both countries, especially in recent years (Figures 1.6 and 1.7). The pricing of seaweed products varies depending on a range of fac-tors aggregated in the statistics (product form, quality, natural disasters, climate variability, and changes in market conditions). Notably, the international price of

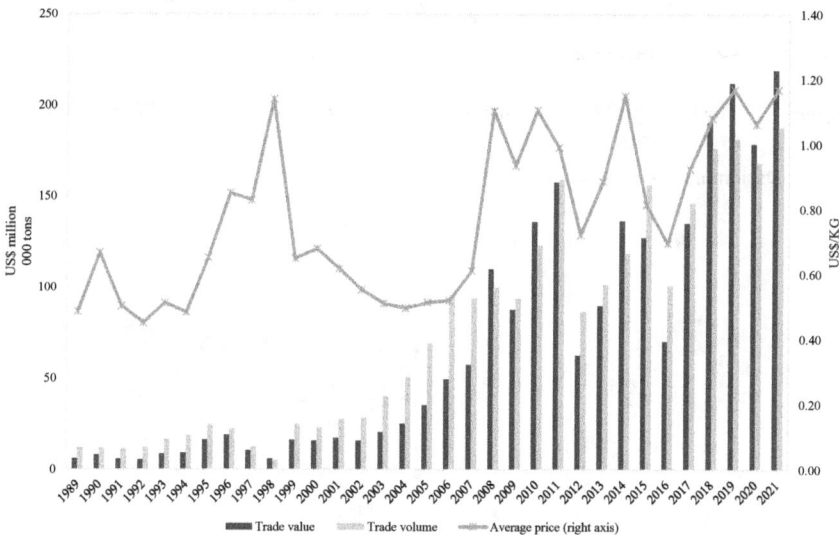

*Figure 1.6* Indonesia carrageenan seaweed export to the world by value, volume, and aver-age price

Source: Data from UN Comtrade (2023).

raw dried carrageenan seaweed from the Philippines has consistently been higher than that from Indonesia, fluctuating between US$0.77 and US$2.33 per kilogram from 2010 to 2022, compared to Indonesia's range of US$0.7–1.17/kg.

The higher prices for seaweed products from the Philippines compared to Indonesia could be due to a different mix of species produced (as spinosum is significantly lower priced than cottonii), lower moisture and dirt content, or different characteristics of the seaweed due to different growing conditions. The destinations for carrageenan seaweed exports vary as shown in Figure 1.8. Indonesia is the largest exporter of carrageenan seaweed, with the overwhelming majority of exports going to China (84 per cent), followed by Vietnam (5 per cent), the Republic

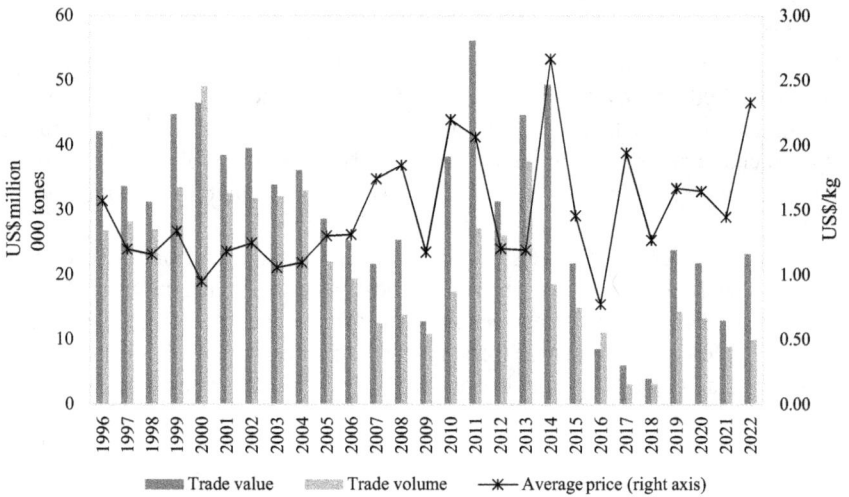

*Figure 1.7*  The Philippines carrageenan seaweed export to the world by value, quantity, and average price

Source: Data from UN Comtrade (2023).

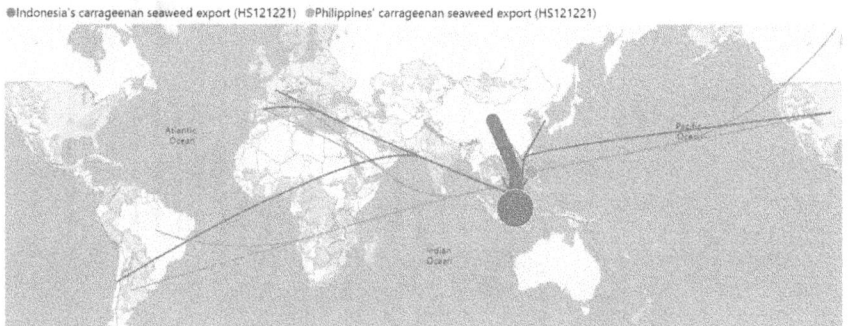

*Figure 1.8*  The major global trade networks of carrageenan seaweed in 2021

Source: Data from UN Comtrade (2023) and China Customs (2023).

of Korea (3 per cent), the United States (2 per cent), and Chile (2 per cent). The Philippines is the second largest exporter of carrageenan seaweed, exports going to the United States (26 per cent) and China (24 per cent). Other significant export destinations include France (17 per cent), Argentina (16 per cent), and Brazil (6 per cent). Overall, China is the largest importer of global carrageenan seaweed. China's demand for carrageenan seaweed is driven by its use in carrageenan production.

## Carrageenan

Despite the absence of precise data regarding the production volume of carrageenan in individual countries, previous industry estimates suggest that the global production capacity of carrageenan ranges between approximately 80,000 million tonnes (MT)/year (Porse and Ladenburg 2015) and 110,000 MT/year (Neish 2015). A more recent estimate from Grand View Research (2023b) suggests a value of US$ 924.7 million for the carrageenan industry with a projected compound annual growth rate of 5.4 per cent from 2023 to 2030. Asia has become a significant producer of carrageenan, with China, Indonesia, and the Philippines ranking among the largest producers. Figure 1.9 provides an illustration of the trade networks of carrageenan among key industry players and their corresponding export destinations in 2021. This finding is supported by Grand View Research (2023b), which highlights Europe as the world's largest consumer market, accounting for 33 per cent of global consumption volume, followed by the Asia Pacific region (30 per cent), and North America (23 per cent). China exported 20,849 tonnes of carrageenan in 2021, with a trade value of US$186 million, shipped to 82 countries and regions. However, 69 countries imported less than 2 per cent of Chinese exports (< 400 tonnes). The majority of China's carrageenan exports (44 per cent) went to the EU, followed by other Asian countries (25 per cent), and Russia and the Ukraine (15 per cent), with very little imported into the United States. Despite the downtrend in seaweed exports, the Philippines remains the leading exporter of carrageenan to the United States (31 per cent of exports) and the EU-27 (27 per cent), with a larger share

*Figure 1.9* The major global trade networks of carrageenan in 2021
Source: Data from UN Comtrade (2023) and China Customs (2023).

than China in these markets. Indonesia's carrageenan exports were mainly within the Asian market, with more than half going to China. Initiatives to build the domestic carrageenan processing sector have seen a tenfold increase in Indonesia's carrageenan export value and a ninefold increase in export volume during the past decade. Values increased from US$9 million in 2010 to 87 million in 2020. Volumes increased from 1,382 tonnes in 2010 to 13,973 tonnes in 2020.

China relies heavily on imported raw materials for carrageenan processing. While most raw dried seaweed was imported from the Philippines before 2000, Indonesia now completely dominates exports to China (Figure 1.10). Other sources such as Chile, the Republic of Korea, and Japan are supplementary sources.

The United States and the EU have significant and longstanding carrageenan processing industries. As shown in Figure 1.11, the market share of the United States, France, and Chile has steadily declined, while several EU countries (Spain, Germany, and the United Kingdom) have maintained a stable market share. One of the challenges facing the US and EU carrageenan industry is competition from low-cost producers, particularly in Asia. This has led to consolidation within the industry, as smaller producers have been acquired by larger companies with the resources to compete on the global market (Bixler 1996, p. 37).

Changes in average export carrageenan price levels in China, the Philippines, and Indonesia are shown in Figure 1.12 and reveal two key dynamics. First, the price of carrageenan in the international market was generally stable in the first half of the 2000s but has increased rapidly since the mid-2000s. Second, carrageenan export prices in China have been consistently higher than Indonesia and the Philippines, reflecting a different mix of high- and low-value carrageenan products (e.g. SRC and RC) and contradictory perceptions that China is a low-cost manufacturing centre.

The price of carrageenan reflects supply considerations as well as demand from carrageenan users. The own-price elasticity of demand for carrageenan (the sensitivity of carrageenan demand to a change in its price) depends on how likely

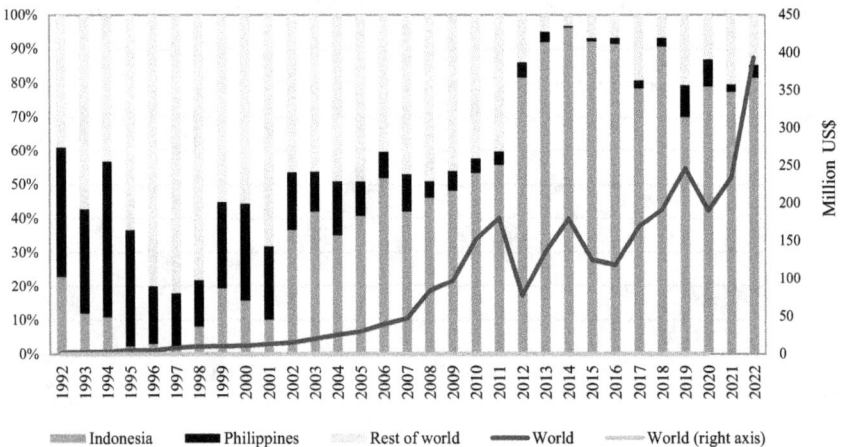

*Figure 1.10* Carrageenan seaweed import by China in trade value

Source: Data from UN Comtrade (2023).

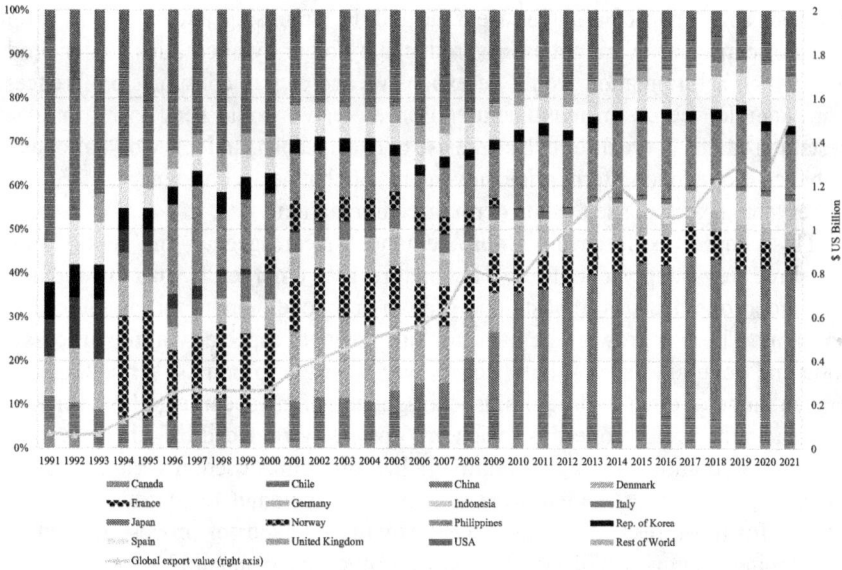

*Figure 1.11* Export of carrageenan from various countries by trade value

Source: Data from UN Comtrade (2023).

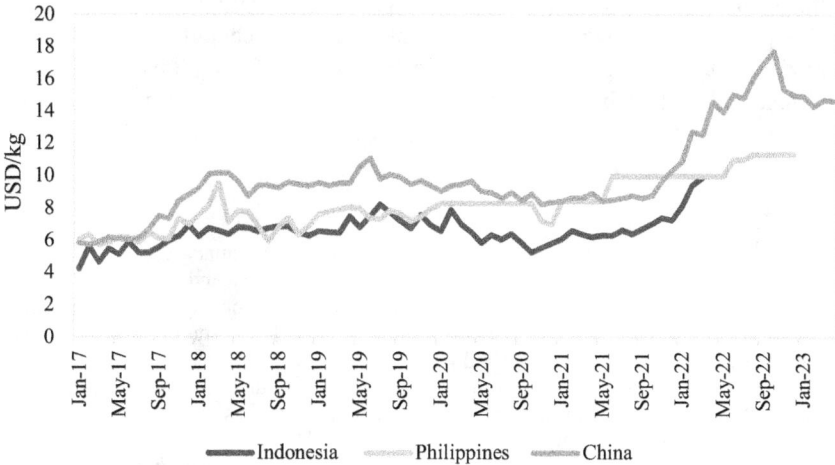

*Figure 1.12* Average monthly unit values of carrageenan exports from selected countries, 2017–23

Source: Conclusion Data from UN Comtrade (2023) and China Customs (2023).

manufacturers of foods and other products are to change their use of carrageenan if prices rise or fall. There is little information to quantify the impact of price fluctuations. However, as carrageenan typically accounts for only a small portion of the cost of an end product, a change in its price may cause a relatively small change in the total cost. Therefore, price changes are likely to have a limited impact on demand. This is particularly the case if the gelling, stabilising, and thickening properties of carrageenan are not easily produced by other hydrocolloids, or for products which reformulation would be expensive and/or risk consumer preferences. For example, ice cream manufacturer Ben & Jerrys would need to reformulate certain products in order to replace the use of carrageenan, and this would probably only occur due to long-term price movements or changes in consumer preferences, rather than as a response to short-term price fluctuations.

This chapter explored the global demand for carrageenan and supply from the three key supply countries. The picture that emerges is that the industry has undergone decades of sustained growth and development that is forecast to continue. There is growing demand for carrageenan, especially in the processed food and beverages sector, but also in a various other products. While there are a wide range of other hydrocolloids, carrageenan offers specific gelling, thickening, and stabilising properties which are not easily replaced by other products. This factor, combined with the relatively low quantities used in various applications, suggests that the own-price sensitivity of demand for carrageenan may be low for the foreseeable future. The growing demand for processed foods in developing countries will also bolster and diversify demand. The shift of processing capacity to Asia would also appear strategically beneficial to Indonesian seaweed growers. These implications suggest that investment and plans to expand exposure to the carrageenan sector by governments, companies, and households within Indonesia appear sound and well aligned with the industry's trends and opportunities. However, it is important to acknowledge that along with the opportunities, challenges regarding industry upgrade, foreign investment, competition, and sustainability will necessitate ongoing adaptation and innovation for long-term success.

## Note

1   These reports draw on data from several sources, including primary research undertaken with industry experts, consultants, manufacturers, distributers and resellers undertaken largely through virtual interviews and online surveys, secondary research including analysis of statistical databases, investor documents and case studies, purchased databases such as Pitchbook, Statista and *Wall Street Journal* which provide information on company financials, industry information and industry journal publications, and third-party perspectives gained through analysis of investor analyst reports, broker reports, government quotes, research institutes and academic centres (Grand View Research 2023a, 2023b). The reports provide a general overview of market dynamics which is used in this chapter to provide insights into the organisation of the industry.

## References

Alves, Vítor D., Nuno Costa, and Isabel M. Coelhoso. 2010. "Barrier Properties of Biodegradable Composite Films Based on Kappa-Carrageenan/pectin Blends and Mica Flakes." *Carbohydrate Polymers* 79, no. 2: 269–276. https://doi.org/10.1016/j.carbpol.2009.08.002

Bixler, Harris J. 1996. "Recent Developments in Manufacturing and Marketing Carrageenan." In *Fifteenth International Seaweed Symposium: Proceedings of the Fifteenth*

*International Seaweed Symposium* held in Valdivia, Chile, in January 1995. 35–57. Dordrecht: Springer.

Bixler, Harris J. 2017. "The Carrageenan Controversy." *Journal of Applied Phycology* 29, no. 5: 2201–2207. https://doi.org/10.1007/s10811-017-1132-4

Bixler, Harris J. and Hans Porse. 2011. "A Decade of Change in the Seaweed Hydrocolloids Industry." *Journal of Applied Phycology* 23, no. 3: 321–335. https://doi.org/10.1007/s10811-010-9529-3

BFAR (Bureau of Fisheries and Aquatic Resources). 2022. "Philippine Seaweed Industry Roadmap 2022–2026." Accessed 31 Jan 2023. www.pcaf.da.gov.ph/wp-content/uploads/2022/06/Philippine-Seaweed-Industry-Roadmap-2022-2026.pdf

Blanchetti-Revelli, Lanfranco. 1997. "Keeping Meat and Dairy Consumers Slim: Philippine Seaweed, American Carrageenan and the USFDA." *Anthropology Today* 13, no. 5: 6–13. https://doi.org/10.2307/2783558

Brenner, Tom, Rando Tuvikene, Yapeng Fang, Shingo Matsukawa, and Katsuyoshi Nishinari. 2015. "Rheology of Highly Elastic Iota-Carrageenan/kappa-Carrageenan/xanthan/konjac Glucomannan Gels." *Food Hydrocolloids* 44: 136–144. https://doi.org/10.1016/j.foodhyd.2014.09.016

Campbell, Ross and Sarah Hotchkiss (2017). "Carrageenan Industry Market Overview." In *Tropical Seaweed Farming Trends, Problems and Opportunities: Focus on Kappaphycus and Eucheuma of Commerce* edited by Anicia Hurtado, Alan Critchley and Iain Neish. 193–205. Cham, Switzerland: Springer Nature. https://doi.org/10.1007/978-3-319-63498-2_13

CBI (Confederation of British Industry). 2019. "Value Chain Analysis Indonesia Seaweed Extracts." Commissioned by the Centre for the Promotion of Imports from Developing Countries. Accessed 6 June 2023. https://www.cbi.eu/sites/default/files/2019_vca_indonesia_seaweed_extracts.pdf

Chan, Gerald, Pak K. Lee, and Lai-Ha Chan. 2011. *China Engages Global Governance: A New World Order in the Making?* Vol. 21. New York: Routledge.

China Customs. 2023. "Customs Statistics Online Query Platform." Accessed 6 June 2023. http://stats.customs.gov.cn

Coenen, Johanna, Simon Bager, Patrick Meyfroidt, Jens Newig, and Edward Challies. 2021. "Environmental Governance of China's Belt and Road Initiative." *Environmental Policy and Governance* 31, no. 1: 3–17. https://doi.org/10.1002/eet.1901

Cohen, Samuel M. and Nobuyuki Ito. 2002. "A Critical Review of the Toxicological Effects of Carrageenan and Processed Eucheuma Seaweed on the Gastrointestinal Tract." *Critical Reviews in Toxicology* 32, no. 5: 413–444. https://doi.org/10.1080/20024091064282

David, Shlomit, Carmit Shani Levi, Lulu Fahoum, Yael Ungar, Esther G. Meyron-Holtz, Avi Shpigelman, and Uri Lesmes. 2018. "Revisiting the Carrageenan Controversy: Do We Really Understand the Digestive Fate and Safety of Carrageenan in our Foods?" *Food & Function* 9, no. 3: 1344–1352. https://doi.org/10.1039/C7FO01721A

Dong, Yan and Chunding Li. 2021. "China and the Reform of International Trade Governance System." *Social Sciences in China* 42, no. 3: 140–164. https://doi.org/10.1080/02529203.2021.1971407

Dwijayanto A. 2018. Delisting dicabut, ekspor rumput laut RI ke Amerika berlanjut. https://industri.kontan.co.id/news/delisting-dicabut-ekspor-rumput-laut-ri-ke-amerika-berlanjut

European Parliament and Council. 1995. "Directive No. 95/2/EC and 96/85EC." Office for Official Publications of the European Communities. https://eur-lex.europa.eu/legal-content/EN/ALL/?uri=CELEX:31996L0085

Grand View Research. 2023a. "Hydrocolloids: Market Estimates and Trend Analysis." Purchased from *Grand View Research*.

Grand View Research. 2023b. "Carrageenan: Market Estimates and Trend Analysis." Purchased from *Grand View Research.*

Hotchkiss, Sarah, Mariel Brooks, Ross Campbell, Kevin Philp, and Angie Trius. 2016. "The Use of Carrageenan in Food." In *Carrageenans: Sources and Extraction Methods, Molecular Structure, Bioactive Properties and Health Effects*, edited by Leonel Pereira. 229–243. New York: Nova Science Publishers. https://www.researchgate.net/publication/318837464_The_use_of_carrageenan_in_food

Huc, D., A. Matignon, P. Barey, M. Desprairies, S. Mauduit, J. M. Sieffermann, and C. Michon. 2014. "Interactions Between Modified Starch and Carrageenan During Pasting." *Food Hydrocolloids* 36: 355–61. https://doi.org/10.1016/j.foodhyd.2013.08.023

Hurtado, Anicia Q., Iain C. Neish, and Alan T. Critchley. 2019. "Phyconomy: The Extensive Cultivation of Seaweeds, Their Sustainability and Economic Value, with Particular Reference to Important Lessons to be Learned and Transferred from the Practice of Eucheumatoid Farming." *Phycologia* 58, no. 5: 472–483. https://doi.org/10.1080/00318884.2019.1625632

ICF (Inner City Fund). 2011. "Technical Evaluation Report for Carrageenan: Handling/Processing." Washington, DC: USDA-AMS-NOP: 1–18. Accessed 6 June 2023. https://www.ams.usda.gov/sites/default/files/media/Carrageenan%20TR%202011.pdf

IFC (International Finance Corporation). 2007. "Chinese Market for Seaweed and Carrageenan Industry." Accessed 10 May 2023. http://marineagronomy.org/sites/default/files/IFC%20-%20Final%20report%20-%20Chinese%20Market%20for%20Seaweed%20and%20Carrageenan%2028-04-07.pdf

JaSuDa. 2022. JaSuDa.net: PT. Jaringan Sumber Daya. Accessed 10 May 2023. https://jasuda.net

JECFA (Joint FAO/WHO Expert Committee on Food Additives). 2015. "Safety Evaluation of Certain Food Additives." *Seventy-ninth Report of the Joint FAO/WHO Expert Committee on Food Additives.* Vol. 79. Geneva: World Health Organization.

Komarek, A., E. R. Cahyadi, J. Zhang, A. Fariyanti, B. Julianto, R. Arsyi, I. Lapong, A. Langford, S. Waldron, and M. Grist. 2023. *Increasing Incomes in Carrageenan Seaweed Value Chains in Takalar, South Sulawesi.* Melbourne, Australia: Australia-Indonesia Centre.

Langford, A., W. Turupadang, and S. Waldron. 2023. "Interventionist Industry Policy to Support Local Value Adding: A Case from the Eastern Indonesia Seaweed Industry." *Marine Policy* 151(105561): 1–11.

Langford, A., S. Waldron, N. Nuryartono, S. Pasaribu, B. Julianto, I. Siradjuddin, R. Ruhon, Z. Zach, I. Lapong, and R. Armis. 2023. *Sustainable Upgrading of the South Sulawesi Seaweed Industry.* Melbourne, Australia: Australia-Indonesia Centre.

Langford, A, S. Waldron, Sulfahri, and H. Saleh. 2021. "Monitoring the COVID-19-Affected Indonesian Seaweed Industry Using Remote Sensing Data." *Marine Policy* 127(104431): 1–10. doi.org/10.1016/j.marpol.2021.104431

Langford, A., J. Zhang, S. Waldron, B. Julianto, I. Siradjuddin, I. Neish, and N. Nuryartono. 2022. "Price Analysis of the Indonesian Carrageenan Seaweed Industry." *Aquaculture* 550(737828). doi.org/10.1016/j.aquaculture.2021.737828

Liu, Zhe, Anthony N. Mutukumira, and Hongjun Chen. 2019. "Food Safety Governance in China: From Supervision to Coregulation." *Food Science & Nutrition* 7, no. 12: 4127–4139. https://doi.org/10.1002/fsn3.1281

Marinalg International. 2023. "Brief History of Regulatory & Scientific Determinations of Carrageenan." Accessed 6 June 2023. http://marinalg.org/news-resources/resources/brief-history-of-regulatory-scientific-determinations-of-carrageenan

Martino, John Vincent, Johan Van Limbergen, and Leah E. Cahill. 2017. "The Role of Carrageenan and Carboxymethylcellulose in the Development of Intestinal Inflammation." *Frontiers in Pediatrics* 5: 96. https://doi.org/10.3389/fped.2017.00096

Matignon, A., P. Barey, M. Desprairies, S. Mauduit, J. M. Sieffermann, and C. Michon. 2014. "Starch/carrageenan Mixed Systems: Penetration in, Adsorption on or Exclusion of Carrageenan Chains by Granules?" *Food Hydrocolloids* 35: 597–605. https://doi.org/10.1016/j.foodhyd.2013.07.028

Matignon, A., G. Moulin, P. Barey, M. Desprairies, S. Mauduit, J. M. Sieffermann, and C. Michon. 2014. "Starch/Carrageenan/Milk Proteins Interactions Studied Using Multiple Staining and Confocal Laser Scanning Microscopy." *Carbohydrate Polymers* 99: 345–355. https://doi.org/10.1016/j.carbpol.2013.09.002

Neish, Iain C. 2015. "A Diagnostic Analysis of Seaweed Value Chains in Sumenep Regency, Madura Indonesia." *UNIDO Project 140140*: 1–66. Accessed 6 June 2023. https://downloads.unido.org/ot/33/70/3370517/A%20diagnostic%20analysis%20of%20seaweed%20value%20chains%20in%20Madura, %20Indonesia.pdf

Neish, Iain C., Miguel Sepulveda, Anicia Q. Hurtado, and Alan T. Critchley. 2017. "Reflections on the Commercial Development of Eucheumatoid Seaweed Farming." In *Tropical Seaweed Farming Trends, Problems and Opportunities: Focus on Kappaphycus and Eucheuma of Commerce*, edited by Anicia Hurtado, Alan Critchley and Iain Neish. 1–27. https://doi.org/10.1007/978-3-319-63498-2_1

Olatunji, O. 2020. "Carrageenans." In *Aquatic Biopolymers: Understanding Their Industrial Significance and Environmental Implications* edited by Ololade OLatunji. 121–144. Cham, Switzerland: Springer. https://doi.org/10.1007/978-3-030-34709-3_6

Palanca-Tan, Rosalina. 2018. "Aquaculture, Poverty and Environment in the Philippines." *The Journal of Social, Political, and Economic Studies* 43, no. 3/4: 294–315. Accessed 10 May 2023. https://www.proquest.com/docview/2119862767?pq-origsite=gscholar&fromopenview=true

Porse, Hans and S. Ladenburg. 2015. "Seaweed Value Chain Final Report." Report submitted to *Smart-fish Indonesia under the UNIDO Seaweed Value Chain Programme*. Jakarta, Indonesia.

Porse, Hans and Brian Rudolph. 2017. "The Seaweed Hydrocolloid Industry: 2016 Updates, Requirements, and Outlook." *Journal of Applied Phycology* 29, no. 5: 2187–2200. https://doi.org/10.1007/s10811-017-1144-0

Rhim, Jong-Whan. 2013. "Effect of PLA Lamination on Performance Characteristics of Agar/κ-Carrageenan/clay Bio-Nanocomposite Film." *Food Research International* 51, no. 2: 714–722. https://doi.org/10.1016/j.foodres.2013.01.050

Rhim, Jong-Whan and Long-Feng Wang. 2013. "Mechanical and Water Barrier Properties of Agar/κ-Carrageenan/konjac Glucomannan Ternary Blend Biohydrogel Films." *Carbohydrate Polymers* 96, no. 1: 71–81. https://doi.org/10.1016/j.carbpol.2013.03.083

Richards-Rajadurai, Nirmala. 1990. "Production, Marketing and Trade of Seaweeds." In *Technical Resource Papers Regional Workshop on the Culture and Utilization of Seaweeds, Volume II*. Rome: Food and Agriculture Organization of the United Nations. Accessed 28 April 2023. https://www.fao.org/3/ab728e/AB728E12.htm

UN (United Nations). 2023. "UN Comtrade Database." Comtrade. Accessed 6 June 2023. https://comtradeplus.un.org

USDA (United States Department of Agriculture). 2018. "National Organic Program: USDA Organic Regulations." *Agricultural Marketing Service*. Accessed 6 June 2023. https://www.federalregister.gov/documents/2018/04/04/2018-06867/national-organic-program-usda-organic-regulations

Waldron, Scott, Nunung Nuryartono, Alexandra Langford, Syamsul Pasaribu, Kustiariyah, Ulfah J. Siregar, Muhammad Farid Dimjati Lusno, Sulfahri, Boedi Sarjana Julianto, Irsyadi Siradjuddin, Radhiyah Ruhon, Zulung Zach Walyandra, Muhammad Imran Lapong, Risya Arsyi Armis, Eugene Sebastian, Helen Brown, Fadhilah Trya Wulandari, Hasnawati Saleh, and Steve Wright. 2022. *Policy Brief: Sustainable Upgrading of the South Sulawesi Seaweed Industry*. Melbourne, Australia: Australia-Indonesia Centre. https://pair. australiaindonesiacentre.org/wp-content/uploads/2022/11/SIP-1-EN-ONLINE.pdf

Wang, Lu, Yiping Cao, Ke Zhang, Yapeng Fang, Katsuyoshi Nishinari, and Glyn O. Phillips. 2015. "Hydrogen Bonding Enhances the Electrostatic Complex Coacervation Between κ-Carrageenan and Gelatin". *Colloids and Surfaces. A, Physicochemical and Engineering Aspects* 482: 604–610. https://doi.org/10.1016/j.colsurfa.2015.07.011

Weenen, H., R. H. Jellema, and R. A. de Wijk. 2005. "Sensory Sub-Attributes of Creamy Mouthfeel in Commercial Mayonnaises, Custard Desserts and Sauces." *Food Quality and Preference* 16, no. 2: 163–170. https://doi.org/10.1016/j.foodqual.2004.04.008

Weiner, Myra L. 2014. "Food Additive Carrageenan: Part II: A Critical Review of Carrageenan in Vivo Safety Studies." *Critical Reviews in Toxicology* 44, no. 3: 244–269. https://doi.org/10.3109/10408444.2013.861798

Zhang, Jing, Scott Waldron, Zannie Langford, Boedi Julianto, and Adam Martin Komarek. 2023. "China's Growing Influence in the Global Carrageenan Industry and Implications for Indonesia." *Journal of Applied Phycology* June: 1–22. https://doi.org/10.1007/ s10811-023-03004-0

Zia, Khalid Mahmood, Shazia Tabasum, Muhammad Nasif, Neelam Sultan, Nosheen Aslam, Aqdas Noreen, and Mohammad Zuber. 2017. "A Review on Synthesis, Properties and Applications of Natural Polymer Based Carrageenan Blends and Composites." *International Journal of Biological Macromolecules* 96: 282–301. https://doi. org/10.1016/j.ijbiomac.2016.11.095

# 2 The Indonesian seaweed industry

*Scott Waldron, Zannie Langford,*
*Syamsul Pasaribu, Nunung Nuryartono,*
*Boedi Julianto, and Irsyadi Siradjuddin*

Given the global context and trade patterns outlined in Chapter 1, this chapter telescopes down to developments in the Indonesian seaweed industry at the national level. It provides industry-wide context, examines the sectors of production, marketing, processing, and end-products, and describes the cross-cutting issues of zoning, investment, and food safety.

The chapter does not provide an exhaustive description of the Indonesian seaweed industry. Many aspects of the industry are already well-documented, including through value chain analyses at various levels,[1] policy or institutional analysis (see Permani et al., 2023), and analysis of the individuals involved (see Neish and Suryanarayan, 2017). These dimensions of the industry are, however, collated in the chapter, with further detail provided in Appendices 2, 3, and 4.

## The industry in context

After gathering momentum globally, the carrageenan seaweed sector found fertile ground for expansion in Indonesia in the late 1970s. The country's aquatic resources, benign weather, and agrarian structures enabled the uptake of new rural activity in parts of the country. In a few short decades, seaweed production burgeoned to over 62,000 households across much of the archipelago (BPS, 2022). From this production base, other sectors of the domestic industry, including seaweed marketing, processing, incorporation into Indonesian foods and exports, evolved.

### Governance structures

Principles of global value chain (GVC) analysis (e.g. Gereffi, 2018) have been applied to the seaweed industry with a focus on the transitions of governance systems (Neish and Suryanarayan, 2017). The principles provide a useful framework for the analysis of the contemporary structures in the Indonesian industry. In the 1980s, the Philippine industry was subject to captive or hierarchical governance systems. A small number of small-to-medium enterprises (SMEs) and later multinational corporations (MNCs) dealt directly with seaweed farmers. These actors aimed to capture returns from expanding cultivation and research and development

DOI: 10.4324/9781003183860-4

activities. Neish and Suryanarayan (2017) argued that these companies explored and helped to establish the Indonesian seaweed production sector. In the 1990s, the development of simple processing technologies (for semi-refined carrageenan (SRC)) led to the proliferation of seaweed processing companies, some of which were founded by previous employees and traders associated with the SMEs and MNCs. Rather than dealing directly with farmers, the companies dealt indirectly through what they called integrated or allied suppliers in modular governance systems. Standards for seaweed farming were set by an international hydrocolloids organisation (MARINALG) and were enforced by the enterprises. By 2000, the Indonesian industry was dominated by market governance systems, where processors had become even more disconnected from farmers, linked by largely autonomous middlemen who conducted spot transactions. Seaweed and seaweed product standards proliferated but were applied unevenly or not at all, and farmers fended for themselves with little company or government support. As elaborated below, there were few cases of relational governance where contracts were used to link seaweed producers and downstream actors.

Neish and Suryanarayan (2017) suggested that industry development in Indonesia was led by economic agents, including farmers, companies, and scientists linked to companies and key individuals.[2] Chapter 3 documents the involvement of Universitas Hasanuddin in South Sulawesi at this early stage. From this base, other actors have entered to formalise the industry development process including industry associations, development agencies, and, of most interest to this chapter, the government. Government interest can be explained by several key factors.

### *Government interest in seaweed*

The seaweed industry has become increasingly economically significant. Like all countries, the share of agriculture, forestry, and fisheries in total GDP has declined and in 2021 stood at 13.28 per cent. Fisheries contributed 22 per cent or Rp. 505,061 billion to GDP at 2021 nominal prices (Bank Indonesia, 2023) and seaweed is a significant contributor to Indonesia's total aquaculture production (BAPPENA, 2021, p. 65).

The industry also provides livelihood activity for a significant number of rural and coastal Indonesian households. Estimates vary largely (see Appendix 1), but suggest that around 62,000 households farm seaweed (BPS, 2022a), with more involved in the seaweed industry in other ways – for example, as casual wage labourers or service providers (see Chapter 8). To provide some context, this compares with 31 million households engaged in agriculture in Indonesia in 2013, 1.6 million of which were engaged in capture fishing, and 985,000 in fish farming (BPS, 2022).

The global carrageenan industry is large with high potential growth prospects (see Chapter 1). Already the dominant global carrageenan seaweed producer, Indonesia, is globally competitive in this sector (Yulisti et al., 2021). Ambitious plans for growth in downstream sectors are thought to provide an opportunity to generate much-needed off-farm employment, investment, and tax revenue. An Omnibus

Law (Job Creation Law, UU Cipta Kerja 11–2020), aims to develop a business-enabling environment and is supported by a large number of subordinate regulations, policies, and activities designed to generate investment.

While during the 1990–2000s the industry developed in a largely organic way, the government is aiming for a more orderly and formalised industry development process. This is thought to require increased policy attention in fields including industry planning, sea-use zoning, research and development, and the coordination of measures to attract investment. This has led to the proliferation of a large number of policies associated with seaweed production. A policy analysis by Permani et al. (2023) of the Indonesian seaweed industry revealed 67 policy documents with a peak of promulgations in 2021 (Figure 2.1).

A major landmark in the evolution of the Indonesian seaweed policy landscape was the issuance of the Presidential Decree 33–2019, Road Map of National Seaweed Industry Development 2018–2021. The Decree is wide in scope and encompasses a large number of other policies and is referred to throughout the chapter. However, there are many other high-level policies[3] which are listed in Appendix 2.

Thus, an extensive institutional web has evolved to support industry development, governance, and service provision. Institutional actors include government administrative line bureaus to design and implement policy, government extension agencies to disseminate technologies, research and development organisations to

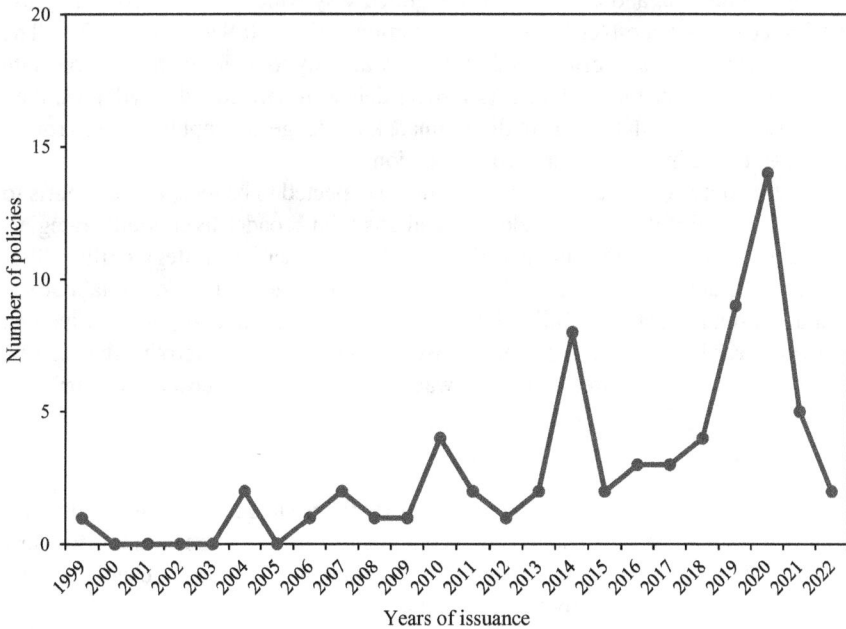

*Figure 2.1* A count of government policies relevant to the Indonesian seaweed industry, 1999–2022

Source: Image reproduced from Permani et al. (2023).

increase technological levels, and associations to represent industry interests. Jurisdiction for the seaweed industry falls mainly under the Ministry of Marine Affairs and Fisheries (KKP) and its line bureaus but intersects with a large number of units in government, research, associations, and international organisations. These are detailed in Appendix 3.

## The production sector

### Early growth period

The carrageenan industry had globalised by the 1970s when MNCs from Western countries invested in seaweed production in the Philippines, to be used for both processing in the country and international export. Exploratory missions for wild seaweed and early efforts in cultivation were attempted in Indonesia in the 1970s, as described in the Preface. Particular staff of companies in the Philippines – Copenhagen and FMC (later to become MCI) sought to ascertain the technical feasibility of seaweed production around Bali and Nusa Dua in the 1980s (Mariño et al. 2019; Iain Neish, personal communication, 10 September 2022). When trials proved successful, seaweed breeding stock and technologies were disseminated to other areas, predominantly through contact with other local coastal communities and by movement between communities. Local businesspeople engaged in retail, trading, wholesale, and credit, in an integrated way which often acted as key extension conduits (Iain Neish, personal communication, 10 September 2022). The entrepreneurs were interested in adding a new activity to their portfolios alongside food and other aquatic products. As traders that were embedded in villages, they had close contact with farmers to disseminate knowledge and inputs (such as ropes, credit, and seedlings) into seaweed production.

Low-income coastal communities could be expected to be receptive to efforts to extend seaweed if the activity adds to or aligns with broader livelihood strategies. New activities may contribute to a livelihood diversification strategy (Ellis, 1998) especially if they are complementary to existing activities in relation to labour demand, seasonality, sustainability, barriers to entry, location, and potential income growth (Reardon, 1997). Seaweed is also a labour-intensive activity that can be expected to develop in areas with low wages and opportunity costs of labour.

### Production structures

Unlike many other agricultural sectors where there are large estates (e.g. palm oil), contract systems (e.g. chickens), or economies of scale (some grains), the seaweed production sector is dominated by individual, autonomous households, with a dearth of examples of corporatised production.

Seaweed cultivation is sensitive to many environmental factors that are highly location-specific, variable, and uncontrollable, as Chapters 6 and 7 discuss (e.g. tides, seasons, rainfall events, disease, and other shocks). To be productive and resilient in these conditions, producers need to have both in-depth knowledge of local conditions and the flexibility and incentive to work around these conditions. For all but the largest households, seaweed farming is not a full-time job so it requires flexibility to allocate labour across a range of other complementary activities.

The predominance of households in seaweed production conforms to theory related to the competitiveness of actors with different scale and governance structures. There can be an inverse relationship between productivity and farm size in agricultural activities where households allocate labour efficiently or endure shocks (Chayanov, 1991). Small farms can be more efficient than large farms when high levels of local knowledge are required (Hazell et al., 2010). When hired labourers are costly to monitor or motivate, the self-supervision function of family farms are more efficient than large farms (Keijiro et al., 2016). Smallholders are also highly responsive to increased access to new technologies and markets (Schultz, 1964).

Alternative structures to autonomous, individual households have been trialled but are yet to be successfully established. A processing company (Widjaya) sought to farm a large area of seaweed using hired labour but encountered problems with production. New technologies and products developed for seaweed farming in deeper waters, including the use of mechanised floating and harvesting methods (e.g. Sea6), are under development but not fully operational. Several processing companies (e.g. Mitsubishi) have, or are seeking to develop, contractual relations with households but these efforts remain at an exploratory stage. While these corporate structures are yet to gain a foothold, conditions that may see some incursion in the future include increases in labour costs (see Chapter 8) and demands for traceability (Kirsten and Sartorius, 2002).

This is not to say, however, that dominant smallholder systems are static. It can take time for news about seaweed to travel and be taken up, especially if the uptake requires substitution out of other activities. Risk-averse households may wait to see the activity ground-tested by other farmers. Households incrementally develop ways to deal with shocks (e.g. weather, disease) and to tinker with systems to increase production or productivity. Data from fieldwork sites in the household sector show the emergence of significant numbers of large-scale household farmers who employ casual wage labourers for some tasks (see Part II of this book and Langford, Waldron et al., 2024).

Household organisational modes also vary. Neish and Suryanarayan (2017) distinguished between two types of seaweed farmers. The first was the traditional nuclear family model, where spouses and their immediate relatives share the work and income from seaweed cultivation. The second is the lead farmer model where one person or a small team manage the farm enterprise and sell the crops, but where labour is bought in for a range of tasks, especially attaching cuttings and drying. Group structures (associations and cooperatives) are also promoted by government. There is wide diversity amongst households in their scale of production, sea space use, and labour use even within the same village (see Part II).

*Policy settings for production*

The Indonesian seaweed industry has grown somewhat organically through the activity of economic agents operating in a conducive biophysical and socio-economic environment. However, government is now playing an increasingly active role in production aspects of the industry through technical extension, research and development, and in seed propagation. Permani et al. (2023) documented 24 policies that relate directly to the production-side aspects of the industry.

An important role of government is to invest in research and development. A wide scope of research has been conducted but production-side aspects feature in all of it.[4] Most research is conducted on seed breeding and supply, but other areas include fish repellents and drying ovens.

Another fundamental role of government in the production sector is in technical extension. With equivalents in the agricultural sector and other aquaculture industries, the Department of Fisheries and Marine Affairs at provincial level (DKP) has a technical extension system charged with developing, testing, adapting, and disseminating new seaweed technologies or practices in coordination with farmers. Staff are located and managed by the DKP at the sub-district level. Similar to extension in other countries, the system is stretched for human resources. In South Sulawesi, the seaweed-intensive sub-districts have just two extension agents responsible for vast distances and large numbers of villages and households. Duties of the staff include a large range of additional administrative duties (e.g. statistics, administration, and certification). Farmers have questioned the effectiveness of the extension system at the local level and developed and disseminated many technologies themselves. However, several key production-side technologies have been derived from the extension system including the para-para drying method, the double-line cultivation method, and mixed-species cultivation. The DKP also disseminates inputs (e.g. ropes and boats), often through group structures that aim to improve production, marketing, and extension.

A final form of government involvement in the production sector is through technical implementation units (TIUs) that produce and disseminate seaweed tissue culture in fresh water, brackish water, and marine aquaculture. Known as seedling gardens (*kebun bibit*) the units come under the jurisdiction of the Ministry of Marine Affairs and Fisheries (Regulation 70–2020) and are located in 18 locations and 20 seaweed villages (*Kampung Budidaya Rumput Laut*). Examples of research laboratories include the Brackish Water Fisheries Aquaculture Centre in Takalar and the Mariculture and Fisheries Centre in Lombok.

The centres were established to increase the quantity and quality of supply of propagules, a key constraint in household production systems particularly at the start of the season (Grist, 2022; Langford, Waldron et al., 2023). The centres aim to propagate seaweed with quality characteristics that include vigour, colour, and branch structure, all of which are largely a function of age. The seedlings are bred in controlled environments using vegetative techniques. Sporulation that would allow increased production is being trialled but is yet to be scaled up for production. To increase production and dissemination, out-grower schemes with selected households are also used.

Despite these efforts, the volumes of propagule material disseminated from TIUs is just a fraction of that produced by households themselves or that traded between households. A relatively small proportion of households directly receive free tissue culture propagules from the programme (Grist, 2022) but with frequent sales and exchanges between households, many may have received the material indirectly.

### Seaweed production

The agro-climatic and socio-economic environment depicted in this discussion is conducive to growth in seaweed production. National production statistics have reported exponential growth in seaweed output since the 2000s with some decline in recent years. Export statistics from Indonesia trend similarly, with rapid growth to 2010 followed by fluctuations in recent years. If accurate, reported declines may have been due to labour transition from seaweed cultivation to other more lucrative or attractive activities such as tourism (Wiratmini, 2018; Keohane, 2016) but the COVID-19 pandemic induced the opposite effect (Langford et al., 2021; Nuryartono et al., 2020). When the tourism sector was severely affected in areas like Bali, affected workers returned to on-farm activities including seaweed production (BBC, 2020; Pratiwi, 2020). Over-use of key production areas may have contributed to recent declines.

Much of the early expansion in seaweed production occurred in key areas including Bali, Nusa Dua, and South Sulawesi. By the 2000s, however, the industry had expanded across the archipelago with some of the expansion occurring informally. For example, members of households, mainly ethnic Buginese, from coastal communities in South Sulawesi, worked on palm plantations in Sabah in Malaysia (Iain Neish, personal communication, 10 September 2022). On route, these workers stopped over in North Kalimantan and observed conditions well suited to seaweed cultivation. The migration of workers and practices from South Sulawesi led to increased seaweed cultivation and has become a significant seaweed producing province.

While much information has been informally passed between those in the sector, more formal government programmes aim to expedite the process. These programmes aim to overcome issues related to dissemination in remote regions and this is reflected in national policies and plans (such as the Presidential Decree). However, provincial and sub-district government have also developed and implement policies. These efforts reflect Indonesia's decentralisation programme implemented in 2000. Provinces hold jurisdiction over major issues such as marine zonation and regional development plans. The spatial distribution of seaweed production in Indonesia in 2020 is shown in Figure 2.2, which shows the importance of Sulawesi (especially South Sulawesi) in national seaweed production.

### The marketing sector

The Indonesian seaweed marketing sector that links the production to processing sectors, comprises of a rich tapestry of actors, transport, and logistics systems, and institutional arrangements. Of particular interest is the role of traders, who play an important role in the organisation of the seaweed industry (Mulyati, 2015; Sutinah et al., 2018). In the earlier stages of industry development, MNCs were one

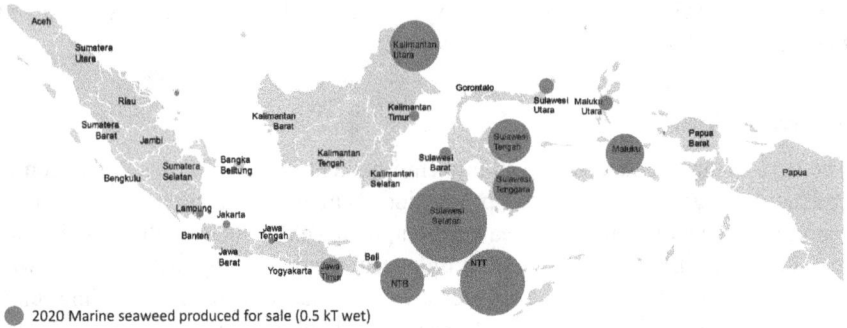

2020 Marine seaweed produced for sale (0.5 kT wet)

*Figure 2.2* Volumes of marine seaweed produced for sale across Indonesia in 2020
*Source:* Data from BPS (2022).

driver of the dissemination of production practices. These companies linked with local businesses with shops in villages or sub-district towns. Local traders were typically integrated with retail, wholesale, trading, and finance activities and were commonly ethnic Chinese Indonesians. With scientists and companies looking to expand seaweed production during the 1980s, these entrepreneurs were the key conduits in organising supply and linkages to farming communities.

These local-level relationships remain as the backbone of the seaweed marketing system, especially at the farmer-market interface. However, with increasing trade volume and demands from buyers, additional intermediaries have entered the industry to form a hierarchy of traders that lead to export markets or processing companies. Neish and Suryanarayan (2017) describe these as market-governed systems run by largely autonomous actors, although there are also remnants of the modular system, where companies and exporters have close relationships with certain buyers.

### The market hierarchy

The structure of the Indonesian seaweed marketing system has evolved to form a hierarchy of actors, linked through the exchange of seaweed for money. Local traders weigh and visually assess local farmers seaweed and buy at an agreed price, usually for cash "on the spot". The relationship is supported by embedded services and backward linkages. For example, traders provide inputs like rope, credit, or seedlings to the households which are paid off on the sale of the seaweed (Neish, 2013). Local traders can deal directly with farmers or, to reduce transaction costs, buy through local-level collectors. Unlike traders, collectors do not take ownership of seaweed but are provided with cash or credit from traders to buy seaweed from households based on their knowledge, contacts, trust, and negotiation, or logistical skills. The collectors might deal with 50–110 farmers and are sometimes heads of the local seaweed associations (Mulyati, 2015). Collectors tend to be more prevalent in the larger seaweed producing and marketing villages, like Laikang, rather than the smaller villages, like Pitu Sunggu (Waldron et al. 2022).

Local-level traders may also dry and clean the seaweed before aggregation with other lots, to then transport to downstream actors, which are larger traders or processors. There are a wide variety of traders in the hierarchy ranging from the village level to intra- and inter-island traders. In addition, there are estimated to be around 100 traders with export licences that supply foreign markets and domestic processors (Hogervorst and Kerver, 2019). This chain of traders can sometimes be shortened by processors that have more modular relations through more stable procurement arrangements with particular buyers. For example, the company Shanghai Brilliant Gum (BLG) sources seaweed through company procurement staff and traders in repetitive, ongoing relationships. These include seven former seaweed exporters[5] based throughout Indonesia but especially in South Sulawesi, North Kalimantan, and the eastern provinces.

### Market characteristics

The market-based governance system of the seaweed marketing system bears close resemblance to that of other commodities in Indonesia and other developing countries. This is especially the case for cash crops like fruit and vegetables, marine products, and some livestock and grain commodities.[6] Indonesian seaweed markets have several characteristics.

The first is that transactions occur in informal spot markets between autonomous actors. Transactions are usually made in cash without compulsory sale, inputs, product specification, or other formal obligations. The relationships can be repetitive between households and the seaweed buyers. Trust and backward linkages of credit, seedlings, or rope are informal and socially bound relationships between the parties. This contrasts with transactions through farmer-buyer contractual systems, where parties are bound by formal legal arrangements. Smallholders are likely to move from spot to contract systems for food products that are differentiated, perishable, or where consumers have food safety concerns (Kirsten and Sartorius, 2002). While these demands are growing in a range of foods, seaweed can be regarded as a bulk commodity where spot markets are generally effective and minimise transaction costs (legal, measurement, and monitoring). Neish and Suryanarayan (2017) describes emerging seaweed technology and chains that may utilise a contractual system.

Second, a language to describe seaweed characteristics is widely accepted and used, but often in a broad, informal, or subjective way. For example, the buyer or seller may claim moisture content of 36–38 per cent and dirt and contamination of 3–5 per cent. Some processors have additional specifications for colour (light) and carrageenan yield (e.g. 25 per cent) linked to pricing schedules. In practice, however, these specifications are not always applied and are rarely measured in farmer-trader transactions. This may raise questions in relation to the accuracy of visual assessments (made by eye) and information asymmetries in the transaction. Notionally, buyers would have a better eye for seaweed characteristics as they buy and sell every day and would have an incentive to discount estimates of grade (moisture and contamination) in order to discount the price. On the other hand, farmers who dry and pack the seaweed presented to the buyer also have an incentive to do so opportunistically (e.g. by putting wet or dirty seaweed at the bottom of sacks) to increase the weight, the measurement unit on which the transaction is made (Stone et al., 2023).

A third feature is that the seaweed chain is relatively long with a large number of actors and stages of transformation (Komarek et al., 2023). This means actors in the early stages of the chain (producers) have no direct contact with downstream actors (e.g. processors) and are unlikely to even recognise the final product. This makes it very difficult to effectively transmit price-grade differentials and buyer preferences down the chain. The indirect signalling of differentials is tested in price analysis discussed below.

Finally, there appears to be a large number of seaweed buyers in the industry which could be expected to create competitive markets. Indeed, in periods of high demand, buyers compete fiercely with each other for supply. Opposingly, however, the organisation of the hierarchy of traders leads to a limited number of end users who can be powerful.[7] For example, the purchasing power of companies like BLG and Greenfresh are known to set prices for the week.[8] The competitiveness of markets is also tested in the price analysis below.

Another feature of the Indonesian seaweed industry is that ethnic Chinese Indonesians dominate the post-production sectors. Chinese Indonesians have traditionally played a major role in seaweed trading and exports and own the majority of domestic Indonesian carrageenan processors, [9] as is the case in the Philippines.[10] This follows structures in agriculture-based trade established by early Chinese diasporas (Skinner, 1963) that have been observed in fisheries (Novaczek et al., 2001) and contemporary local-level business activities (Chiang and Cheng, 2017). The literature has documented ethnic Chinese business networks that form alliances with elites and may expedite business (McVey, 1992). These alliances can extend to mainland China through trade and investment flows (Ren and Liu, 2022). As established in Chapter 1, the vast majority of Indonesian seaweed is exported to China.

*Price analysis*

Prices provide valuable insights into the functioning of marketing systems. They signal the interplay between supply and demand, show patterns of change over time, and the degree of integration in time, space, and product attributes. Langford, Zhang et al. (2022) conducted a price analysis for Indonesian seaweed based on fortnightly price observations collected by Jaringan Sumber Daya (JaSuDa) in 13 locations across Indonesia. The data has been collected since 2005, but a sub-set from 2011 to 2021 were used. The prices have been updated to May 2023 and are presented in Figure 2.3.

The price data provides several insights. In January 2015 there was a large price decline that coincided with the announcement of a ban on raw seaweed exports as part of a broader industry policy to stimulate domestic processing. In September 2017, there were rapid price increases that coincided with the start of operations at Indonesia's largest processor, BLG. From mid-2019 to mid-2021, prices declined

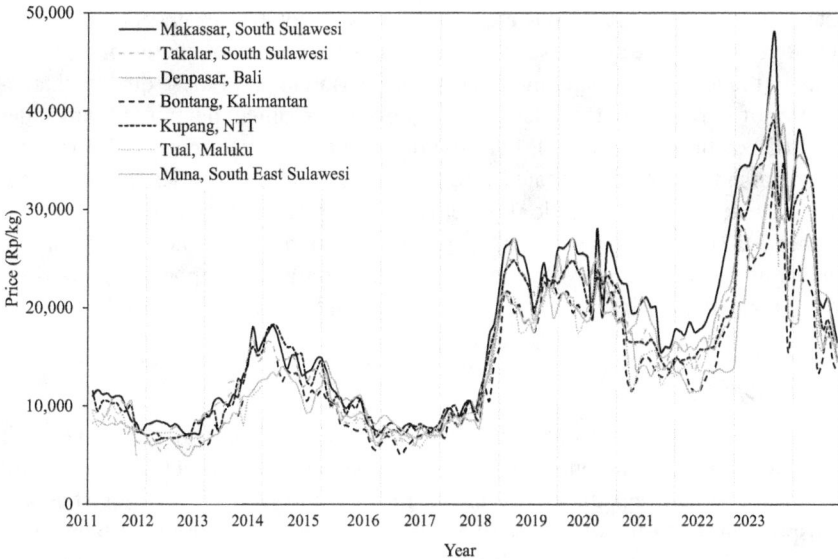

*Figure 2.3* Nominal seaweed prices in seven locations in Indonesia, 2011–23

Source: Data provided by Jasuda in 2023.

and stagnated, aligning with the most severe disruptions from the COVID-19 pandemic. After the price analysis by Langford, Zhang et al. (2022), there were rapid and sustained price increases through the second half 2021 and first half of 2022 (Langford, Waldron et al. 2023). It is important to note that the prices are notional, but seaweed price increases outstripped inflation. The historically high prices were good for seaweed farmers but placed a strain on the capital stocks and margins of downstream actors. Price levels have since corrected but appear to remain above pre-COVID levels.

Another finding of the price analysis is that prices are (spatially) co-integrated between regions (Figure 2.3). This is an indicator of a competitive and generally well-functioning market, underpinned by competition and flow of information. However, prices however became less integrated after the rapid increases in 2017, possibly as a result of BLG purchasing, and into the 2020–21 COVID pandemic. Transport and supply chain disruptions meant that prices were relatively lower in more remote areas (Palopo in South Sulawesi, Tual in Maluku, and Bontong and Nunukan in Kalimantan) compared to areas closer to major trading hubs (Makassar). Model results also show that the area closest to Makassar (Takalar) leads prices in other regions (Langford, Zhang et al., 2022).

The data collected includes seaweed prices and the basic attributes of moisture content and contaminant levels (sand and salt). Regression analysis found a low correlation between these variables, suggesting low transmission of the value for quality characteristics of seaweed and narrow price-grade differentials. This may be a function of inaccurate (subjective) measurement or incentives by all parties to not reveal characteristics (Stone et al., 2023). There would also appear to be limited incentives to produce quality seaweed defined by the content or quality of the carrageenan (the length of the gelling molecule). Measurement of these quality characteristics of seaweed are feasible in a laboratory environment but not at farm-trader level (Cozzolino et al., 2023). Furthermore, the downstream chemical processes are so harsh that they can negate carrageenan quality characteristics, especially for SRC, which is a more generic product (Neish, personal communication, 10 September 2022). In these conditions, subjective trading based on knowledge – where traders associate quality with particular production areas – appears appropriate. Measures to upgrade the marketing system should be examined critically.

### Marketing policy

The trade of seaweed is conducted mainly on an informal and subjective basis but is nevertheless underpinned by broader, cross-sectoral market regulations. This includes business registration, access to finance, infrastructure for logistics and transport, and industry standards. In addition, government has issued a large number of regulations, standards, and certification schemes that are directly relevant to seaweed and its products (see Permani et al., 2023 and Appendix 2). Most of the standards relate to food safety, particularly given international contestation in the organic status and safety of carrageenan as a food additive. Several standards apply to the production sector (e.g. SNI 8228.2: 2–15 Good Fish Farming Practices: Seaweed) while others apply to downstream products (e.g. SNI 8391.1: 2017 (Refined Carrageenan) – Part 1: Kappa carrageenan – quality requirements and processing). These national standards are designed to harmonise with international standards (e.g. Halal, ISO 22000 Food Safety Management System Certification, Good Manufacturing Practice, FDA registration).

There are also a large number of standards that apply to dry seaweed for seaweed marketing from the farmer to processor level. For example, SNI 2690:2018 Dried Seaweed specifies that seaweed must meet requirements for its origins (unpolluted areas, a harvest time of at least 45 days), characteristics (clean and free from decomposition), appearance (clean, with large thallus, and, for *Kappaphycus alvarezii*, ivory yellow, green, light/dark brown), texture (not easily broken between the stem and the branch), and must fall below a threshold of metal contamination. The standard applies to a range of seaweeds.[11]

While standards such as these provide minimum thresholds and a common language for industry actors, the standards are not widely or strictly used in farmergate transactions, partly because they are difficult to enforce. It is also important to note that processors have their own standards (for seaweed inputs) that deviate from national standards. Hogervorst and Kerver (2019) reported that Indonesian processors perform quality checks but lack written specifications (p. 24). They also claim that processors ignore their own standards when demand for seaweed is high, seldom invest in traceability or chain development, and do not provide incentives or feedback to traders and farmers to produce or maintain quality differentials. Issues in marketing practices at local level are discussed further in Chapter 9.

Another marketing initiative relevant to seaweed is the Warehouse Receipt System (*Sistem Resi Gudang*) under the jurisdiction of the Ministry of Trade and also endorsed in the Presidential Decree (33–2019). Under the system, receipts received by farmers or traders after storing their produce in a warehouse can be used as a collateral, including to buy more seaweed (BAPPEBTI, 2020). Organisations (such as the Indonesian Merdeka Workers' Cooperative) and companies (such as PT Asia Sejahtera Mina) provide loans to seaweed chain actors on this basis. Regulations of the Ministry of Trade 33–2020 (amended by Regulation of the Ministry of Trade 14–2021) included seaweed in the list of 18 commodities that can use the warehouse receipt system. The Presidential Decree (33–2019) aimed to open warehouses in Nunukan, Wakatobi, and Makassar. Despite these efforts, transaction volumes in the Warehouse Receipt System remain low, reportedly because it is difficult to find professional warehouse management and farmers still have low awareness of the system and its benefits (Permani et al., 2023).

Another marketing initiative conducted by the Ministry of Trade was to develop three physical seaweed auction markets in South Sulawesi, Nusa Tenggara Barat, and East Java. There is also a plan to develop an online market platform to facilitate seaweed trade auctions (Permani et al., 2023). Initiative to develop apps that could be used by farmers to sell seaweed are still under development.

## The processing sector

Scientists and industry explored Indonesia as a location for seaweed cultivation for export to processing facilities elsewhere, including the Philippines. While the majority of Indonesian seaweed is still exported, a combination of market forces and policy attention has seen the Indonesian carrageenan processing sector grow to become the third largest in the world.

Seaweed industrialisation started in Indonesia in 1976 when PT Bantimurung Indah was founded and began producing alkaline treated carrageenan (ATC) in Makassar. In 1988, PT Galic Artabahari was launched as an SRC company in Bekasi, West Java (Mulyati, 2015). From the 1990s conditions were conducive to growth. Seaweed cultivation was burgeoning and more accessible SRCs allowed for the proliferation of enterprises with lower barriers to entry. SRC was initially used for pet food, but food grade SRC was developed and became a substitute for refined carrageenan (RC) in the 1990s (Stanley, 1987). By this time there were a considerable number of companies producing semi-refined and refined carrageenan.[12]

Despite the development of the processing sector, the industry remained based on the export of raw dried seaweed (RDS). In 2015, the KKP reported that only one-third of Indonesia's seaweed was partially or fully processed, while 70 per cent was traded in a raw dried form (Patutie, 2015; Porse and Ladenburg, 2015). To Indonesian policy makers, seaweed was an under-utilised resource that could otherwise be used to advance the national interest through further processing, thereby value-adding and generating investment, employment, and tax revenue. In 2014, Indonesia announced that it planned to process at least half of the seaweed produced in the country by 2020 (Harrison-Dunn, 2014). A reconciliation of statistical sources indicates that in 2022, the proportion had reached 35 per cent (for *Kappaphycus alvarezii, Kappaphycus striatus* and *Eucheuma denticulatum*) (Zhang et al., 2023).[13]

The plans for greater domestic processing were pursued through industry policy on two fronts.[14] First, was the mooted introduction of export restrictions and tariffs. This instrument was first announced in 2011 by the KKP. The export restrictions were not implemented, due largely to concerns that they would reduce prices and hurt farmer interests (Harrison-Dunn, 2015) including those in remote locations (Neish, 2013). Nevertheless, a complete ban of seaweed exports by 2018 was announced by President Widodo as part of Indonesia's Blue Economy programme. Again, the ban was not implemented (Wright, 2017).

A concomitant measure was to promote and encourage the development of domestic industrial capacity through investment policy. At a general level, national industrial policies (e.g. Presidential decrees 28–2008 and 2–2018) encourage districts to develop industry capacity.[15] Indonesia's Blue Economy programme launched in 2013 and called specifically for the development of seaweed processing capacity. To fast-track investment for seaweed in a more interventionist way, the KKP developed seaweed processing plants in the major production areas of Bulukumba and Takalar (established from 2003 to 2005), and Saumlaki (2010). The plants were operated by a regional state-owned enterprise (Badan Usaha Milik Daerah (BUMD)) but are now closed for reasons that include inadequate access to water, electricity, and human resources. In a renewed programme of the KKP in 2016 (called Downstreaming, or *hilirisasi*) saw further state investment in at least nine seaweed processing plants, mainly on Sulawesi Island.[16] The plants were barely operating at the time of writing.

Even if the interventionist industry policies did successfully establish a group of operational processors, questions arise as to the effects. Export restrictions are likely to put downward pressure on prices, which effectively constitute a transfer from the revenues of seaweed farmers to processors (see Langford, Turupadang, and Waldron, 2023) for the case of provincial restrictions in Nusa Tenggara Timur (NTT)). Distortionary policies also cause price volatility that harms industry by making it risky to enter into longer-term purchase or supply contracts. The implication is that a more market-conforming approach where exporters compete with domestic processors (both Indonesian and foreign-invested) is more conducive to sustainable industry development and the interests of seaweed farmers. This is widely acknowledged within the industry (Seaweed Association of Indonesia (ARLI), personal communication, 10 September 2022).

### Foreign investment in processing

While attempts at industry-building through domestic investment and ownership have faltered, one of the outcomes of Indonesia's enthusiasm for processing within Indonesia was an injection of direct foreign investment, virtually all of which came from China. This includes investments from at least six Chinese companies, including the three largest carrageenan processors in China. Greenfresh invested in Indonesia (Hongxin) in 2012 with plans to expand, followed by BLG in 2015, and Longrun Newstar in 2016 (Zhang et al., 2023). The processing capacity of these three investments is approximately 18,000 tonnes per year, with an average capacity utilisation of 60 per cent (higher than most domestic plants), and amounts to an estimated actual production of 10,000 tonnes for all carrageenan products (ATC, SRC, and RC). The Ministry of Industry estimates total domestic production of 25,057 tonnes of carrageenan ATC, SRC, and RC (Kemenperin, 2022), suggesting that the three Chinese-invested plants account for 40 per cent of total domestic production. This figure would increase well above 50 per cent if other Chinese plants were included in the estimates and the planned investment from Greenfresh is actualised.

The investments are driven by several forces. The companies already import the majority of their seaweed inputs from Indonesia but locating closer to production sites can reduce exposure to the risks of restrictive Indonesian export policies, increases security of supply, and generates competitive advantages over competitors in China and the Philippines. In addition, the investors were able to leverage Indonesia's enthusiasm for investment to negotiate favourable terms. For example, BLG is located in a bonded zone (Pinrang) and Longrun Newstar in a special economic zone (Kendal) and are therefore eligible for preferential treatment for the import of inputs (machinery and chemicals) and exemptions from import duties, VAT, and excise tax.[17] Indonesian companies and industry associations have complained that they do not receive the same treatment (ARLI, personal communication, 10 September 2022).

### Current processing structures

Market and policy forces have caused an ebb and flow of companies in the Indonesian carrageenan processing sector, with companies opening, closing, merging, and changing names. The Ministry of Industry provided a snapshot of the sector in 2021 (Kemenperin, 2022) which listed 41 seaweed processing companies in Indonesia.[18] The reported production of the companies in 2017, 2021, and planned production under the Presidential Roadmap (year not defined) are shown in Table 2.1. Of the total seaweed extract production in 2021, 86 per cent is carrageenan. Of total carrageenan production, 56 per cent is SRC, 34 per cent is ATC, and 10 per cent is RC. The Presidential Roadmap plans to increase the proportion of RC to 28 per cent.

Astruli also collected details from member companies, which was collected and supplemented with further information from other major non-members. Details on a total of 32 processors (name, locations, capacity, capacity utilisation, production, and country of investment) are provided in Appendix 4, locations are mapped in Figure 2.4.

### Capacity utilisation in the processing sector

The Ministry of Industry also reported that the sector has a production capacity of 68,604 tonnes. With an actual production of 28,968 tonnes, this suggests a 42 per cent utilisation rate. The Presidential Roadmap aims for a 72 per cent utilisation rate for

*Table 2.1* Seaweed production processing output in Indonesia, 2017, 2022, and forecast

| Product | 2017 (tonnes) | 2021 (tonnes) | Target for Roadmap (tonnes) |
|---|---|---|---|
| Carrageenan | 13,116 | 25,057 | 27,838 |
| ATC | 2,352 | 8,531 | 3,850 |
| RC | 2,618 | 2,423 | 7,814 |
| SRC | 8,146 | 14,102 | 16,174 |
| Agar-agar | 4,140 | 3,911 | 10,393 |
| Total | 17,256 | 28,968 | 38,231 |

Source: Data from Kemenperin (2022).

*Table 2.2* Carrageenan production, capacity, and capacity utilisation of ASTRULI member companies, 2018–22

| | 2018 | 2019 | 2020 | 2021 | January–July 2022 |
|---|---|---|---|---|---|
| Carrageenan production (tonnes) | 30,314 | 26,977 | 17,988 | 25,655 | 5,000 |
| Production capacity (tonnes) | 40,230 | 40,230 | 40,230 | 52,299 | 26,149 |
| Capacity utilisation rate (%) | 75% | 67% | 45% | 49% | 19% |

Source: Data from ASTRULI (unpublished).

the sector as a whole. In another data series, the Indonesia Seaweed Industry Association (ASTRULI) collected data on carrageenan production, capacity, and capacity utilisation from member companies (Table 2.2). Utilisation rates were high in 2018 (75 per cent) but reduced over the 2020–2021 period, possibly due to the COVID-19 pandemic. The rate was especially low in the first six months of 2022 (19 per cent) and recovered to 45 per cent in the second half of 2022 (August to December).

ASTRULI and its members report that the stark declines in capacity utilisation from January to July 2022 were caused by very high prices in the second half of 2021 and first half of 2022 (Figure 2.3). When processors (particularly smaller processors) had met their orders, many companies re-assessed input-output prices and reduced or closed operations. The companies resumed operations and utilised more capacity when prices began dropping in the second half of 2022.

## Seaweed products

With a seaweed production and processing base established, in recent years the government of Indonesia has turned its attention to the final demand end of the industry with at least two objectives: to grow aggregate demand for seaweed; and to generate opportunities for value adding. While the downstream sectors may seem a world away from village life (Part II of this book), development in the downstream sectors impacts on the livelihoods of farmers through derived demand, prices, and potentially the form in which seaweed is cultivated and sold.

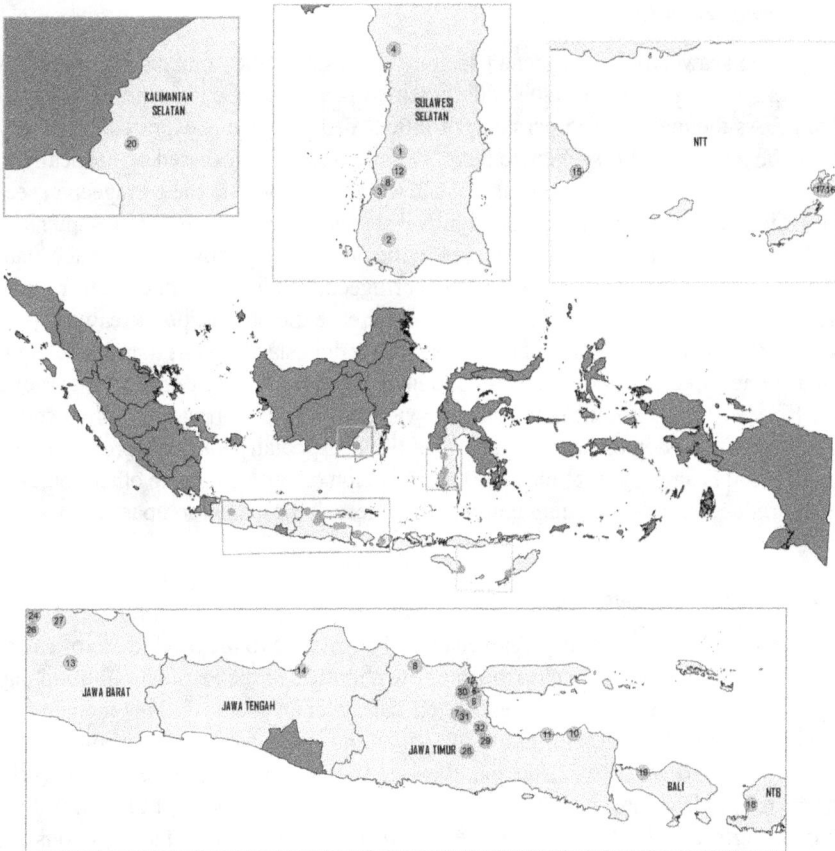

**Sulawesi Selatan**
(1) Anugerah Mapan Jaya - Maros
(2) Giwang Citra Laut - Makassar
(3) Wahyu Putra Bimasakti - Makassar
(4) Biota Ganggang Laut - Pinrang
(21) Cahaya Cemerlang - Makassar
(22) Bantimurung Indah - Makassar
**Nusa Tenggara Timor**
(15) Algae Sumba Timur Lestari - Sumba
(16) Rote Karaginan Nusantara - Kupang
(17) Agara Kembang - Kupang
**Nusa Tenggara Barat**
(18) Pheonix Mas - Lombok
**Bali**
(19) Sea6 Energy Indonesia - Buleleng
**Kalimantan Selatan**
(20) Batu Licin Algae Perdana - Tanah Bumbu
**Jawa**
5) Centram – Pasuruan (Jawa Timur)

(6) Karaginan Indo Mandiri – Blitar (Jawa Timur)
(7) Ocean Carrageenan Indonesia – Mojokerto (Jawa Timur)
(8) HW Marine – Situbondo (Jawa Timur)
(9) GreenOne Biotechnology – Surabaya (Jawa Timur)
(10) Hongxin Algae International – Situbondo (Jawa Timur)
(11) Fuyuan Biologi Teknologi – Situbondo (Jawa Timur)
(12) Sansawita (Jawa Timur)
(13) Segoro Algae – Bandung (Jawa Barat)
(24) Galic Bina Mada – Cikarang (Jawa Barat)
(25) Galic Arta Bahari – Bekasi (Jawa Barat)
(26) Hydrocolloid Indonesia – Cibinona (Jawa Barat)
(27) Buanatama Fajar Abadi – Karawang (Jawa Barat)
(14) Karaginan – Semarang (Jawa Tengah)
(28) Indoflora Cipta Keria Mandiri – Malang (Jawa Timur)
(29) Indonusa Algaemas Prima – Malang (Jawa Timur)
(30) Amara Carragenan Indonesia (Jawa Timur)
(31) Algalindo Perda – Pasuruan (Jawa Timur)
(32) Kappa Carragenan Nusantara – Pasuruan (Jawa Timur)
(23) Gumindo Perkasa Industri – Cilegon (Banten)

*Figure 2.4* Location of major Indonesian carrageenan processors
Source: Location of major Indonesian carrageenan processors

*Carrageenan products*

Within the seaweed-to-carrageenan sector, Indonesian policy makers pursue multiple forms of upgrading or value-adding, which is a major focus of the Presidential Decree. As the majority (65 per cent) of Indonesian seaweed is exported in raw dry form, the government is seeking to increase the proportion processed domestically to 50 per cent. It also aims to make the country a global leader in the carrageenan sector, which appears sound given the high and sustained forecast growth (Chapter 1).

Another form of upgrade can be seen in the processing pathways of carrageenan. The majority (56 per cent) of Indonesian carrageenan is in the form of semi-refined product. The Presidential Roadmap aims to increase the proportion of seaweed processed into RC from 10 per cent to 18 per cent. Indonesia also aims to produce high-value niche carrageenan products for certain markets, including Europe (Hogervorst and Kerver, 2019). Within Indonesia, the vast majority of carrageenan produced is exported or used domestically as a food additive, especially for drink products. The Presidential Decree plans to increase the utilisation of carrageenan in other domestically produced foods including coffee, milk, meat, jellies, and toothpaste.

*Non-carrageenan products*

It is likely that the Indonesian seaweed will be oriented to the production of carrageenan for some time. However much of the attention of the Presidential Roadmap is concerned with the development of non-carrageenan products. This includes direct food products (Adharini et al., 2019), animal feed, fertilisers (biostimulants, liquid and solid fertilisers, and planting media), cosmetics (e.g. capsules, pills, toothpaste, hair cream, soap), and bioethanol (Sulfahri, Husain et al., 2020; Sulfahri, Langford et al., 2020). The use of seaweed in a wide range of applications has generated attention world-wide, partly due to perceived environmental benefits, especially for *Asparagopsis* (Kinley et al., 2020; Ball et al., 2022). However, many of the technologies and applications are in the early stages of development and face logistic or commercial challenges. Further development should not be based only on technological development but should be subject to a full cost-benefit analysis to ascertain economic viability.

In line with potential and ambition, Indonesia has invested significantly in organisations to conduct research and development into new seaweed projects. These include: the Centre of Excellence for Seaweed at Hasanuddin University; Badan Pengkajian dan Penerapan Teknologi (BPPT) (Agency for the Assessment and Application of Technology) with a seaweed-based capsules programme; Seaweed-based Capsule Shell Teaching Industry facility at Universitas Airlangga; and the Department of Aquatic Product Technology, Faculty of Fisheries and Marine Sciences, IPB University in Bogor.

## Cross-cutting policies

The previous discussion overviewed the seaweed industry by reviewing on a sector-by-sector basis. However, several aspects of the industry cut across industry sectors. This includes food safety and the regulation of the use of carrageenan in organic food, which is overviewed in Chapter 1. Other cross-cutting issues such as marine zoning and investment are reviewed below.

### Zoning

The rules and norms that govern use of sea space have been developed by communities and households themselves through the development of informal institutions (see Chapter 5). However, the expansion and intensification of seaweed production may involve an increased role for government to mediate competing interests between seaweed farmers, other aquacultural activities, use of boat lanes, and marine protection zones. Zoning may have a role in reducing conflict, protecting public goods, allocating resources, and attracting investment (Permani et al., 2023). While land-based property rights are more established in Indonesia, the government in recent years has turned its attention to zonation in marine areas, which may impact on seaweed cultivation.

The national government has a stake in zoning, but jurisdiction lies at the provincial level. Law 7–2007 on the management of coastal areas and small islands provided a mandate for provincial governments to apply RZWP3K (*Rencana Zonasi Wilayah Pesisirdan Pulau-pulau Kecil* – 'Coastal Area and Small Island Zone Planning')to respective regulations. This mandate was amended by Law 23–2014 on regional governments, which requires each province to issue a provincial regulation to govern RZWP3K, and more recently Law 11–2020 which stipulates the integration of RZWP3K into RTRWP (*Rencana Tata Ruang Wilayah*, 'Spatial Planning'). As a provincial issue, the issue of zoning is discussed further in Chapter 3.

### Investment

The government is very interested in attracting and promoting inward investment. Investment is a driver of industry growth and development with associated public benefits including employment. The government can also generate revenue and taxes from the involvement in projects. In line with industry development objectives, investment in seaweed processing plants and farms is a priority for the government.

Investment is promoted through a large number of activities, forums, and trade shows at an international and local level. An underlying aspect of investment promotion is to provide an investor-friendly or a business-enabling environment including areas relevant to business processes (e.g. registration), preferential policies (e.g. tax treatment), and the clarification and harmonisation of laws.

To draw together these disparate and sometimes controversial objectives and mechanisms, China has used an Omnibus instrument to guide investment in Indonesia, known as the Job Creation Law 11–2020 (UU *Cipta Kerja*), enacted by the Indonesian president in November 2020. The Law aims to attract investment, generate employment, and stimulate the Indonesian economy by simplifying the licensing process and harmonising various laws and regulations. While the Omnibus

Law provides a high-level, over-arching framework that transcends specific industries, it amended 76 laws that related to seaweed (Permani et al., 2023). This includes Law 32–2014 on the sea, Law 18–2012 on food, Law 31–2004 on fisheries, Law 23–2014 on regional governments, Law 7–2007 (amended by Law 1–2014) on the management of coastal areas and small islands, and Law 33–2014 on Halal assurance. In 2021, the Indonesian government enacted 49 implementing regulations to the Omnibus Law (Permani et al., 2023).

## Conclusion

This chapter outlined the historical development of the seaweed industry that has led to current industry structures. While it is argued that most developments have occurred in a bottom-up way led by economic agents, it also outlines the plans and measures that industry and government actors are making to meet future challenges and objectives. While these measures seem significant at the (national) level of analysis, they may be unrecognisable at the local levels which are the subject of subsequent chapters.

## Notes

1  See Kalimajari (2016), Neish (2015), Mulyati et al. (2020), Porse and Rudolph (2017), Wright (2017), Neish (2007), Zamroni and Yamao (2012), Suadi and Kusano (2019), Yulisti et al. (2021), Sutinah et al. (2018), Hogervorst and Kerver (2019); Porse and Ladenburg (2015).
2  Several individuals were instrumental in the development of the industry. These included foreign scientist-entrepreneurs (Hans Porse and Iain Neish) and a founder of the Indonesian industry known as the seaweed politician, Sulfahri Aziz (also known as Sulfahri Hussain).
3  Other strategic plan documents from ministries include: Ministry of Marine Affairs and Fisheries Strategic Plan 2020–2024, Ministry of Industry Strategic Plan 2020–2024, and the Coordinating Ministry for Maritime and Investments Affairs Strategic Plan 2020–2024. Prior to the 2018 Presidential Decree, notable documents include the Revitalization Program for Agriculture, Fisheries, and Forestry (Program Revitalisasi Pertanian, Perikanan, dan Kehutanan) initiated by the President of Indonesia in 2005, the Acceleration of Fisheries Industry Program (Inpres No. 7 Tahun 2016, Percepatan Industri Perikanan), and Presidential Regulation No. 3 of 2017, which focuses on the development of non-food industries using seaweed as a raw material.
4  Key research centres supported by the central government include the Research Institute for Seaweed Culture in the Gorontalo Province, the Agency for the Assessment and Application of Technology (BPTP), Institute for Marine Socio-Economic and Fisheries Research, the Indonesian Institute of Science and the Indonesian Institute of Science Centre for Oceanography Research. International centres include the Southeast Asian Regional Centre for Tropical Biology and the Tropical Seaweed Innovation Network. Other research centres are located in Lombok, Bali, and South Sulawesi. The latter includes the Centre of Excellence for Seaweed at Hasanuddin University and a Public Agricultural Polytechnic in Pangkep.
5  These are CV Jala Ganggang, PT Sindo Serene International, PT Mega Citra Karya, PT Rika Rayhan Mandiri, CV Mitra Sejahtera, PT Central Pulau Laut, and CV Guna Bahari.

6  These differ from estate crops (cocoa, rubber) or staple crops (wheat, rice) where there are centralised, corporatised, or state-led marketing systems. For detailed analysis of market structures for six different commodities see InterCAFE (2018).
7  The InterCAFE (2018) found all six commodities studied to be characterised by oligopsony or oligopoly structures.
8  For example, an industry association seeking market intelligence asked a trader on a Friday what their prices would be next week. The answer was "I don't know, the Chinese haven't bought yet." The trader was waiting until then, because it would be too risky to set up a purchase order in case prices moved against them.
9  For example, Chinese Indonesians have established seaweed processing plants in West Java (Gumindo, Galic Artha Bahari, Hydrocolloid Indonesia), East Java (Algalindo Perdana, Seatech Carrageenan, Amarta Carrageenan) and South Sulawesi (Cahaya Cemerlang, Giwang Citra Laut, Wahyu Putra Bimasakti, Anugerah Mapan Jaya Hydrocolloid).
10  For an account of a well-known Chinese-Filipino entrepreneur see Gargan (1995).
11  For carrageenan seaweeds, moisture content must be a maximum of 38 per cent with a clean anhydrous weed (CAW) yield of 50 per cent minimum and a maximum of 3 per cent impurities. The standard goes on set requirements for proper handling techniques in harvesting, drying, packaging, labelling, and storage.
12  They include Banti Murung Indah, Algalindo Perdana, Amarta Carrageenan, Centram, Cahaya Cemerlang, Galic Artha Bahari, and Gumindo
13  Reconciliations are based on trade statistics from UN Comtrade for raw dried seaweed (HS code 121221), thickeners not confined to but dominated by carrageenan (HS code 130239), and domestic processing statistics of the Ministry of Industry (Kemenperin, 2022). Conversion from RDS to a generic carrageenan product (average for ATC, SRC, and RC) is based on a coefficient of 4:1.
14  The policy and outcomes for seaweed industry policy resemble those used in the Indonesian cocoa industry, where in 2010 exports taxes were imposed to encourage domestic processing and resulted in investment from MNCs (Harrison-Dunn, 2015). Another parallel is Russia's industry policy settings which discourage the export of raw timber (export tariffs) in order to encourage domestic wood processing (Ekstrom, 2014). The provinces use similar industry policies, Nusa Tenggara Timur (NTT) province for beef cattle (Waldron et al., 2016) seaweed (Langford, Turupadang et al., 2022; Langford, Turupadang, and Waldron, 2023).
15  The Ministry of Industry (Kemenperin, 2022) list policies to encourage downstream processing: tax allowances, deductions for research and development, and vocational training costs, exemptions for machinery for industrial development and using commodity balance sheets to expedite export and import approvals.
16  These are in the South Sulawesi Province (Luwu Timur, Janeponto, Bone), South East Sulawesi (Bombana, Buton, Buton Tengah), Gorontalo, North Kalimantan (Tarakan), and Maluku Utara.
17  See, for example, information on the Kendal Industry Park/Special Economic Zone. https://www.kendalindustrialpark.co.id/page/index/17/special-economic-zone?p=1
18  A list of 52 seaweed processors and exporters in South Sulawesi were reported by the Makassar Agricultural Quarantine Agency (2023).

## References

Adharini, Ratih Ida, Eko Agus SuyonoSuadi, Anes Dwi Jayanti, and Arief Rahmat Setyawan. 2019. "A Comparison of Nutritional Values of *Kappaphycus alvarezii, Kappaphycus striatum,* and *Kappaphycus spinosum* from the Farming Sites in Gorontalo Province, Sulawesi, Indonesia." *Journal of Applied Phycology* 31 (1): 725–730. doi:10.1007/s10811-018-1540-0

Ball, Alex, Scott Williams, and Russel Pattinson. 2022. *Scoping Study of the Capital Requirements for Commercial Production of* Asparagopsis *for Methane Reduction in Cattle*. Agrifutures Australia: Emerging Industries and Commonwealth Bank. https://agrifutures.com.au/product/scoping-study-of-the-capital-requirements-for-commercial-production-of-asparagopsis-for-methane-reduction-in-cattle

Bank Indonesia. 2023. *Statistik Ekonomi dan Keuangan Indonesia (Indonesian Economic and Financial Statistics)*. Accessed 10 March 2023. https://www.bi.go.id/SEKI/tabel/TABEL7_1.pdf

BAPPEBTI. 2020. *Annual Report of Performance Accountability of Commodity Futures Trading Regulatory Agency*, Jakarta.

BAPPENAS. 2021. *Blue Economy Development Framework for Indonesia's Economic Transformation*, Ministry of National Development Planning/National Development Planning Agency. https://perpustakaan.bappenas.go.id/e-library/file_upload/koleksi/dokumenbappenas/file/Blue%20Economy%20Development%20Framework%20for%20Indonesias%20Economic%20Transformation.pdf. Accessed 10 March 2023.

BBC (British Broadcasting Corporation). 2020. "Covid-19: Pariwisata Bali terpuruk, rumput laut menyelamatkan warga Nusa Lembongan." *BBC News: Indonesia*, 16 September. https://www.bbc.com/indonesia/majalah-54172281

BPS (*Badan Pusat Statistik*). 2022. Hasil Survei Komoditas Perikanan Potensi Rumput Laut 2021 Seri 2. *Badan Pusat Statistik*. https://www.bps.go.id/publication/2022/08/29/269de33babc6e3d52bbae5b6/hasil-survei-komoditas-perikanan-potensi-rumput-laut-2021-seri-2.html

Chayanov, Alexander. 1991. *The Theory of Peasant Co-operatives*. Translated by D. W. Benn, and introduction by Viktor Danilov. Columbus: Ohio State University Press.

Chiang, Bien and Jean Chih-yin Cheng. 2017. "Ethnic Chinese Enterprises in Indonesia: A Case Study of West Kalimantan." In *Chinese Capitalism in Southeast Asia* edited by Yos Santasombat, pp.135–153. London: Palgrave.

Cozzolino, Daniel, Mohamad Rafi, Wahyu Ramadhan, and Rudi Heryanto. 2023. *Technology to Improve Seaweed Marketing, Prices and Small-holder Incomes*. Report to the Australia-Indonesia Centre, forthcoming.

Ekstrom, Hakan. 2014. "Increases in Russian Lumber Exports Have not Compensated for the Decline in Log Exports that Came as the Result of Higher Log Export Tariffs." *ResourceWise* 1, 17 March https://news.cision.com/resourcewise/r/increases-in-russian-lumber-exports-have-not-compensated-for-the-decline-in-log-exports-that-came-as, c9552752

Ellis, Frank. 1998. "Household Strategies and Rural Livelihood Diversification." *The Journal of Development Studies* 35 (1): 1–38.

Gargan, Edward. 1995. "Mandaue Journal; They Plow the Waves for the Squire of Seaweed." *New York Times*, 5 June. www.nytimes.com/1995/06/05/world/mandaue-journal-they-plow-the-waves-for-the-squire-of-seaweed.html

Gereffi, Gary. 2018. *Global Value Chains and Development: Redefining the Contours of 21st Century Capitalism*. Cambridge: Cambridge University Press.

Grand View Research. 2023. *Carrageenan Market Size, Share & Trends Analysis Report by Processing Technology (Semi-refined, Gel Press, Alcohol Precipitation), by Function, by Product Type, by Application, by Region, and Segment Forecasts, 2020–2028*. Accessed 6 January 2023. https://www.grandviewresearch.com/industry-analysis/carrageenan-market/toc

Grist, Madeleine. (2022). "Propagule Distribution Systems of Tropical Carrageenan Seaweed in South Sulawesi: Value Chain and Social Media Analyses". Honours Thesis, University of Queensland. Supervised by Scott Waldron and Zannie Langford.Harrison-Dunn, Annie-Rose. 2014. "'Dramatic' Carrageenan Supply Chain Shake Up as Indonesia Eyes

Domestic Processing." *Food Navigator Europe*, 8 December. https://www.foodnavigator. com/Article/2014/12/08/Carrageenan-supply-shake-up-as-Indonesia-eyes-domestic-processing#

Harrison-Dunn, Annie-Rose. 2015. "What Will It Take to Make Indonesian Seaweed Competitive?" *Food Navigator Europe*, 2 June. www.foodnavigator.com/Policy/What-will-it-take-to-make-Indonesianseaweed-competitive

Hazell, Peter, Colin Poulton, Steve Wiggins, and Andrew Dorward. 2010. "The Future of Small Farms: Trajectories and Policy Priorities." *World Development* 38 (10): 1349–1361.

Hogervorst, Robbie and Kasper Kerver. 2019. *CBI Value Chain Analysis: Seaweed Extracts – Indonesia*. Centre for the Promotion of Imports from Developing Countries. https://www. cbi.eu/sites/default/files/2019_vca_indonesia_seaweed_extracts.pdf

InterCAFE. 2018. *Market Study on Food Sector in Indonesia*. International Center for Applied Finance and Economics (InterCAFE) https://kppu.go.id/wp-content/uploads/2019/09/ Market_Study_Report_JICA.pdf

Kalimajari. 2016. *Seaweed Sub-sector Growth Strategy in East Nusa Tenggara, West Nusa Tenggara, West Papua*. AIP-PRISMA (Australia-Indonesia Partnership for Promoting Rural Income through Support for Markets and in Agriculture). https://cms.aip-prisma. or.id/uploaded_file/2018-03-10_08-45-33am_40._41._42._GSD_Seaweed_report_ PRISMA_FINAL_-NTT_NTB_West_Papua.pdf

Keijiro, Otsuka. 2015. "Viability of Small-scale Farms in Asia." In *Handbook on Food: Demand, Supply, Sustainability and Security*, edited by R. Jha, R. Gaiha, and A. Deolalikar. Cheltenham: Edward Elgar.

Keijiro, Otsuka, Yanyan Liu, and Futoshi Yamauchi. 2016. "Growing Advantage of Large Farms in Asia and Its Implications for Global Food Security." *Global Food Security* 11 (December): 5–10.

Kemenperin. 2022. *Kebijakan Industri Pengolahan Rumput Laut*. Jakarta: Kemenperin.

Keohane, Aaron. 2016. *A Look into the Carrageenan Industry: How Tourism, Markets and Demand Affect the Seaweed Farmers of Bali Indonesia*. University of California San Diego: Center for Marine Biodiversity and Conservation. https://escholarship.org/uc/ item/9pw032tn

Kinley, Robert D., Gonzalo Martinez-Fernandez, Melissa K. Matthews, Rocky de Nys, Marie Magnusson, and Nigel W. Tomkins. 2020. "Mitigating the Carbon Footprint and Improving Productivity of Ruminant Livestock Agriculture Using a Red Seaweed." *Journal of Cleaner Production*, 259: 120836. doi: 10.1016/j. jclepro.2020.120836

Kirsten, Johann and Kurt Sartorius. 2002. "Linking Agribusiness and Small-scale Farmers in Developing Countries: Is There a New Role for Contract Farming?" *Development Southern Africa* 19 (4): 503–529.

Komarek, A., E. R. Cahyadi, J. Zhang, A. Fariyanti, B. Julianto, R. Arsyi, I. Lapong, A. Langford, S. Waldron, and M. Grist. 2023. *Increasing Incomes in Carrageenan Seaweed Value Chains in Takalar, South Sulawesi*. Melbourne, Australia: Australia-Indonesia Centre.

Kusano, Suadi and Eiichi Kusano. 2019, "Indonesian Seafood Supply Chain." In *Food Value Chain in ASEAN: Case Studies Focusing on Local Producers* edited by Eiichi Kusano, 134–163. ERIA Research Project Report FY2018, No. 5. Jakarta: ERIA.

Langford, Alexandra, Welem Turupadang, Marcelien Djublina, Ratoe Oedjoe, Christian Liufeto, Berty Bire, Marskel Yohanis, Yulius Suni, and Scott Waldron. 2022. *Seaweed Farmer Resilience in Eastern Indonesia after COVID-19*. ANU Indonesia Project.

Langford, Alexandra, Welem Turupadang, and Scott Waldron. 2023. "Interventionist Industry Policy to Support Local Value Adding: Evidence from the Eastern Indonesia Seaweed Industry." *Marine Policy* 151, 1–11. https://doi.org/10.1016/j.marpol.2023.105561

Langford, Alexandra., Scott Waldron, Nunung Nuryartono, Syamsul Pasaribu, Boedi Julianto, Irsyadi Siradjuddin, Radhiyah Ruhon., Zulung Zach Walyandra., Imran Lapong,
and Risya Arsyi Armis. 2023. *Sustainable Upgrading of the South Sulawesi Seaweed
Industry*. Melbourne, Australia: Australia-Indonesia Centre.
Langford, Alexandra, S. Waldron, Sulfahri, and H. Saleh. 2021. "Monitoring the COVID-
19-Affected Indonesian Seaweed Industry Using Remote Sensing Data." *Marine Policy*
127 (104431): 1–10. doi.org/10.1016/j.marpol.2021.104431
Langford, Alexandra, Scott Waldron, Jing Zhang, Radhiyah Ruhon, Zulung Zach Waly-
andra, Risya Arsyi Armis, Imran Lapong, Boedi Julianto, Irsyadi Siradjuddin, Syamsul
Pasaribu, and Nunung Nuryartono. (2024). "Diverse Seaweed Farming Livelihoods in
Two Indonesian Villages." In *Tropical Seaweed Cultivation – Phyconomy: Proceedings
of the Tropical Phyconomy Coalition Development, TPCD 1, Held at UNHAS University,
Makassar, Indonesia. July 7th and 8th* edited by A. Hurtado, A. Critchley, and I. Neish. 00-
00. New York: Cham, Switzerland: Springer Nature.
Langford, Alexandra, Jing Zhang, Scott Waldron, Boedi Julianto, Irsyadi Siradjuddin, Iain
Neish, and Nunung Nuryartono. 2022. "Price Analysis of the Indonesian Carrageenan
Seaweed Industry." *Aquaculture* 550: 737828: 1–13.
Makassar Agricultural Quarantine Agency. 2023. *Annual Report, 2022*. https://makassar.
karantina.pertanian.go.id/po-content/uploads/Laptah2022.pdf
Mariño, Mónica, Annette Breckwoldt, Mirta Teichberg, Alfred Kase, and Hauke Reuter.
2019. "Livelihood Aspects of Seaweed Farming in Rote Island, Indonesia." *Marine Pol-
icy* 107 (103600). doi.org/10.1016/j.marpol.2019.103600
McVey, Ruth. 1992. *Southeast Asian Capitalists (No. 9)*. Ithaca, NY: SEAP Publications.
Mulyati, Heti. 2015. "Supply Chain Risk Management Study of the Indonesian Seaweed
Industry." PhD thesis. Universität Göttingen. Göttingen. https://d-nb.info/1074758560/34
Mulyati, Heti., Jutta Geldermann, and Tridoyo Kusumastanto. 2020. *Carrageenan Supply
Chains in Indonesia*. In *IOP Conference Series, Earth and Environmental Science* 414,
012013. IOP Publishing. doi 10.1088/1755-1315/414/1/012013
Neish, Iain. 2007. "Helping Indonesia to Grow." *Assessment of the Seaweed Value Chain in
Indonesia*. Washington, DC: US Agency for International Development.
Neish, Iain. 2013. "Social and Economic Dimensions of Carrageenan Seaweed Farming
in Indonesia." In *Social and Economic Dimensions of Carrageenan Seaweed Farming*
edited by Diego Valderrama, Junning Cai, Nathanael Hishamunda, and Neil Ridler. Fish-
eries and Aquaculture Technical Paper No. 580: 61–89. Rome: FAO.
Neish, Iain C. 2015. *A Diagnostic Analysis of Seaweed Value Chains in Sumenep Regency,
Madura Indonesia*. Report submitted for UNIDO Project: 140140.
Neish, Iain C. and Shrikumar Suryanarayan. 2017. "Development of Eucheumatoid Seaweed
Value-Chains Through Carrageenan and Beyond." In *Tropical Seaweed Farming Trends,
Problems and Opportunities: Focus on Kappaphycus and Eucheuma of Commerce*, edited
by Iain Neish, Alan Critchley, and Anicia Hurtado. Cham, Switzerland: Springer.
Novaczek, Irene, Ingvild Harkes, Sopacua Marcus, and M. D. D. Tatuhey. 2001. "An
Institutional Analysis of Sasi Laut in Maluku, Indonesia." *ICLARM Tech. Rep.* (59): 327
WorldFish (worldfishcenter.org). Penang, Malaysia.
Nuryartono, N., S. Waldron, A. Langford, Sulfahri, K. Tarman, S. H. Pasaribu, U. J. Siregar,
and M. F. D. Lusno. 2020. *A Diagnostic Analysis of the South Sulawesi Seaweed Industry*.
Melbourne, Australia: Australia-Indonesia Centre.
Patutie, Rahmat. 2015. "KKP Minta Investor Buka Pabrik Rumput Laut Di Indonesia."
*Tribun News*, 21 May. http://www.tribunnews.com/bisnis/2015/05/21/kkp-minta-investor-
buka-pabrik-rumputlaut-di-indonesia

Permani, Risti, Yanti N. Muflikh, Fikri F. Sjahruddin, Nunung Nuryartono, Scott Waldron, Alexandra Langford, and Syamsul Pasaribu. 2023. "The Policy Landscape and Supply Chain Governance of the Indonesian Seaweed Industry: A Focus on South Sulawesi." Melbourne, Australia: Australia-Indonesia Centre. https://pair.australiaindonesiacentre. org/wp-content/uploads/2023/06/FINAL-REPORT_ENG_TWP-3_The-policy-landscape-and-supply-chain-governance-of-the-Indonesian-seaweed-industry_-A-focus-on-South-Sulawesi-2.pdf

Porse, Hans and S. Ladenburg. 2015. *Seaweed Value Chain Final Report*. Report submitted to Smart-Fish Indonesia under the UNIDO Seaweed Value Chain Programme.

Porse, Hans and Brian Rudolph. 2017. "The Seaweed Hydrocolloid Industry: 2016 Updates, Requirements, and Outlook." *Journal of Applied Phycology* 29 (5): 2187–2200. https://doi.org/10.1007/s10811-017-1144-0

Pratiwi, Fuji. 2020. "Warga Nusa Lemongan Beralih Jadi ke Budidaya Rumput Laut. Republika." 20 August. https://republika.co.id/berita/qfddfp457/warga-nusa-lembongan-beralih-jadi-ke-budidaya-rumput-laut

Reardon, Thomas. 1997. "Using Evidence of Household Income Diversification to Inform Study of the Rural Nonfarm Labour Market in Africa." *World Development* 25 (5): 735–747.

Ren, Na and Liu, Hong. 2022. "Southeast Asian Chinese Engage a Rising China: Business Associations, Institutionalised Transnationalism, and the Networked State. *Journal of Ethnic and Migration Studies* 48 (4), 873–893.

Schultz, T. W. 1964. *Transforming Traditional Agriculture*. New Haven, CT: Yale University Press.

Seaweed Insights. n.d. "*Eucheumatoids.*" https://seaweedinsights.com/global-production-eucheumatoids

Skinner, G. William. 1963. "The Chinese Minority." In *Indonesia, Survey of World Cultures*, edited by Rith McVey, 97–117. New Haven, CT: Yale University Southeast Asia Studies.

Soethoudt, J. M (Han), Heike Axmann, and Melanie G. Kok. 2022. *Indonesian Seaweed Supply Chain Analysis and Opportunities*. Wageningen Food & Biobased Research. Report 2309, Wageningen. https://research.wur.nl/en/publications/indonesian-seaweed-supply-chain-analysis-and-opportunities

Stanley, N. 1987. "Chapter 3 – Production, Properties and Uses of Carrageenan." In *Production and Utilization of Products from Commercial Seaweeds*, edited by Dennis J. McHugh. FAO Fisheries Technical Paper. Rome: FAO. https://www.fao.org/3/X5822E/x5822e05.htm

Stone, S., A. Langford, R. Arsyi, I. Lapong, Z. Zach, R. Ruhon, B. Julianto, I. Siradjudding, A. Wong, S. Waldron. 2023. "Technology Adoption by Smallholder Farmers: The Case of Drying Technology in the Indonesian Seaweed Industry." *Journal of Agribusiness in Developing and Emerging Economies*. https://doi.org/10.1108/JADEE-01-2023-0011

Sulfahri, Siti Mushlihah, Dirayah R. Husain, Alexandra Langford, Asmi Citra Malina, and A. R. Tassakka. 2020. "Fungal Pretreatment as a Sustainable and Low Cost Option for Bioethanol Production from Marine Algae." *Journal of Cleaner Production* 265: 121763. doi: https://doi.org/10.1016/j.jclepro.2020.121763

Sulfahri, Siti Mushlihah, Alexandra Langford, Asmi Citra Malina, and A. R. Tassakka. 2020. "Ozonolysis as an Effective Pretreatment Strategy for Bioethanol Production from Marine Algae." *BioEnergy Research* 13: 1269–1279. doi: 10.1007/s12155-020-10131-w

Sutinah, Hamzah Tahang, M. Chasyim Hasani, and Nurmaena. 2018. "Study of Seaweed (*Eucheuma cottoni*) Marketing in Makassar Industrial Area (PT. Kima) as Seaweed Marketing Center in Eastern Indonesia." *International Journal of Advances in Science Engineering and Technology* 6 (2): 80–84.

Waldron, S., J. Ngongo, S. Kusuma Putri Utami, M. Halliday, T. Panjaitan, B. Tutik Yuliana, Shelton M. Dahlanuddin, J. Nulik, and D. Nulik. 2016. *Economic Analysis of Cattle Fattening Systems Based on Forage Tree Legume Diets in Eastern Indonesia*. Report for ACIAR Project Report LPS-2014-034.

Waldron, Scott, Nunung Nuryartono, Alexandra Langford, Syamsul Pasaribu, Tarman, Kustiariyah Siregar, J. Ulfah, Muhammad Farid Dimjati Lusno, Julianto Sulfahri, Sarjana Boedi Siradjuddin, Ruhon Irsyadi, Walyandra Radhiyah, Zulung Zach Walyandra, Muhammad Imran Lapong, Risya Arsyi Armis, Eugene Sebastian, Helen Brown, Fadhilah Trya Wulandari, Hasnawati Saleh, and Steve Wright. 2022. *Policy Brief: Sustainable Upgrading of the South Sulawesi Seaweed Industry.* Melbourne, Australia: Australia-Indonesia Centre.

Wiratmini, Ni Putu Eka. 2018. "Produksi Rumput Laut di Bali Anjlok 99% (Seaweed Production in Bali Plummets 99%)." *Ekonomi*, 8 March. Accessed 2 February 2023. https://ekonomi.bisnis.com/read/20180308/99/747619/produksi-rumput-laut-di-bali-anjlok-99

Wright, Emily. 2017. "The Upshot of Upgrading: Seaweed Farming and Value Chain Development in Indonesia (Order No. 10656708)." Doctoral dissertation, University of Hawai'i at Mānoa. Available from ProQuest Dissertations & Theses Global (1954699097). Accessed 9 September 2022. https://www.proquest.com/dissertations-theses/upshot-upgrading-seaweed-farming-value-chain/docview/1954699097/se-2

Yulisti, Maharani, Estu Sri Luhur, Freshty Yulia Arthatiani, and Irwan Mulyawan. 2021. *Competitiveness Analysis of Indonesian Seaweeds in Global Market, IOP Conference Series: Earth and Environmental Science*, 860 012061. doi: 10.1088/1755-1315/414/1/012013

Zamroni, Achmad and Masahiro Yamao. 2012. "An Assessment of Farm-to-market Link of Indonesian Dried Seaweeds: Contribution of Middlemen toward Sustainable Livelihood of Small-scale Fishermen in Laikang Bay." *African Journal of Agricultural Research* 7 (30): 4198–4208.

Zhang, Jing, Scott Waldron, Alexandra Langford, Boedi Julianto, and Adam Martin Komarek. 2023. "China's Growing Influence in the Global Carrageenan Industry and Implications for Indonesia." *Journal of Applied Phyconomy.* https://doi.org/10.1007/s10811-023-03004-0

# 3 The South Sulawesi seaweed industry

*Radhiyah Ruhon, Scott Waldron, Zannie Langford, Adam Komarek, Jing Zhang, and Eko Ruddy Cahyadi*

Chapter 2 explored how Indonesia has worked to support the development of the seaweed industry through national investment in production, marketing, processing, and research. This chapter focuses on the provincial level, on South Sulawesi Province. South Sulawesi is the largest seaweed-producing province in Indonesia and the home of the case study villages analysed in Part II of the book. This chapter outlines the features of the South Sulawesi seaweed industry, including production, trade, processing, and export.

## Establishment of the seaweed industry in South Sulawesi

South Sulawesi has a long history as a hub of maritime routes. Several ports along the west coast of South Sulawesi have serviced international trade since the 16th century (Hadrawi 2018). Seaweed was exported from South Sulawesi from the 17th through to the 19th century. Makassar Port continued as a major export hub throughout the 20th century (Naval Intelligence Division 1944; Soegiarto and Sulustijo 1981). These exports included wild seaweed stock from Makassar waters and the other islands around Sulawesi (Brugman 1882) and as far as Australia (Pelras 1996). China and Japan were direct export destination countries at that time, while Singapore and Hong Kong became transit countries for products exported to the United States and a number of European countries. Increasing global demand for agar led to the expansion of harvesting of agar-bearing seaweeds in Indonesia, however this was interrupted by the outbreak of World War II (Soegiarto and Sulustijo 1981). In the 1960s and 1970s, four ports in Indonesia exported significant volumes of wild-harvested seaweed, including Makassar (Figure 3.1).

Farming of *Kappaphycus alvarezii* (known colloquially as 'cottonii') was successfully achieved in Indonesia for the first time in the 1980s (as described in the Preface). This was followed by investments in seaweed production and processing in several sites in Indonesia (Chapter 2). However, these initiatives did not initially identify South Sulawesi as a central location for the development of the seaweed industry (Hatta and Purnomo 1994; Soegiarto and Sulustijo 1981). Only 10–20 per cent of the total volume of (wild-harvested) seaweed exported from Ujung Pandang in Makassar Port originated from the waters of South Sulawesi (Mubarak 1980).

DOI: 10.4324/9781003183860-5

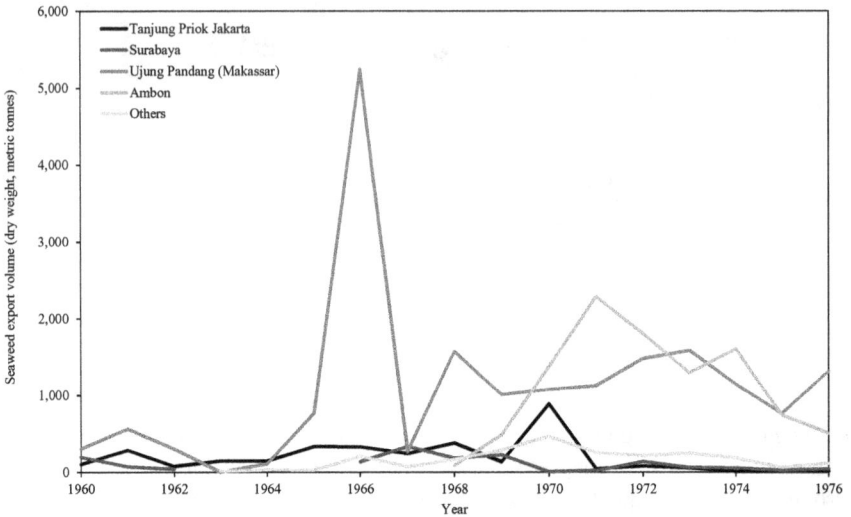

*Figure 3.1* Annual dried seaweed exports 1960–1976 from four major ports in Indonesia

Source: Data from The Indonesian Directorate General of Fisheries reported in Soegiarto and Sulustijo (1981).

There are no detailed records of exactly when and how seaweed farming was introduced to South Sulawesi, but farming was regarded as a step forward in development as it departed from a reliance on wild stock (Hatta and Purnomo 1994). In 1988, South Sulawesi was still an emerging seaweed farming location, with 2,171 marine cultivation plots established and Bali being the dominant producing area with 111,104 plots (Firdausy and Tisdell 1989).

Encouraged by growing conditions in South Sulawesi and overseas demand from Europe and Japan, exploration for a broader range of seaweed expanded. Exploration was carried out by Universitas Hasanuddin from 1988–1990 in the waters of the Spermonde Islands, South Sulawesi, through the Buginesia III project (Verheij and van Reine 1993). In the 1980s, a small number of fishermen in the Spermonde Archipelago were known to have cultivated species of *Eucheuma* and *Kappaphycus* by tying wild propagules from the sea to a monoline (Verheij and van Reine 1993). By the year 2000, the practice was increasingly widely adopted among the coastal communities in South Sulawesi (Briggs 2003). Today South Sulawesi is the largest seaweed producing province in Indonesia (BPS 2022). The widespread uptake of seaweed cultivation in South Sulawesi has been driven by a number of factors: maritime history, the skills and culture of local people, favourable marine conditions for seaweed, the easy-to-adopt methods, and the low cost of seaweed production (Doty 1973). Easy access to markets and ports (Fougeres 2005), government economic programmes (Firdausy and Tisdell 1989), the banning of destructive fishing practices (Briggs 2003), and commodity intensification (Fougeres 2005) also played a role.

## South Sulawesi seaweed supply chains

The seaweed-to-carrageenan supply chain involves many stages of transformation by several actors (Figure 3.2).

### Inputs

Production begins in the coastal regions where it dominates as a household industry. Farmers source planting material primarily from previous harvests (72 per cent of SulSel farmers) or other farmers (25 per cent of SulSel farmers) (BPS 2022). There is also some input of tissue cultured propagules provided by the Technical Implementation Units (Unit Pelaksana Teknis KKP), which can be requested periodically through farmer groups.

### Production

The first step in seaweed farming is to attach seaweed propagules to farming ropes. This is by far the most time-consuming step, and as a result many farmers rely on paid labour to support this stage (discussed in Part II). During intensifying seaweed production through 2021, some areas paid labourers from further afield, leading to the emergence of intermediary binding service providers who transported seaweed from farms to inland villages for binding. After binding, ropes with the propagules attached were transported to farming sites, where they were installed for a growth period of 45–50 days, or 25–35 days for seaweed destined to be reused as propagules (MKPRI 2019) (see Chapters 6 and 7). The seaweed is then harvested, dried, and cleaned, packed into bags, and stored in preparation for sale. Storage may be for short or extended time periods depending on farmer preferences.

### Distribution

Seaweed is then sold to local traders, who may redry and repack the seaweed if necessary to meet quality requirements, or to reduce tax liability (which is charged per sack rather than per kg). Local traders may also purchase seaweed from other regions and mix it with locally produced seaweed. They then either on-sell the seaweed to larger local traders or transport the seaweed to Makassar warehouses or processors for sale (see Chapter 9). Makassar warehouses may repack and redry seaweed if necessary for export or sale to processors.

### Processing and marketing

Regional traders then either package seaweed for export as raw dried seaweed (RDS) or sell it to seaweed processing companies. These companies produce a range of products including alkali treated cottonii (ATC), semi-refined carrageenan

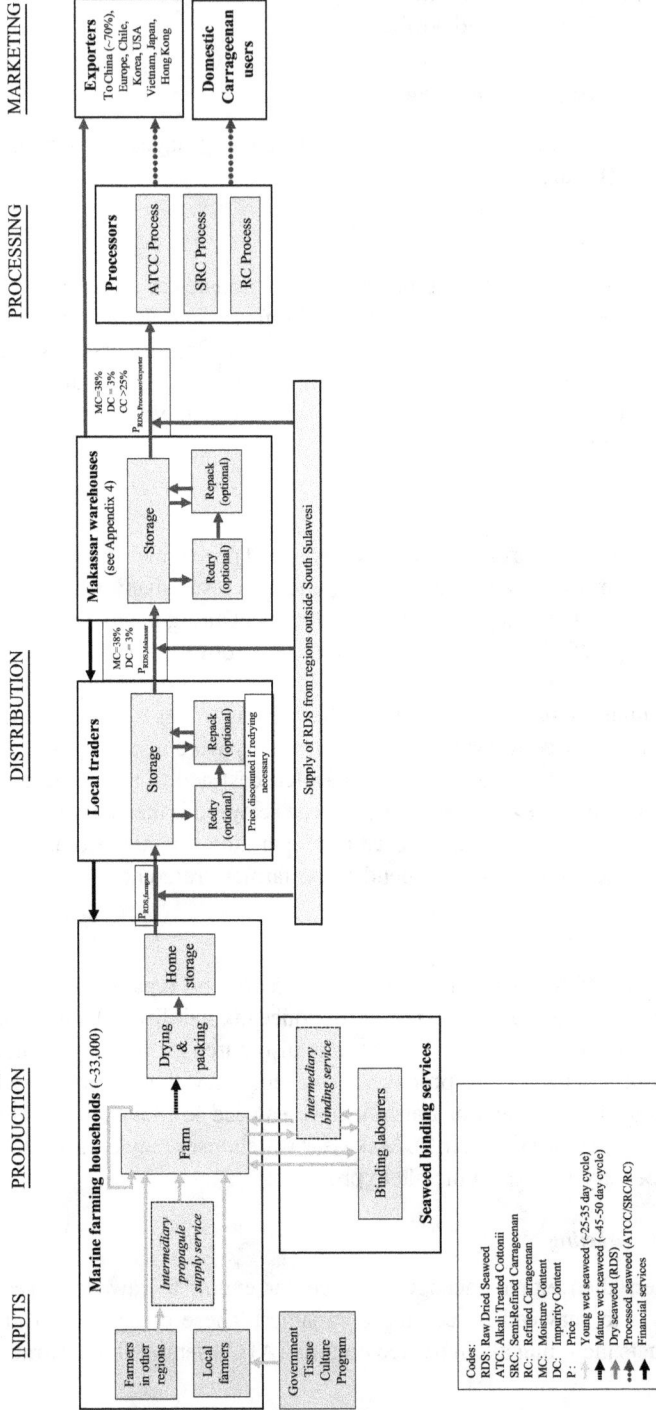

*Figure 3.2* South Sulawesi carrageenan seaweed supply chain

Source: Authors' schematic.

(SRC), and refined carrageenan (RC). The products are then either exported or consumed domestically.

The actors in the South Sulawesi seaweed-to-carrageenan supply chain (Figure 3.2) are connected through the exchange of product and money. At each stage of the chain the actors encounter different costs and prices. These costs and prices reflect the costs of transforming seaweed into carrageenan as the raw seaweed product changes over time, space, and form, and the profits obtained by each actor. Table 3.1 provides a specific example of these costs and prices from August 2022 based on seaweed being produced in the Laikang village of the Takalar Regency in South Sulawesi. The costs include transport to regional traders and processors in Takalar and Makassar. The data reported in Table 3.1 are averages taken from face-to-face interviews with: (a) 5 seedling suppliers, 15 farmers, and 12 local traders in Laikang village of Takalar Regency; and (b) 1 regional trader, 1 exporter, and 2 processors in Takalar and Makassar.

**Propagule inputs**

Propagules are a crucial material in seaweed cultivation practices, in terms of quality and quantity (supply). Moreover, as propagules are the major variable cost encountered by seaweed farmers, propagules are a key factor influencing farmer profitability (Table 3.1). At the beginning of the growing season, farmers will try to get the propagules from local traders who usually obtain the seed from other regions in South Sulawesi. However, when farmer demand for propagules cannot be met by the local collectors, some will travel to another seaweed district to get the seed directly from other farmers (see Chapters 6 and 7). Based on the annual survey results conducted by BPS (2022), approximately 34 per cent of all seaweed produced in South Sulawesi is used to propagate new cycles, with a further 1.2 per cent sold to other farmers as propagules.

When mass disease attacks or crop failure occurs due to bad weather or rising temperatures, which affects a significant number of farms, this can lead to a sudden

*Table 3.1* Buying costs and selling prices for actors along the seaweed-to-carrageenan value chain in Takalar, South Sulawesi in the 2022 high season

| Actor | Commodity purchased | Commodity sold | Average cost to buy (Rp/kg) | Average price to sell (Rp/kg) |
|---|---|---|---|---|
| Seedling supplier | NA | Seedling | NA | 5,273 |
| Farmer | Seedling | Raw dried seaweed | 5,200 | 34,500 |
| Local trader | Raw dried seaweed | Raw dried seaweed | 37,150 | 43,320 |
| Regional trader | Raw dried seaweed | Raw dried seaweed | 43,000 | 48,500 |
| Processor | Raw dried seaweed | alkali treated cottonii | 45,000 | 133,391 |
| Processor | Raw dried seaweed | semi-refined carrageenan | 45,000 | 185,265 |

Source: Data from Komarek et al. 2023, p. 8.
Note: Alkali treated cottonii and SRC sold in US$, average exchange rate of US$ to Rupiah was 14821.22 in August 2022 (interview time). The carrageenan content of dried seaweed was around 30 per cent.

increase in demand for propagules. Under these conditions, farmers usually try to get access to local propagules from fellow farmers. According to Grist (2022), there are technical and socio-economic factors considered by the farmer before obtaining propagules from other farmers. For example, the technical factors relate to: (a) the information availability of the seeds – the source, quantity, and quality of the propagules; and (b) the proximity and timely availability – is the person located close enough and will the propagules be ready when needed. Socio-economic considerations include: (a) the anticipated benefits; and (b) whether the propagule is reasonably priced (Grist 2022) (quality is discussed in Chapters 6 and 7). To overcome problems of seed quality and availability, SEAMEO BIOTROP, a stakeholder in the seaweed industry has developed seaweed breeding technology through a tissue culture system (SEAMEO BIOTROP 2022). This initiative is supported by the government Ministry of Marine and Fisheries Affairs (KKP). The ministry supported the industry with the implementation of the Kebun Bibit (propagules farms) programme in seaweed producing districts in South Sulawesi, namely Takalar, Luwu, and Bantaeng (Permani et al. 2023). This programme sourced the propagules from tissue culture technique facilitated by the relevant UPT (technical implementation unit), which then worked together with the local farmers' collectives involved in the Kebun Bibit programme. Platelets were grown before being shared with other districts.

## Production

South Sulawesi is reported to be the largest seaweed producer in Indonesia which produces at least 30 per cent of national production (BPS 2022). Carrageenan seaweed is one of South Sulawesi's leading commodities (DKP Prov. SulSel 2021), and marine seaweed farming in South Sulawesi is dominated by coastal households. Carrageenan producing seaweeds are the most common type cultivated in 18 seaweed producing areas in South Sulawesi (DKISP Provinsi Sulawesi Selatan 2017; DKP Prov. SulSel 2022a). This mostly consists of farming of cottonii and *Kappaphycus striatus* (known colloquially as 'sacol'), with *Eucheuma denticulatum* (known colloquially as 'spinosum') and *Gracilaria* also produced in smaller quantities.

This section has discussed trends in seaweed production primarily through official statistics reported by the Dinas Kelautan dan Perikanan (Marine and Fisheries Agency) DKP Provinsi Sulsel. SulSel's DKP production volume and area estimates appear to be over-stated but estimates of households engaged in production appear to be realistic (see Appendix 1). The statistics remain the only source of time-series data on the development of the South Sulawesi seaweed industry and therefore are subsequently used in this discussion. Figure 3.3. compares the three estimates collected by DKP and demonstrates some correlation, although production estimates diverge from the estimates of production areas and number of households over the last ten years.

Seaweed production in South Sulawesi has grown significantly in recent years (Figure 3.4). A period of significant increase occurred from 2007–2017, where the

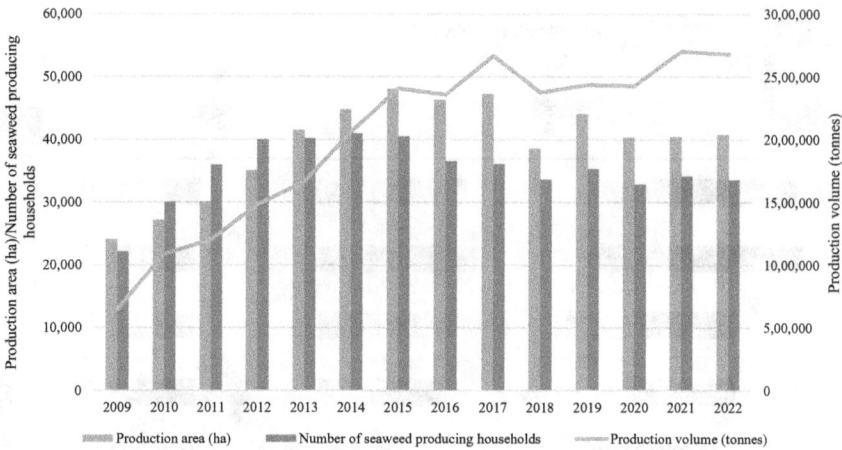

*Figure 3.3* Seaweed production volume, area, and number of marine farming households in South Sulawesi

Source: Annual Statistical Reports of DKP Provinsi Sulawesi Selatan 2006–2021 (published) and 2022 (unpublished).

increase in production was more than 80 per cent in ten years. Figure 3.4 demonstrates that the most rapid production increase occurred in the 2009–2010 period, where seaweed production increased by almost 40 per cent in one year. The figure also suggests that Takalar and Luwu contributed significantly to the increase in the total production. The same graph also suggests that there was a significant decline in production between 2017–2018, when Takalar experienced a very drastic decline in production. Despite the overall upward trend, there were fluctuations in the yearly values. Even though there was a decline in price during the COVID-19 pandemic, the price of seaweed bounced back in 2021. There was a slight decrease in volume from 2021 to the year after and the price increased rapidly in 2022. A milestone was recorded in August 2022 when the farm-gate price of dried seaweed reached Rp. 48,500/kg in Pitu Sunggu and Laikang villages. In line with this increase in prices from time to time, the incomes of seaweed cultivators have also increased. The DKP estimates that in 2021 the income of seaweed farmers was around Rp. 50,770,242 per season (DKP Prov. SulSel 2022b).

In 2022, seaweed production in South Sulawesi spanned across 17 districts (DKP Prov. SulSel 2023a). Of these districts, Jeneponto, Pangkep, Takalar, Bantaeng, Bulukumba, Wajo, and Bone had the largest number of households involved in the sector, together comprising almost 84 per cent of the total number of seaweed farming households (Figure 3.5). According to data from the Centre for Maritime and Fishery Socioeconomic Research, the national cultivation area experienced a growth of over 260 per cent in the period 2013–2015 (BBRSEKP 2019). However, the corresponding increase in seaweed production during the same period was only around 60–70 per cent. As shown in Figure 3.3, the number of seaweed farming households (RTP) increased significantly, starting at approximately 22,152 RTP in

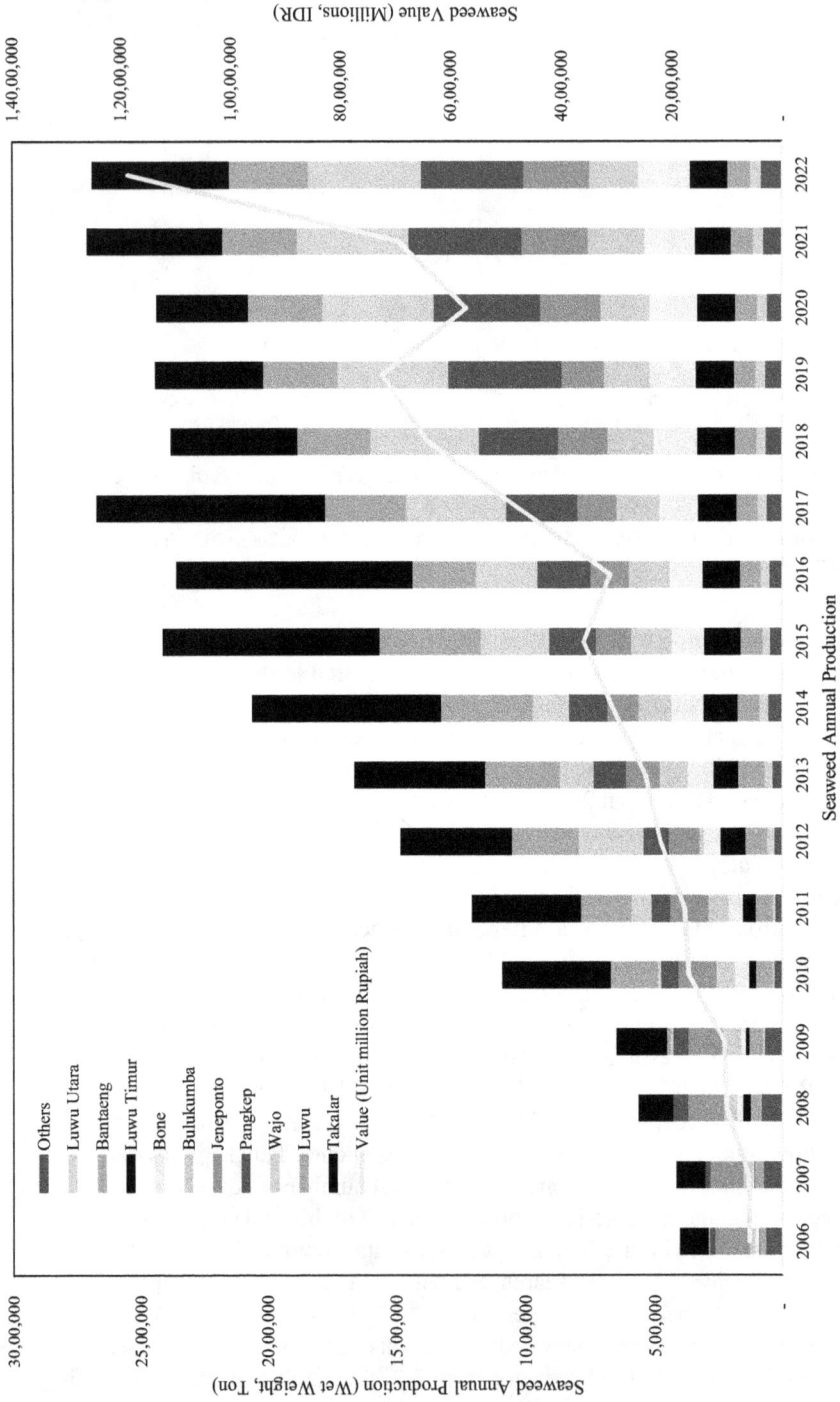

*Figure 3.4* Seaweed production from top ten seaweed-producing districts in South Sulawesi from 2006–2022

All data from DKP Prov. SulSel.

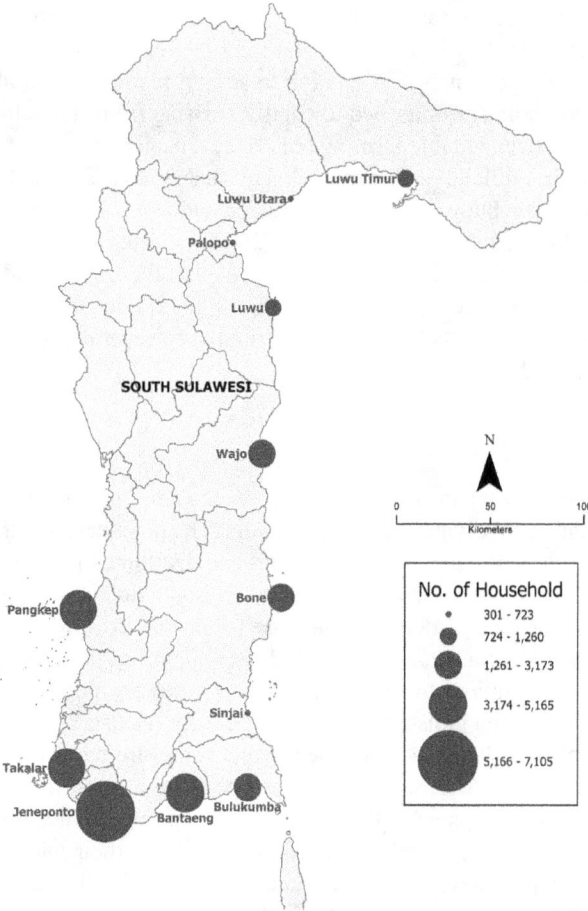

*Figure 3.5* Map of household participation in marine farming in South Sulawesi
Source: Authors' schematic. Data provided by DKP Sulsel 2023.

2009 and reaching 40,947 RTP in 2014. These numbers fluctuated and decreased to 33,589 in 2022.

As an indication of profitability, Komarek et al. (2023) indicates that farmer average total costs (including variable costs, cash fixed costs, and non-cash fixed costs) were Rp. 6,867/kg of raw dried seaweed. Farmer operating profit was Rp. 27,633/kg of raw dried seaweed. Seedling costs were the main cost incurred by seaweed farmers.

The experiences of farmers in Pangkep highlight the challenges of extreme weather conditions and marine space conflicts faced in seaweed production. Extreme weather conditions such as high rainfall (which may be exacerbated by climate change) frequently result in production losses. The farmers expressed concern regarding the adverse impact of these environmental factors that alter temperature and precipitation and have a severe impact on cultivation (see Chapter 6). Another challenge identified by farmers is related to sea space conflicts between farmers and other stakeholders. As seaweed farming has expanded extensively, there is a lack of formal governance in relation to zoning regulation. An absence of clear guidelines and boundaries has led to conflicts arising from competing interests and disagreements over suitable farming areas (see Chapter 5).

Overall, the official statistics indicate that seaweed production in South Sulawesi has been buoyant but has fluctuated and levelled off in recent years. The remarkable farm-gate price increases of 2022 impacted not only the livelihoods of the farmers, but also the motivation to develop the trading and distribution of seaweed products. Discussion will now turn to the trading of seaweed products in South Sulawesi, where the seaweed is distributed by several actors prior to entering the domestic and international market.

## Distribution

The seaweed industry in South Sulawesi operates through a complex network of trading channels involving various key actors. The main actors at this stage are the collectors/traders who act as intermediaries and distribute the seaweed from the locals to the domestic or international market. In South Sulawesi, like many other areas in Indonesia, local traders will purchase the raw dried seaweed from the farmers and aggregate it into a certain volume before selling it on to regional traders. Local traders may be required to redry the seaweed when the purchased seaweed does not meet the required moisture level of 38 per cent (Badan Standardisasi Nasional 2018). This usually happens when the product comes from other regions or another province. The local traders operate by maintaining a close relationship with the farmers by providing them with inputs (seeds, ropes, to reduce operational costs) and technical support. In return, the farmer will sell their harvest to the trader. The relationship between the farmers and the local traders is considered mutual as the traders provide a steady market and the farmers provide the local traders with a consistent supply. The relationship between the key actors may promote trust, reputation, and long-standing connections (Zamroni 2021; Langford et al. 2024; Waldron et al. 2022).

According to BPS (2022), around 94.78 per cent of seaweed households sell their crops to local traders/collectors, while the rest is sold to other farmers (2.48 per cent), exporters (1.77 per cent), and others (1.41 per cent). Others includes selling to processing industries, restaurants, cooperatives, or directly to other parties. Most of the transactions between farmers are in the form of propagules (Langford et al. 2024). The percentage shown by BPS also indicates that the direct purchase of seaweed products from farmers by local processing industries appears to be limited. This trend is similar to that in a study by Neish in 2007–2008, which suggested that in South Sulawesi, almost all farmers interviewed chose to sell to local collectors (Neish 2013). The most common reasons were price (100 per cent) and kinship (88 per cent). Results obtained by PAIR, suggest the main reason (60 per cent) farmers prefer to sell to collectors is that collectors often provide credit (Langford et al. 2024).

Another key actor in the trading network is the broker/middleman, a level of local traders who work as an extension of large traders or companies, either as local traders or as an individual who goes to the village door-to-door making trading offers to the farmers (Komarek et al. 2023). This type of actor was common in Laikang village, which may be due to the vastness of the area and the high number of seaweed farmers in the region. In Pitu Sunggu, the farmers mentioned that a number of local traders/collectors had established branches or worked together with small collectors in adjacent villages. These alliances mean they are able to purchase more local produce, preventing it from falling into the hands of competitors who also operate in the same villages. This extension of the network is usually a member of the trader's family who lives in the next village.

In the context of South Sulawesi, 2022 witnessed a significant increase in the farm-gate price of dried seaweed. The increased price motivated an increase in farmers and traders in several districts, including Pangkep. In Pitu Sunggu, the farmers associated the higher price with the new traders as it encouraged competition between new and established traders. It was also assumed the new traders would promote more transparent pricing.

The 2022 farm-gate price of dried seaweed may, however, have presented opportunities and challenges for seaweed farmers in South Sulawesi. Although farmers benefitted from higher income and profit, the increasing demand during this period led to unsustainable practices such as premature harvesting or overexploitation of environmental resources. These practices resulted in a market oversupply and was followed by a price crash.

Regional collectors act as middlemen between local traders and the larger traders/exporters or processing companies. They play a vital role in ensuring that seaweed products make their way smoothly along the distribution line. In South Sulawesi, regional traders mostly reside in Makassar, Maros, and Takalar City, close to the big port in Makassar. Similar to local traders, the regional traders may also conduct repacking and redrying (if necessary) and storing. Because of the many sources/channels of RDS products, quality control is a critical part of the operation of the regional trader who is the last gate for the product before entering the export market. If the seaweed does not meet quality standards, the regional trader may negotiate a lower price or reject the batch altogether. Regional traders/collectors are also aware of the different requirements and preferences of export destinations. For instance, certain countries may require specific product quality standards or have specific packaging regulations (Komarek et al. 2023).

In South Sulawesi, many of the regional traders/collectors have a large amount of capital and may trade in other marine-derived products in addition to seaweed. Where the main trading product is seaweed, the regional collector will usually trade in a number of seaweed species such as cottonii, *Gracilaria*, sacol, and spinosum. The RDS is also collected from Kalimantan and several regions in Eastern Indonesia (Komarek et al. 2023). Regional traders will organise the shipment of goods to customers either in Indonesia or overseas. Some of these regional traders are individuals who are affiliated to processing companies (Zamroni 2021). In 2022 local traders purchased seaweed from farmers for Rp. 37,150/kg and on sold the seaweed for Rp. 43,320/kg (Table 3.1). A margin of Rp. 6,170/kg for the local traders included Rp. 1,098/kg in costs related to payment to collectors, labour, and transportation.

## Processors

The seaweed processing industry plays a vital role in transforming RDS into various value-added products for domestic consumption and international trade. The RDS is supplied by local and regional collectors in South Sulawesi as well as by other traders from different provinces. Similar to the relationships that are built between farmers and local traders, processing companies also prefer to buy products from trading partners. Trust and a common understanding have been built in relation to the quality of the seaweed needed by the processing companies. If the collector is negligent or intentionally engages in harmful marketing practices (such as mixing seaweed with salt or other foreign matter), the collector will be blacklisted by the company. During the course of this research, cases of fraudulent collectors were common in Jeneponto and Takalar. It was also a topic of conversation among local traders in Pitu Sunggu where they expressed concern that these practices would undermine prices and could impact the trust relationships. Product quality relates to impurity, humidity, and gel content which follows Standar Nasiona Indonesia (SNI) standards (SNI 2690: 2018) that have been set by the government.

At this stage, RDS is then processed into ATC, SRC, or RC. These products, together with unprocessed RDS, are then exported or used in domestic industries that utilise carrageenan. It is estimated that nationally around 80 per cent of raw dried seaweed is exported (Ratnawati et al. 2020; Anggadiredja 2017), and the rest of the product is processed domestically before being exported or used in domestic processing (Anggadiredja 2017).

As of 2023, there are around seven seaweed processing companies in South Sulawesi (DKP Prov. SulSel 2022c, 2023c; BBKP, 2023). These processors mostly specialise in the production of agar or carrageenan (Rimmer et al. 2021). One of the main companies in South Sulawesi is PT. Biota Laut Ganggang (BLG) in Pinrang, which specialises in carrageenan production, positioning itself as the largest seaweed company in the world in that segment (Kukarpaper 2022). The company has a long history and has experience in seaweed research and development that can be traced back to 1996. Since its establishment in Pinrang, the company has expanded its production quantities (Pratiwi 2022). In 2019, the company had a production capacity of 100 tonnes per day for powdered seaweed. However, the available supply of seaweed as a raw material was only 50 tonnes per day (sourced from South Sulawesi and East Kalimantan), which resulted in a shortage of raw materials for BLG's production (Pemprov Sulsel 2019; Alfarizi 2019). At the time, BLG employed around 510 employees, with more than half of those employed being locally recruited (Sulfiani 2022). As of 2023, the number of employees at BLG is 735 people, and raw material requirements have increased to 150–200 tonnes per day. In 2022, around 70 per cent of this daily quota was being met, leaving up to 30 per cent of the quota unfilled (Kukarpaper 2022). The company exports its powdered seaweed to countries such as China, the United States, Europe, and Malaysia.

For the development of the South Sulawesi seaweed industry, the presence of BLG is considered to have been of benefit as it appears to have encouraged an increase in the seaweed price (Langford et al. 2022). It has also given rise to increased job opportunities for the local residents (Sulfiani 2022), improvements to roads and other infrastructure adjacent to the industry, guaranteed adherence to price standards, and an end to local buyers manipulating prices (Sulsel 2020). The challenges faced by BLG are related to a limited supply of RDS which has been a problem for other processing companies in previous studies investigating the challenges and threats faced by processors (van der Heijden et al. 2022; Soethoudt et al. 2022). Other problems include the inconsistent supply and low-quality of raw materials, which affects the overall quality of the end product. The factories also struggle to operate at full capacity. High transportation costs between islands and long waiting times also pose difficulties. There is strong competition from Chinese buyers who also purchase dried seaweed, and from hydrocolloid producers in other countries. The processors also may have limited experience in exportation to high-end markets and struggle to meet the specific requirements of those markets. Most of the seaweed exported still consists of raw materials (Sesditjen DJPB KKP n.d.). This means that the economic value adding from seaweed processing is relatively low (although value adding through improved drying practices is undertaken by some companies (Langford, Turupadang et al. 2023).

As a comparison, PT. Bantimuring Indah (PT. BI) is a smaller processing company based in Maros which was established in 1985 as a cracker factory using shrimps as raw material. In 1986, it started processing *Gracilaria* product in collaboration with Japan. By 1989, the company switched to processing cottonii to produce ATC and SRC. In 2017, it briefly produced *Gracilaria*-based products for a year. More recently it has focused on the production of cottonii-based products (SRC), having a production capacity of 4.8 tonnes per day. PT. BI primarily trades two containers of SRC (50 tonnes) per month on the international market, mostly to the UK, but has also supplied Russia, Argentina, and Chile in the past. The company has specifications for its raw materials, including a maximum water content of 37 per cent and impurities limited to 3 per cent. The origin of the raw material is crucial for the processing plant as buyers often require traceability to ensure quality. Buyers may even demand specific origins such as Nusa Tenggara Timur (NTT), Nusa Tenggara Barat (NTB), and Maluku.

In an interview with the Head of Plant on March 2023, it was disclosed that PT BI purchased seaweed from Jeneponto and Bone at a price of Rp. 34,000/kg. When sourcing from outside South Sulawesi, the price increased to Rp. 35,000–36,000/kg, including transportation costs. Surprisingly, only 30 per cent of their raw materials currently derive from within the province (Jeneponto, Bone, Pangkep, and Maros), with the remaining 70 per cent sourced from elsewhere. The company has identified a significant decline in seaweed quality within South Sulawesi, especially in relation to gel strength. The minimum gel strength required is 700g/cm$^2$, but seaweed from South Sulawesi typically averages between 500g/cm$^2$ and 600g/cm$^2$. The decline in quality can be attributed to various factors such as when farmers harvest seaweed earlier than is optimal, especially when there is high demand for raw material and during changing oceanic conditions (Langford, Waldron et al. 2023). Some exporters prioritise dry levels over gel strength and are willing to purchase any seaweed that meets the dry-level requirement. Additionally, poor handling practices by collectors,

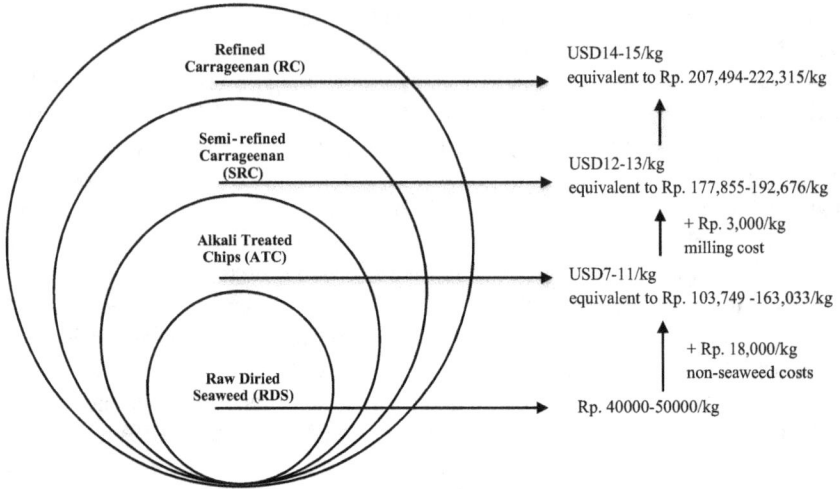

*Figure 3.6* Value adding of domestic carrageenan seaweed processing
Source: Authors' schematic. Data from Komarek et al. (2023).

who mix seaweed from different origins and harvest ages, contribute to the decline in quality. In some cases, collectors even use salt to expedite the drying process, further deteriorating the quality of the raw material.

At the processing level, processing techniques and the quality of seaweed influence prices and costs; however, a detailed decomposition of margins for processors is unavailable. Figure 3.6 provides an indication of the value of products produced at different stages of the carrageenan supply chain. Margins are composed of both operating costs and profit. Operating costs probably contribute to the size of the margin due to the fixed and variable costs of seaweed processing (rather than the physical movement of the product along the chain).

The yield of carrageenan (ATC, SRC, and RC) extracted from carrageenan seaweed (RDS) is shown in Figure 3.7, however this can be influenced by the processing technique and the quality of the raw materials. For instance, processing 1,200kg of RDS would result in a production output of 400kg of ATC, 300kg of SRC, or 240kg of RC. In terms of production costs, raw materials account for approximately 75–80 per cent of the total cost, followed by labour costs at 10–15 per cent, with chemical and energy costs each accounting for 5–10 per cent. Profit margins, as estimated by the processors interviewed, ranged between 10 and 20 per cent, with an average of 15 per cent.

### Provincial export

There are an estimated 52 seaweed exporters in South Sulawesi Province (DKP Prov. SulSel 2021; BBKP 2023). The export of seaweed from South Sulawesi has generally increased over the last decade (DKP Prov. SulSel 2021). Figure 3.8 reports statistics on the weight of RDS exported over the decade 2012 to 2022 and

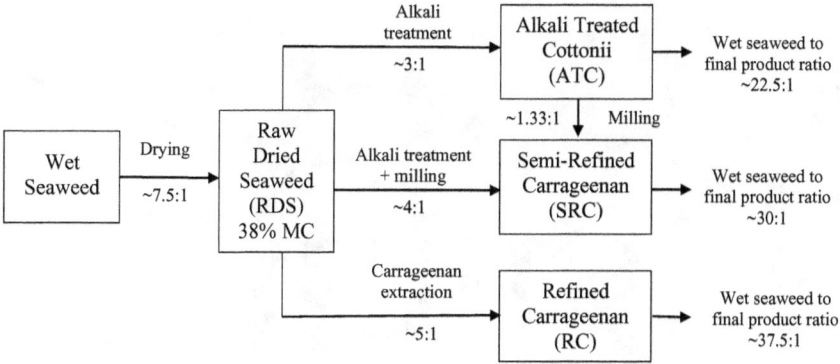

*Figure 3.7* Rates of conversion of seaweed into other products

Source: Authors' schematic. Data from Komarek et al. (2023).

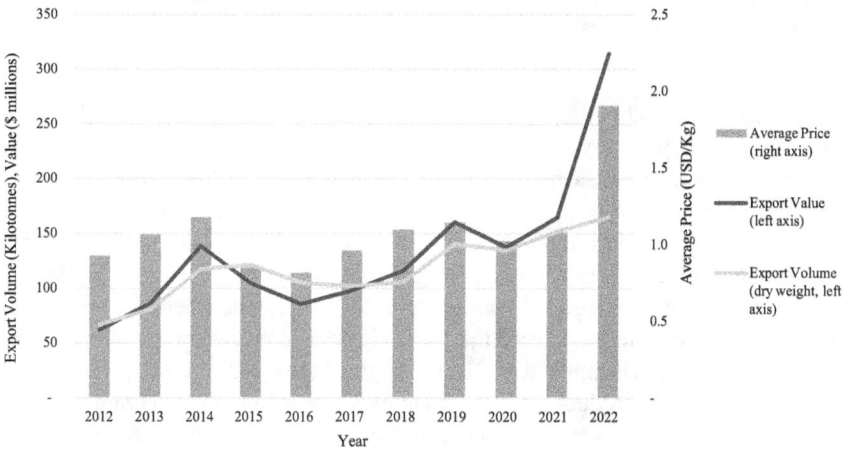

*Figure 3.8* Weight, value, and average price of all types of seaweed exports from South Sulawesi, 2012–2022

Source: Realisasi Pembangunan Perikanan Sulawesi Selatan Tahun 2012–2021. pdf and Data Expor Perikanan 2022 DKP.pdf, released by DKP Prov. Sulsel

Note: total export volume representative for all types of seaweed products, accounted in dry weight (different from the released data for the production which is represented in wet weight, personal communication, April 2023)

average prices. The weight of exports trended up over the period with a slight decline from 2015 to 2018, a spike in 2019, and a levelling-out over the COVID-19 pandemic years of 2020–2021. While weights increased slightly in 2022, prices surged, which led to a large increase in value exported. Figure 3.9 presents the same indicators (weight, value, and average price) but on a monthly basis to show variation in a single year (2022). There were slight fluctuations in the weight of seaweed exported. Average derived prices varied more over the year, with a peak in August 2022 which accurately reflects reports from the field where the farm-gate

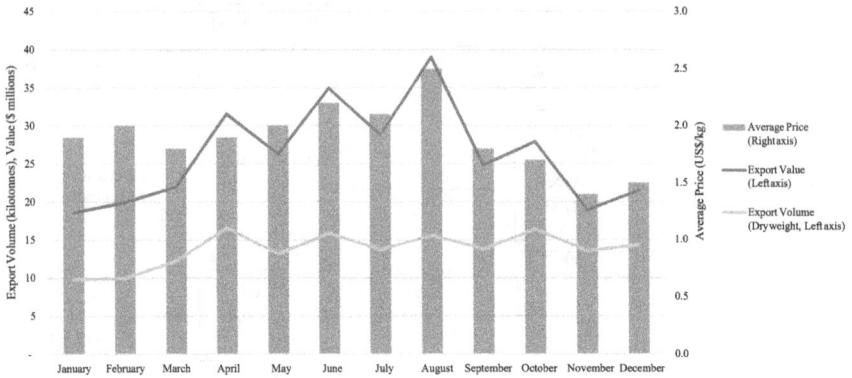

*Figure 3.9* Weight, value, and average price of all types of seaweed exports from South
Sulawesi by month

Source: Data from Trade Agency and BKIPM Sulsel and prepared by DKP Sulsel. Average prices are
derived from total values and weights.

price reached Rp. 48,000/kg (prices in Makassar peaked at Rp. 50,000/kg), the
highest in industry history. Prices then declined at the end of 2022 and into 2023.

For the last 20 years, China has been the major export destination for South
Sulawesi's seaweed products. According to BKIPM, more than 70 per cent of the
total exported seaweed from South Sulawesi has gone to China each year since 2018
(BKIPM Statistik 2023). Previous studies have highlighted that the price paid for seaweed in Indonesia was largely driven by demand from China, where the bulk of the
processing occurs. This has caused the domestic price to be generally lower than the
average global price (Rimmer et al. 2021; Langford et al. 2022). With the increasing
demand for seaweed product in the global market, the Indonesian government is trying to dominate this market by encouraging the development of the industry through
several policies and targets being set for the South Sulawesi seaweed industry.

**South Sulawesi seaweed policy**

The national government has enacted a wide range of policies aimed at the seaweed industry (see Chapter 2 and Appendix 1). Provincial governments also play
a role in policy-making by enacting higher-level policy and developing policies
within their jurisdiction. These are outlined briefly below.

South Sulawesi Province does not yet have an over-arching equivalent of the
national government "Roadmap" for the provincial seaweed industry, however,
industry policies are collated in policy documents. For example, South Sulawesi
Province states that provincial programmes (in 18 districts/cities) should support
seaweed production through inputs (superior propagule production through tissue culture programmes) to post-harvest handling (DKP Prov. SulSel 2022d). To
promote collaboration within the provincial government, the KKP established the
seaweed aquaculture fishing village in Laikang village, located in Takalar Regency
of South Sulawesi (*ANTARA News Makassar* 2022).

Further downstream, there are also programmes to support the establishment of new seaweed processing plants in South Sulawesi, in Bone, Jeneponto, and Luwu Timur. These plants will be required to obtain processing feasibility certification, Hazard Analysis and Critical Control Point (HACCP) certificates for health, and other certification for export or domestic distribution (Ningsih 2020). The government is also involved in various programmes to support exports. In 2016 the Sea Toll programme was developed to facilitate inter-regional distribution of seaweed within Indonesia, including supplies from South Sulawesi (Saputra 2023; Sesditjen DJPB KKP n.d.). Furthermore, the provincial government's commitment to supporting the industry is evident through several strategies, for example by adopting the recommendation of Omnibus Law No. 11 of 2020 in the South Sulawesi Provincial Spatial Plan (RTRW) for 2022–2041. These recommendations are aimed at easing business and investment license registration to attract foreign investment. The government has also partnered with the Ministry of Investment (BKPM) to promote South Sulawesi's seaweed industry as one of the targeted development projects in the Investment Opportunity Map (PPI) 2022 (Nooca 2022).

Amongst all the policy areas relevant to seaweed, provincial government has jurisdiction over zoning. The management of coastal areas and small islands is governed by Law 7/2007, which gives provincial governments the authority to implement regulations for coastal areas and small island zone planning (Permani et al. 2023). As such, the South Sulawesi provincial government issued two regulations: the Spatial Planning Plan (RTRW) 2009–2029 and the Coastal and Small Island Zoning Plan (RZWP3K) 2019–2039. These were later superseded by a new regulation (Regional Regulation No. 3 of 2022 concerning Regional Spatial Plans, 2022–2041).[1]

## Conclusion

Seaweed collection and exports have been practised in South Sulawesi since ancient times when South Sulawesi served as a global trading hub. Seaweed cultivation began in the early 1980s and increased during the 21st century. The organic, bottom-up growth has led to a sophisticated web of interactions at the community level and an atomised supply chain populated by a large number of actors. Industry growth has attracted the attention of corporate actors (processors and export companies) and government. This is reflected in the enactment of several policy initiatives, although policy attention can be expected to grow as the industry encounters new growth-induced challenges and opportunities. This includes availability of seedlings, volatile prices, labour shortages, ecological problems, conflicts over sea space, and the objectives of local government to promote value-adding and employment. While higher levels of government play a role in addressing these issues, it is imperative that policy is based on detailed and robust information from the local level such as the data on costs and prices encountered by supply chain actors presented in this chapter. It is suggested that decisions that require knowledge of the local and commercial environment should be vested in the local-level agents. The importance of understanding local context is explored in subsequent chapters.

## Note

1 Perda Nomor 3 Tahun 2022 Tentang Rencana Tata Ruang Wilayah (RTRW) Provinsi Sulawesi Selatan Tahun 2022–2041.

## References

Alfarizi, Rasid. 2019. "BLG Butuh 100 Ton Rumput Laut Per Hari." FAJAR.co.id. Accessed 27 May 2023. https://fajar.co.id/2019/10/25/blg-butuh-100-ton-rumput-laut-per-hari.

Anggadiredja, J. T. 2017. *Peningkatan Produktivitas dan Mutu Dalam Usaha Budidaya Rumput Laut Jenis Euchema sp.* SMART-Fish Indonesia.

*ANTARA News Makassar.* 2022. "KKP bangun kampung budi daya rumput laut di Takalar Sulsel." *ANTARA News Makassar.* Accessed May 2023. https://makassar.antaranews.com/berita/379473/kkp-bangun-kampung-budi-daya-rumput-laut-di-takalar-sulsel.

Badan Standardisasi Nasional. 2018. *SNI 2690–2018 Rumput Laut Kering.* Jakarta: Badan Standardisasi Nasional.

BBKP. 2023. *Laporan Tahunan 2022 BBKP. Balai Besar Karantina Pertanian.* Makassar: Balai Besar Karantina Pertanian.

BBRSEKP. 2019. *Potensi Sumberdaya kelautan dan Perikanan WPPNRI 713.* Edited by Sonny Koeshendrajana, I. Wayan Rusastra, and Purwito Martosubroto. Jakarta: AMAFRAD Press.

BKIPM Statistik. 2023." Operasional Volume Ekspor Produk Perikanan Mati: Rumput Laut. Badan Karantina Ikan, Pengendalian Mutu dan Keanaman Hasil Perikanan KKP 28 April 2023." http://bkipm.kkp.go.id/bkipmnew_rubah/?r=stats/#_ops_volume/E/Kg/m/0/2020/0/nm_umum/Rumput%20Laut.

BPS (*Badan Pusat Statistik*). 2022. Hasil Survei Komoditas Perikanan Potensi Rumput Laut 2021 Seri 2. *Badan Pusat Statistik.* https://www.bps.go.id/publication/2022/08/29/269de33babc6e3d52bbae5b6/hasil-survei-komoditas-perikanan-potensi-rumput-laut-2021-seri-2.html.

Briggs, Matthew R. P. 2003. *Destructive Fishing Practices in South Sulawesi Island, East Indonesia and the Role of Aquaculture as a Potential Alternative Livelihood.* APEC Secretariat. Bangkok: APEC. https://enaca.org/?id=783.

Brugman, B. Ch. 1882. "Zeevisscheeijen langs de kusten der eilanden van Netherlandsch indie celebes." *Makassaarsch Handels-Blad*, 26 April 2023.

DKISP Provinsi Sulawesi Selatan. 2017. "Data Komoditas Unggulan Provinsi Sulawesi Selatan Tahun 2014 Sampai Dengan Tahun 2016." Dinas Komunikasi, Informatika, Statistik dan Persandian Provinsi Sulawesi Selatan. panel.sulselprov.go.id/pages/komoditas-unggulan-rumput-laut.

DKP Prov. SulSel. 2021. *Rencana Strategis (Renstra Perubahan) Tahun 2018–2023.* Makassar: Dinas Kelautan dan Perikanan Provinsi Sulawesi Selatan.

DKP Prov. SulSel. 2022a. *Data Sementara Produksi Produksi Komoditas Perikanan Budidaya Semester I 2022.* Dinas Kelautan dan Perikanan Provinsi Sulawesi Selatan.

DKP Prov. SulSel. 2022b. *Laporan Tahunan Dinas Kelautan dan Perikanan Provinsi Sulawesi Selatan 2021.* Dinas Kelautan dan Perikanan Provinsi Sulawesi Selatan. Makassar: Dinas Kelautan dan Perikanan Provinsi Sulawesi Selatan. https://dkp.sulselprov.go.id/page/info/22/laporan-tahunan.

DKP Prov. SulSel. 2022c. *Rekapitulasi Penerbitan Rekomendasi Tahun 2022.* Makassar: Dinas Kelautan dan Perikanan Provinsi Sulawesi Selatan.

DKP Prov. SulSel. 2022d. *Rencana Kerja Dinas Kelautan dan Perikanan Tahun 2022.* Makassar: Dinas Kelautan dan Perikanan Provinsi Sulawesi Selatan.

DKP Prov. SulSel. 2023a. *Data Sementara Produksi Komoditas Unggulan Perikana Budidaya 2022*. Dinas Kelautan dan Perikanan Provinsi Sulawesi Selatan.

DKP Prov. SulSel. 2023b. *Jumlah RTP dan Tenaga Kerja di Laut Kategori Besarnya Usaha Menurut Kabupaten/Kota, Tahun 2022*. Dinas Kelautan dan Perikanan Provinsi Sulawesi Selatan.

DKP Prov. SulSel. 2023c. *Rekapitulasi Penerbitan Sertifikat Kelayakan Pengolahan Tahun 2023*. Makassar: Dinas Kelautan dan Perikanan Provinsi Sulawesi Selatan.

DKP Prov. SulSel. 2007. *Laporan Statistik Perikanan Sulawesi Selatan 2006*. Makassar: Dinas Kelautan dan Perikanan Provinsi Sulawesi Selatan.

DKP Prov. SulSel. 2008. *Laporan Statistik Perikanan Sulawesi Selatan 2007*. Makassar: Dinas Kelautan dan Perikanan Provinsi Sulawesi Selatan.

DKP Prov. SulSel. 2009. *Laporan Statistik Perikanan Sulawesi Selatan 2008*. Makassar: Dinas Kelautan dan Perikanan Provinsi Sulawesi Selatan.

DKP Prov. SulSel. 2010. *Laporan Statistik Perikanan Sulawesi Selatan 2009*. Makassar: Dinas Kelautan dan Perikanan Provinsi Sulawesi Selatan.

DKP Prov. SulSel. 2011. *Laporan Statistik Perikanan Sulawesi Selatan 2010.*(Makassar: Dinas Kelautan dan Perikanan Provinsi Sulawesi Selatan.

DKP Prov. SulSel. 2012. *Laporan Statistik Perikanan Sulawesi Selatan 2011*. Makassar: Dinas Kelautan dan Perikanan Provinsi Sulawesi Selatan.

DKP Prov. SulSel. 2013. *Laporan Statistik Perikanan Sulawesi Selatan 2012*. Makassar: Dinas Kelautan dan Perikanan Provinsi Sulawesi Selatan.

DKP Prov. SulSel. 2014. *Laporan Statistik Perikanan Sulawesi Selatan 2013*. Makassar: Dinas Kelautan dan Perikanan Provinsi Sulawesi Selatan.

DKP Prov. SulSel. 2015. *Laporan Statistik Perikanan Sulawesi Selatan 2014*. Makassar: Dinas Kelautan dan Perikanan Provinsi Sulawesi Selatan.

DKP Prov. SulSel. 2016. *Laporan Statistik Perikanan Sulawesi Selatan 2015*. Makassar: Dinas Kelautan dan Perikanan Provinsi Sulawesi Selatan.

DKP Prov. SulSel. 2017. *Laporan Statistik Perikanan Sulawesi Selatan 2016*. Makassar: Dinas Kelautan dan Perikanan Provinsi Sulawesi Selatan.

DKP Prov. SulSel. 2018. *Laporan Statistik Perikanan Sulawesi Selatan 2017*. Makassar: Dinas Kelautan dan Perikanan Provinsi Sulawesi Selatan.

DKP Prov. SulSel. 2019. *Laporan Statistik Perikanan Sulawesi Selatan 2018*. Makassar: Dinas Kelautan dan Perikanan Provinsi Sulawesi Selatan).

DKP Prov. SulSel. 2020. *Laporan Statistik Perikanan Sulawesi Selatan 2019*. Makassar: Dinas Kelautan dan Perikanan Provinsi Sulawesi Selatan.

DKP Prov. SulSel. 2021. *Laporan Statistik Perikanan Sulawesi Selatan 2020*. Makassar: Dinas Kelautan dan Perikanan Provinsi Sulawesi Selatan.

DKP Prov. SulSel. 2022. *Laporan Statistik Perikanan Sulawesi Selatan 2021*. Makassar: Dinas Kelautan dan Perikanan Provinsi Sulawesi Selatan.

Doty, Maxwell. 1973. "Farming the Red Seaweed, Eucheuma, for Carrageenans." *Micronesia* 9, no. 1: 59–73.

Firdausy, Carunia and Clem Tisdell. 1989. *Seafarming as a Part of Indonesia's Economic Development Strategy – Seaweed and Giant Clam Mariculture as Cases*. The University of Queensland. Brisbane, Australia: University of Queensland the Department of Economics.

Fougeres, Dorian 2005. "Aquarian Capitalism and Transition in Indonesia." PhD Thesis. Graduate Division University of California, University of California.

Grist, Madelaine. 2022. "Propagule Distribution Systems of Tropical Carrageenan Seaweed in South Sulawesi: Value Chain and Social Media Analyses." Bachelor Degree Research Thesis, School of Agriculture and Food Science, University of Queensland (AGRC6001).

Hadrawi, Muhlis. 2018. "Sea Voyages and Occupancies of Malayan Peoples at the West Coast of South Sulawesi." *International Journal of Malay-Nusantara Studies* 1, no. 1: 80–95.

Hatta, Agus M. and Agus H. Purnomo. 1994. "Economic Seaweed Resources and Their Management in Eastern Indonesia." *Naga, the ICLARM Quarterly* 17, no. 2: 10–12. https://aquadocs.org/handle/1834/25844.

Jayo, Muhamad Iqbal. 2018. "Peluncuran Eksport Perdana ATG Gracilaria di Bone." Puslatluh KKP. Accessed May 2023. https://kkp.go.id/brsdm/puslatluh/artikel/3863-peluncuran-eksport-perdana-atg-gracilaria-di-bone.

Komarek, Adam, Eko R. Cahyadi, Jing Zhang, Anna Fariyanti, Boedi Julianto, Risya Arsyi Armis, Imran Lapong, Alexandra Langford, Scott Waldron and Madeleine Grist. 2023. *Increasing Incomes in Carrageenan Seaweed Value Chains in Takalar, South Sulawesi.* Australia-Indonesia Centre.

Kukarpaper. 2022. "Wabup Melihat Langsung Pengolahan Rumput Laut PT BLG Pinrang." Pemerintah Kabupaten Kutai Kartanegara. Last modified 3 August 2022. Accessed May 2023. https://kukarpaper.com/wabup-melihat-langsung-pengolahan-rumput-laut-pt-blg-pinrang.

https://kukarpaper.com/wabup-melihat-langsung-pengolahan-rumput-laut-pt-blg-pinrang.

Langford, Alexandra, Irsyadi Siradjuddin, Jing Zhang, Nunung Nuryartono, and Scott Waldron. 2022. "Indonesia is the World's Largest Seaweed Producer but Why Are Prices so Volatile." *The Conversation* January 4, 2022. Accessed May 2023. https://theconversation.com/indonesia-is-the-worlds-largest-seaweed-producer-but-why-are-prices-so-volatile-171961.

Langford, Alexandra, Welem Turupadang, and Scott Waldron. 2023. "Interventionist Industry Policy to Support Local Value Adding: A Case from the Eastern Indonesia Seaweed Industry." *Marine Policy* 151 (105561): 1–11.

Langford, Alexandra, Scott Waldron, Nunung Nuryartono, Syamsul Pasaribu, Boedi Julianto, Irsyadi Siradjuddin, Radhiyah Ruhon, Zulung Zach Walyandra, Imran Lapong, Risya Arsyi Armis. 2023. *Sustainable Upgrading of the South Sulawesi Seaweed Industry.* Melbourne, Australia: Australia-Indonesia Centre.

Langford, Alexandra, Scott Waldron, Jing Zhang, Radhiyah Ruhon, Zulung Zach Walyandra, Risya Arsyi Armis, Imran Lapong, Boedi Julianto, Irsyadi Siradjuddin, Syamsul Pasaribu, and Nunung Nuryartono. (2024). "Diverse Seaweed Farming Livelihoods in Two Indonesian Villages." In *Tropical Seaweed Cultivation – Phyconomy: Proceedings of the Tropical Phyconomy Coalition Development, TPCD 1, Held at UNHAS University, Makassar, Indonesia. July 7th and 8th* edited by A. Hurtado, A. Critchley, and I. Neish. 00-00. New York: Cham, Switzerland: Springer Nature.

Langford, Alexandra, Jing Zhang, Scott Waldron, Boedi Julianto, Irsyadi Siradjuddin, Iain C. Neish, and Nunung Nuryartono. 2022. "Price Analysis of the Indonesian Carrageenan Seaweed Industry." *Aquaculture* 550 (737828). doi.org/10.1016/j.aquaculture.2021.737828

Marhawati, Muhammad Rakib, Agus Syam, and Muhammad Imam Ma'ruf. 2020. "Analysis of the Feasibility of Seaweed Farming in Pangkep District." *Proceedings from the International Conference on Science and Advanced Technology* (ICSAT), pp. 1486–1492.

MKPRI (Menteri Kelautan dan Perikanan Republik Indonesia). 2019. *Keputusan Menteri Kelautan dan Perikanan Republik Indonesia Nomor 1 / Kepmen-KP/2019 Tentange Pedoman umum pembudidayaan rumput laut.*

Mubarak, Hasan. 1980. "Indonesian Seaweeds, Its Resources and Culture." Marine Fisheries Research Institute, Jakarta, Indonesia. https://www.fao.org/3/bm800e/bm800e.pdf

Naval Intelligence Division. 1944. *Geographical Handbook Series: Netherlands East Indies, Volume II*. Edited by H. C. Darby. Great Britain: H.M. Stationery Office.

Neish, Iain C. 2013. "Social and Economic Dimensions of Carrageenan Seaweed Farming in Indonesia." In *Social and Economic Dimensions of Carrageenan Seaweed Farming,* edited by Diego Valderrama, Junning Cai, Nathanael Hishamunda and Neil Ridler, 66–89. Rome: Food and Agriculture Organization of the United Nations.

Ningsih, Trisna. 2020. *Progress Roadmap Pengembangan Industri Rumput Laut Nasional Tahun 2018–2021.* Jakarta, Indonesia: Dirjen Penguatan Daya Saing Produk Kelautan dan Perikanan KKP.

Nooca, Dhafintya. 2022. "BKPM Nilai Rumput Laut Masuk Peta Peluang Investasi 2022." Suara Surabaya.com. Accessed June 2023. https://www.suarasurabaya.net/ ekonomibisnis/2022/bkpm-nilai-rumput-laut-masuk-peta-peluang-investasi-2022.

Pelras, Christian. 1996. *The Bugis.* Cambridge, MA: Wiley-Blackwell.

Pemerintah Republik Indonesia. 2019. *Peraturan Presiden Nomor 33 Tahun 2019 Tentang Peta Panduan (Roadmap) Pengembangan Industri Rumput Laut Nasional Tahun 2018– 2021.* Jakarta: Sekretariat kabinet RI Deputi Bidang Kemaritiman.

Pemprov Sulsel. 2019. "PT BLG Kekurangan 50 Ton Pasokan Rumput Laut per Hari." Humas Pemerintah Provinsi Sulawesi Selatan. Accessed May 2023. https://humas.sulselprov. go.id/index.php/2019/10/25/pt-blg-kekurangan-50-ton-pasokan-rumput-laut-per-hari.

Permani, Risti, Yanti Nuraeni Muflikh, Nunung Nuryartono, Scott Waldron, Alexandra Langford., Syamsul H. Pasaribu, and Fikri Sjahruddin. 2023. *The Policy Landscape and Supply Chain Governance of the Indonesian Seaweed Industry: A Focus on South Sulawesi.* Melbourne, Australia: Australia-Indonesia Centre. https://pair. australiaindonesiacentre.org/wp-content/uploads/2023/06/FINAL-REPORT_ ENG_TWP-3_The-policy-landscape-and-supply-chain-governance-of-the-Indonesian-seaweed-industry_-A-focus-on-South-Sulawesi-2.pdf.

Pratiwi, I. I. 2022. "Laporan Kuliah Kerja Praktik PT. Biota Laut Ganggang Pinrang." Undergraduate fieldwork report, Politeknik ATI Makassar Kementerian Perindustrian Republik Indonesia.

Ratnawati, Pustika, Nova Francisca Simatupang, Petrus Rani Pong-Masak, Nicholas A. Paul, and Giuseppe C. Zuccarello. 2020. "Genetic Diversity Analysis of Cultivated Kappaphycus in Indonesian Seaweed Farms using COI Gene." *Squalen Bulletin of Marine and Fisheries Postharvest and Biotech* 15, no. 2: 65–72.

Rimmer, Michael A., Silva Larson, Imran Lapong, Agus Heri Purnomo, Petrus Rani Pong-Masak, Libby Swanepoel, and Nicholas A. Paul. 2021. "Seaweed Aquaculture in Indonesia Contributes to Social and Economic Aspects of Livelihoods and Community Wellbeing." *Sustainability* 131, 10946. https://doi.org/10.3390/su131910946. https:// mdpi-res.com/d_attachment/sustainability/sustainability-13-10946/article_deploy/ sustainability-13-10946-v2.pdf?version=1633934576.

Saputra, Dany. 2023. "Pakar Maritim Sebut Kendala Tol Laut yang Sudah Jalan 7 Tahun." Bisnis Indonesia.com. Accessed May 2023. https://ekonomi.bisnis.com/read/20230105/98/1615559/ pakar-maritim-sebut-kendala-tol-laut-yang-sudah-jalan-7-tahun.

SEAMEO BIOTROP. 2022. *SEAMEO BIOTROP Introduces Agricultural Commodities Produced by Using Tissue Culture Technique in Panen Raya Nusantara 2022.* Accessed May 2023. https://biotrop.org/news/seameo-biotrop-introduces-agricultural-commodi-ties-produced-by-using-tissue-culture-technique-in-panen-raya-nusantara-2022.

Sesditjen DJPB KKP. n.d. "KKP Pacu Pengembangan Daya Saing Rumput Laut Na-sional." Direktorat Jenderal Perikanan Budidaya. Accessed May 2023. https://www. djpb.kkp.go.id/index.php/arsip/c/651/KKP-PACU-PENGEMBANGAN-DAYA-SAING-RUMPUT-LAUT-NASIONAL/?category_id=8.

Soegiarto, Aprilaini and Sulustijo. 1981. "Utilization and Farming of Seaweeds in Indonesia." In *Proceedings of the Symposium on Culture and Utilization of Algae in Southeast Asia*, edited by I. J Dogma, G. C. Trono, and R. A. Tabbada. Iloilo, Philippines: Tigbauan.

Soethoudt, J. M. (Han), Heike Axmann, and Melanie G. Kok. 2022. *Indonesian Seaweed Supply Chain Analysis and Opportunities.* Wageningen Food & Biobased Research. Wageningen: Wageningen Food & Biobased Research.

Sulfiani. 2022. "Peran PT. Biota Laut Ganggang Dalam menanggulangi Pengangguran di Desa Polewali Kecamatan Suppa Kabupaten Pinrang." Undergraduate Thesis. Fakultas Ushuluddin, Adab dan Dakwah. Institut Agama Islam Negeri Parepare. Parepare. Indonesia.

Sulsel, Humas Pemprov. 2020. "Di Tengah Pandemi, Perusahaan Pengolahan Rumput Laut Pinrang Tetap Berproduksi dan Tidak PHK Karyawan – Humas Pemerintah Provinsi Sulawesi." Pemerintah Sulawesi Selatan. https://humas.sulselprov.go.id/index.php/2020/11/16/di-tengah-pandemi-perusahaan-pengolahan-rumput-laut-pinrang-tetap-berproduksi-dan-tidak-phk-karyawan.

TRIBUNLUTIM. 2021. "Koperasi Adi Luwung Ekspor Perdana 100 Ton Rumput Laut ke China." *Tribun Timur*. 8 August 2021. Accessed May 2023. https://makassar.tribunnews.com/2021/08/08/koperasi-adi-luwung-ekspor-perdana-100-ton-rumput-laut-ke-china.

Valderrama, D. 2012. "Social and Economic Dimensions of Seaweed Farming: A Global Review". In *Visible Possibilities: The Economics of Sustainable Fisheries, Aquaculture and Seafood Trade.* Proceedings of the Sixteenth Biennial Conference of the International Institute of Fisheries Economics and Trade, July 16–20, Dar es Salaam, Tanzania. Edited by Ann L. Shriver. International Institute of Fisheries Economics and Trade (IIFET), Corvallis: Curran Associates Inc. Proceedings.

van der Heijden, Peter G. M., Romy Lansbergen, Axmann Heike, Han Soethoudt, Gemma Tacken, Jos van den Puttelaar, and Nita Rukminasari. 2022. *Seaweed in Indonesia: Farming, Utilization and Research.* Wageningen Centre for Development Innovation. Wageningen: Wageningen University & Research.

Verheij, E. and W. F. P van Reine. 1993. "Seaweeds of the Spermonde Archipelago, SW Sulawesi, Indonesia." *Blumea: Biodiversity, Evolution and Biogeography of Plants* 37, no. 2: 385–510.

Waldron, Scott, Nunung Nuryartono, Alexandra Langford, Syamsul Pasaribu, Kustiariyah Tarman, Ulfah J. Siregar, Muhammad Farid Dimjati Lusno, Sulfahri, Boedi Sarjana Julianto, Irsyadi Siradjuddin, Radhiyah Ruhon, Zulung Zach Walyandra, Muhammad Imran Lapong, Risya Arsyi Armis, Eugene Sebastian, Helen Brown, Fadhilah Trya Wulandari, Hasnawati Saleh, and Steve Wright. (2022). *Policy Brief: Sustainable Upgrading of the South Sulawesi Seaweed Industry.* Melbourne, Australia: Australia-Indonesia Centre. https://pair.australiaindonesiacentre.org/wp-content/uploads/2022/11/SIP-1-EN-ONLINE.pdf

Zamroni, A. 2021. "Sustainable Seaweed Farming and Its Contribution to Livelihoods in Eastern Indonesia." *Earth and Environmental Science*. The 3rd International Conference on Fisheries and Marine Sciences, Surabaya, Indonesia. IOP Conference Series: 718.

# PART II

# Livelihood transformations

# 4 Export commodity frontiers and the transformation of village life

*Zannie Langford, Radhiyah Ruhon, Zulung Zach Walyandra, and Risya Arsyi Armis*

## Agrarian change in Pitu Sunggu

On a sunny day in Pitu Sunggu, we meet with Pak Cakra at his home near the coastline. The walls are freshly painted a sky blue, ornate furniture decorates the home and an impressive array of biscuits and sweets sit on the table. There are new frilly green curtains hanging over the windows and a fish tank with several goldfish. The house is not new, however. It was built with a large attic to fill with rice after the harvest. But nobody here along the coastline farms rice anymore. The rice fields were converted into shrimp ponds in the 1980s and 90s when tiger prawn prices were high, and now house milkfish and whiteleg shrimp, and most people make their living from the sea farming seaweed. There are signs of wealth all around the hamlet, with new and ornately decorated houses, new motorbikes, and many people doing well. Pak Cakra explains how he started farming seaweed when the price was just Rp. 8,000/kg, and watched it go up – 11,000, 12,000, 13,000, 14,000, up and up, 22,000, 23,000 … now, he says, the price is Rp. 33,000/kg – so times are good for seaweed farmers.[1]

Pitu Sunggu is a place that has seen rapid and widespread transformation of its landscape at several points in time. Some people have done very well out of these transformations, adapting entrepreneurially to new circumstances. Others have had to move around to find work, selling their labour to landowners in the village and its surrounds or moving to other islands. The last century has seen landscapes transformed from mixed agriculture to rice fields, from rice fields to shrimp ponds, and over the last decade, the sea from communal fishing grounds to private seaweed farms. These rapid transformations, pulled along by broader developments in global value chains, have changed livelihoods in agrarian households. They have led to shifting diets, education levels, gender roles, housing, infrastructure and transport, migration and labour use.

This chapter describes the history of agrarian transformation in Pitu Sunggu from the perspective of village residents. Information gained through oral histories with older residents of the villages is triangulated with historical maps, statistical information, and published literature to describe the history of export-orientated agrarian change in the village. This history illustrates the series of agricultural transitions and changes in land ownership which enabled some residents to enthusiastically take up seaweed farming and claim extensive rights to the sea in the early

DOI: 10.4324/9781003183860-7

2000s. Others persisted with established livelihood activities and were left behind on this unexpected new frontier of sea space enclosure. These patterns shaped the way that the current industry is organised and the different livelihoods that people draw from it.

### Pitu Sunggu village

This section of the book focuses on our case study location of Pitu Sunggu, a small village on the coastline of Pangkajene dan Kepulauan Regency, a few hours' drive north of the provincial capital of Makassar. The population of Pitu Sunggu at the end of 2022 was around 2,108, divided into 613 households, 181 of which farm seaweed. The village is divided into three hamlets (Figure 4.1): the coastal hamlet of Pungkalawaki, where the majority of seaweed is farmed; the inland hamlet of Bonto Sunggu, where residents farm little seaweed but own large areas of agricultural land; and the central hamlet of Kampung Baru, a settlement built in the 1970s between the Sidenreng River and the road, whose residents have traditionally worked as traders and who are now heavily engaged in seaweed farming (Table 4.1). Our survey of 96 out of a reported 136 seaweed farmers (at the commencement of the research) in the village found that 55 were located in Pungkalawaki, 36 in Kampung Baru, and 3 in Bonto Sunggu. Few are located in Bonto Sunggu, due to both the inland location of this hamlet and the larger landholdings of this cohort.

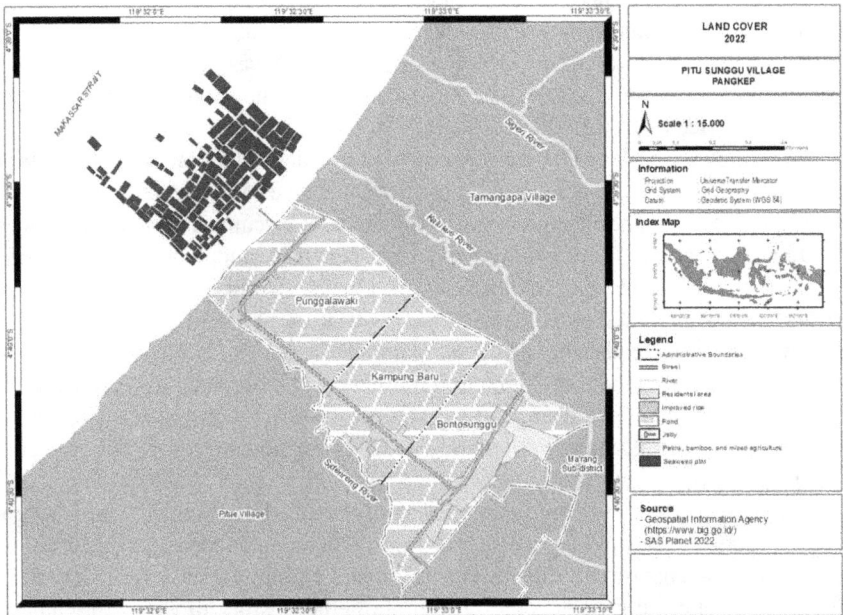

*Figure 4.1* Layout of the village with three hamlets

Source: Map created by Risya Arsyi Armis using ArcMap. Land cover based on maps of village available in office of village head, combined with analysis of satellite imagery and triangulated with published information.

*Table 4.1* Overview of Pitu Sunggu hamlets

| | |
|---|---|
| Pungkalawaki | In the coastal hamlet of Pungkalawaki, residents live along a road running parallel to the coastline and the riverbank close to the river mouth. Most residents are seaweed farmers or fishermen, and some also have brackishwater ponds in which they grow shrimp and fish. Pungkalawaki was historically a remote hamlet until roads were improved in the 1970s. This area was converted to brackishwater ponds as rice fields near the coast salinised. Most seaweed (67 per cent) produced in Pitu Sunggu is from Pungkalawaki farmers, who access the sea via a coastal pier. Before commencing seaweed farming, 67 per cent of Pungkalawaki seaweed farmers had ponds and 80 per cent undertook marine fishing. Pungkalawaki people have a history of working at sea. |
| Bonto Sunggu | The inland hamlet of Bonto Sunggu is located nearest to a main road and fresh water source, and the residents grow rice and vegetables. It has historically been the central settlement in the village as a result of its proximity to the main road, and residents here have historically owned large areas of the agricultural land in the village. It was the site of an Oxfam project to develop organic vegetable and rice farming in the region from 2010–2015, and these agricultural activities continue today in a small area of Bonto Sunggu (see Muchtar, 2017). Bonto Sunggu has a fresh water source and land use is a combination of shrimp and fish ponds, rice fields, and forest. Very little seaweed farming is currently undertaken by Bonto Sunggu residents. Residents typically own agricultural land which they use primarily for pond farming. We located only seven seaweed farmers from this hamlet during our long-term fieldwork, and just three of these appear in the household survey. |
| Kampung Baru | Kampung Baru (literally translated as 'new village') is the most recent hamlet in the village. It is located between Pungkalawaki and Bonto Sunggu and on the banks of the Sidenreng River. Residents of this village moved to this location from their previous location amidst the rice fields which was known as 'Bonto Baddo'. Bonto Baddo had limited access to roads, rivers, or sea, and residents report that poor roads made transporting goods in and out difficult. Government road projects in the 1970s improved the roads in Pitu Sunggu and Bonto Baddo residents abandoned their previous site and moved to live along the new road, forming the new hamlet of Kampung Baru. With access to the sea via the river, many residents took up work as fishermen or seaweed farmers. They typically hold limited land areas and were traditionally seen to have poorer access to education and employment than the residents of the other hamlets, but have recently experienced growth in wealth as a result of seaweed farming. Around 26 per cent of the seaweed produced in Pitu Sunggu is produced by Kampung Baru residents. Before commencing seaweed farming, just 11 per cent of Kampung Baru seaweed farmers had ponds, 64 per cent undertook ocean fishing, crab netting, or clam collecting, and 33 per cent undertook neither of these activities, instead working as traders of fisheries products or undertaking off-farm work. The lack of agricultural land holdings by people in this village may have increased uptake of seaweed farming by people in Kampung Baru, and as a result of their early adoption, they now own large areas of sea space and have rapidly increased their wealth over time. |

*Land use change in Pitu Sunggu*

Over the last century, Pitu Sunggu village has undergone a series of landscape transformations which reflect broad changes that have occurred along much of the coastline in the area. A series of historical maps combined with recent satellite imagery (Figure 4.2) shows the repeated transformation of land use in the region over the last century. Historical maps illustrate land use in the village at four points in time: in 1917 prior to large scale clearing of the land for rice fields the landscape was probably used for mixed agriculture including palm trees and bamboo; in 1925 when the landscape had been transformed into rice fields; in 1981 when the conversion of the land to brackishwater ponds began along the coastline but did not at that point reach inland areas; and in 2022 when the area was almost entirely used for brackishwater pond farming and seaweed farming (Figure 4.1). These historical maps reflect several distinct periods of agricultural land use in the village which are described in the following sections.

## The early history of Pitu Sunggu

Pitu Sunggu sits within the regency of *Pangkajene dan Kepulauan* (hereafter 'Pangkep'), as outlined in the Introduction. The area which is now Pangkep was previously part of the Kingdom of Siang, founded around 1112 AD and persisting until around 1544 (Zainal and Aprasing, 2014). The kingdom was one of the oldest and most influential kingdoms of South Sulawesi (Muhaeminah and Makmur, 2016; Zainal and Aprasing, 2014) and was the first kingdom in Sulawesi to establish trade relations with Europe (Hadrawi et al., 2019), benefitting from the use of Pangkajene River as a strategic port for trading. The kingdom declined in the mid-16th century partly as a result of the siltation of the Pangkajene River such that it was no longer suitable for anchorage for merchant ships (Hadrawi, 2018; Pelras, 1996). At the same time the nearby Gowa Kingdom was growing in strength.

After the decline of the Siang Kingdom, several smaller kingdoms, including the Barasa Kingdom of Pangkajene, grew in the void, but these fell under the rule of the Makassarese Kingdom of Gowa (Makkulau, 2008). In the 1660s some of these kingdoms changed their affiliation to fall under the Buginese Kingdom of Bone, which eventually defeated the Kingdom of Gowa in the 18th century. This kingdom was subsequently governed by the Dutch East India Company (Verenigde Oostindische Compagnie (VOC)) in the 18th century. Throughout the period of VOC control through the 17th and 18th centuries, many Bugis-Makassarese people travelled to other parts of Indonesia and abroad and settled there to avoid the rule of the Kingdom of Bone and the trade monopoly of the Dutch VOC (Cribb, 2000; Drakeley, 2005). In the 19th century the decline of the VOC saw Pangkep come under the direct administration of the Netherlands government (Van Gorsel, 2022). The Dutch rule maintained local systems of power in the region, which both supported its control of the territories while simultaneously fuelling desire for independence (Cribb, 2000). In 1942 with the Japanese occupation, independence

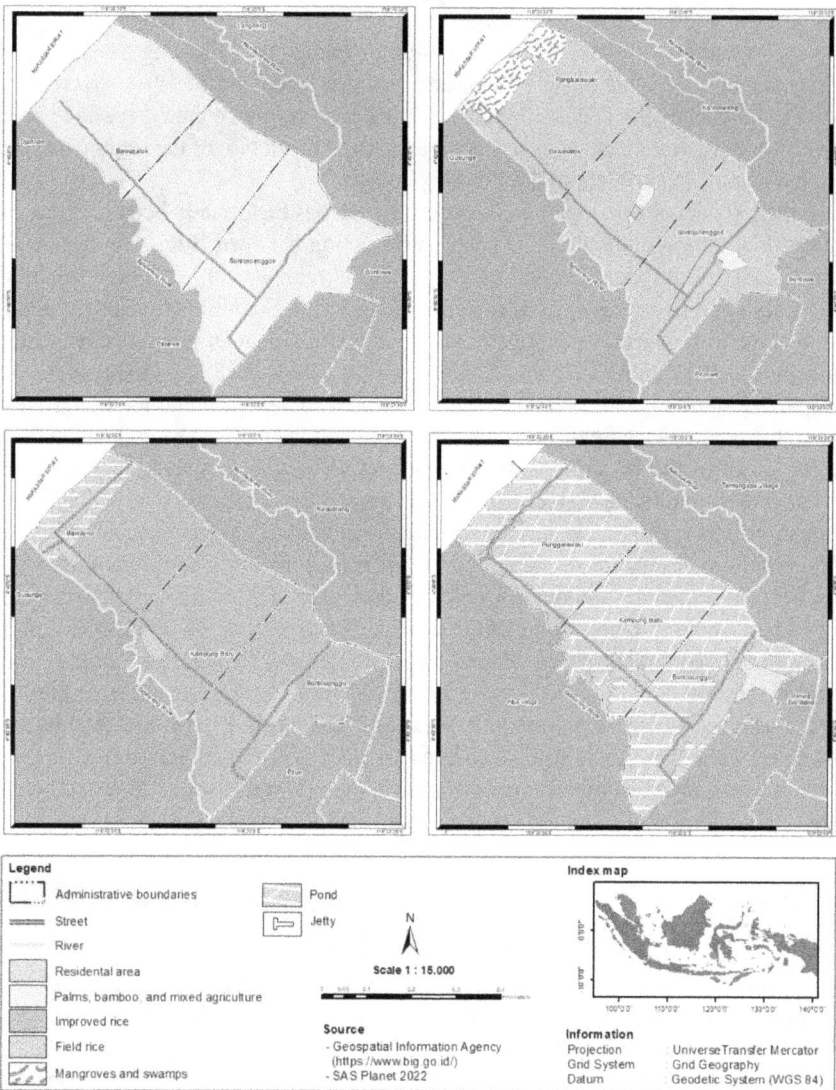

*Figure 4.2* Land use in Pitu Sunggu in 1917 (top left); 1925 (top right); 1981 (bottom left); 2022 (bottom right)

Source: Maps created by Risya Arsyi Armis using ArcMap. Land use data extracted from historical maps. Figure 4.2a is based on a historical map produced in 1943 from aerial photographs taken in 1917–1918 (United States Army Map Service 1943). Figure 4.2b is based on a Dutch map produced based on aerial photographs taken in 1925. (Topografische Inrichting & Topografische Dienst 1927). Figure 4.2c is based on maps available from the Indonesian government showing the landscale in 1981–2 (Bakosurtanal 1991). Figure 4.2d was created based on triangulation of maps on display in the Pitu Sunggu village office, satellite imagery from 2023, and primary data. Administrative boundaries shown are those currently in use.

struggles in Indonesia intensified (Cribb, 2000), and resistance movements grew in several parts of South Sulawesi, including Pangkep. Following the Japanese surrender in 1945, Indonesia proclaimed independence and in 1949 sovereignty was recognised by the Dutch after an extended process. Since then there have been several administrative changes to the organisational structure of the region, and the village names and borders in the Pangkep region.

Pangkep is therefore located in a region which has historically been of strategic and economic importance. It sits at the intersection between Bugis and Makassar ethnic groups, and identification of people in the region with these two groups is often fluid and overlapping. The area that is now Pitu Sunggu was sparsely populated at the turn of the 20th century, and the first available map of land cover which informs the 1917 map in Figure 4.2 shows that in 1917–1918, the village area was primarily 'bamboo' and 'palm' (coconut), with 'woodland' (including mangroves) along the coasts. Rice fields were already visible on the maps in an area about 5km to the south of the village but were not identified within Pitu Sunggu. This does not necessarily indicate that no rice was grown in the area, rather that the land was probably used for mixed cropping activities which are not finely differentiated in the colonial maps. Most residents did not directly recall this period as they were not born at that time, although a few of them offered short reflections on what they knew about it. One respondent noted that 'I heard from the old people that in colonial times, people here consumed sweet potatoes or bananas as a staple food, I heard that my grandfather ate those foods'. Several unpaved roads were present in the village, but residents reported that until the 1970s these were muddy and provided limited accessibility. According to historical maps, between 1917 and 1925 the land area of Pitu Sunggu was almost entirely transformed into rice fields, beginning a period of more than fifty years of rice farming.

## Rice farming

Older residents of Pitu Sunggu today recall the rice farming era which began in approximately the 1920s and continued until the rice fields were converted into shrimp and fishponds in the 1980s and 1990s. This era was divided into two distinct periods – the period of field rice farming from the 1920s to the 1970s, in which a low yielding variety was grown with minimal inputs and no irrigation yielding just one harvest per year, and an intensified period which occurred after the Green Revolution in the 1970s, in which farmers grew a new variety of rice using chemical fertilisers.

### *Field rice farming (1920s–1970s)*

For most of the rice farming era of Pitu Sunggu (until the Green Revolution in the 1970s), the type of rice grown was a variety of field rice known as *Ase lapang*. Residents recalled the flavour of this rice favourably, describing the grains as 'big and fragrant' with a pleasant taste. As one resident recalls, the rice 'had hair, was tall, had a long growing period, was harvested using [a wooden tool known as] *ani-ani*, and yielded little'. The rice was grown in fields without irrigation, and as a result of the reliance on rainwater, cultivation for most of the fields was limited to a single crop over a three-month period in the rainy season. Buffalo were used to work the fields and their manure was used as fertiliser.

Residents' recollections of this period vary depending on their landownership status. Rice fields were unequally distributed amongst the villagers, and some of the land was owned by non-residents of the village, including residents of nearby villages and residents of offshore islands. As one resident recalls, 'there were people from outside [the village] who owned land [here]. It depends, if people here were wealthy they might have had land, but if not, even though they lived here they didn't have a pond or a rice field.' Landownership was dominated by a small group of 'rural elite' who were able to accumulate land over time, and who were mostly residents of the inland hamlet Bonto Sunggu. The majority of residents did not have significant land holdings, and generally recall this as a difficult period when food was scarce. Residents reported mixing their rice with low-cost vegetables as the supply of rice alone was insufficient to meet household needs. As one resident recalled,

> We had corn and sweet potato which we cut into small pieces and then cooked. There was no other food at that time. The food was purchased at the Segeri Market and was brought to the village by carrying it on your shoulders and walking on foot. There were no vehicles at that time ... all the land in the village was used for growing rice. Corn and sweet potato could not grow ... [The rice fields] belonged to the villagers, but there was still not enough food. It used to be very hard, people suffered. Even the fine corn husks were eaten. There were often people who did not cook because they didn't have any food. I sometimes feel sad when I think about the suffering of the people in the past. I will always remember that time.

Small fish supplemented these starchy foods and were reportedly relatively cheap during that period. They could be obtained from rice fields where, as one resident described, 'at that time there were no pesticides, so usually when we were working in the fields, we also caught fish'. Fish, shrimp, crabs, and clams were also caught in the sea using traditional boats and fishing methods, and these were sold at local markets.

Some residents of Pungkalawaki and Bonto Baddo (the former residence of Kampung Baru people) owned rice fields, but many worked as labourers in the rice fields of others – a practice known as *ma'sangki*. This occurred under a profit-sharing system in which the labourers would receive a portion of the harvested rice. As one respondent recalled:

> My father was a farm labourer [on a] rice field [owned by] a resident of an island. The owner could not come to the village so he asked my father to work on his rice fields and the yield was divided in two ... some [of the rice fields] are owned by people outside the village, that's why a lot of people here are

labourers. There are also villagers who own rice fields but there are not many of them, only certain people in the village do.

It was common for women, men, and children alike to undertake *ma'sangki*, both within the village and outside it. One respondent described working with her parents as a child:

> I would work on other people's land, in remote places, too, because we helped our parents. Because I had many younger siblings ... after graduating from junior high school I went to Pinrang and Sidrap for farm work. Because I wasn't going to school anymore – a money problem. Usually we would go somewhere for about 15 or 20 days. When the harvest in our village had finished we would move to another area for farm work, and when the work was finished there, we would move again, to another place.

Around one third of Kampung Baru seaweed farmers report that prior to commencing seaweed farming they undertook off-farm work or migrated to look for work. Migration for off-farm work has been common in Pitu Sunggu historically, and many residents reported periods of 'wandering' to Makassar, Kalimantan, and Papua for work. During the field rice farming era, seasonal migration for work was common due to the low productivity of the field rice, as one resident described:

> Before, after the rice was planted, the farmers would leave it unattended and go to the cities to earn another living ... many people would work elsewhere, such as being pedicab drivers in Makassar. They would return to the village when it entered the harvest season. Fortunately, at that time there were no planthopper pests.

This movement was made both necessary and possible by the non-intensive nature of the field rice farming system, which was single-crop, unirrigated, with low inputs. Farmers report that they did not experience significant issues with pests at that time and attribute pest issues to increasing use of insecticides and chemical fertilisers – possibly because such inputs interfere with natural biological control mechanisms (Settle et al., 1996; Heong et al., 2014). Farmers had to move around to earn their living during this period, as one resident described:

> The problem was that rice fields cannot be planted all year round, only during the west monsoon season, and at that time there was no irrigation, so when the west monsoon season ended and the harvest was over, there was no other work, so I would go to Makassar.

Farming activities in the period were therefore dominated by non-intensive rice farming supplemented by marine fishing and off-farm work. Rice field inputs of labour were low, fertilisers were limited to buffalo manure, irrigation was rainfed only, and rice fields once planted required little additional work until harvesting.

Fish were obtained through marine fishing and fishing in rice fields, but starch needs were often unfulfilled and many residents recall this as a difficult and 'hungry' time. This situation continued until the Green Revolution entered the village in the 1970s.

### The Green Revolution (1970s)

From the early 1970s, the Indonesian government encouraged the implementation of 'Green Revolution' technologies. The five-pronged approach included the introduction of high yielding rice varieties, chemical fertilisers, synthetic pesticides, irrigation and improved planting methods, stimulating a transition from low-input, low-productivity field rice production to high-input, high-productivity (IR8, locally known as *citarung*) rice production (Rahmi et al., 2020). These methods supported substantial increases in Indonesian rice production over time (Figure 4.3).

In line with broader national programmes (e.g. Hansen, 1972), residents recall being 'pressured' by the government to carry out the directives of the Green Revolution, but also that the yields of farmers who adopted them were so much higher that soon many farmers took them up of their own volition. As one farmer described,

We were not *forced*, but the community did not want to accept something which they did not know about … [but] after it had been proven that the new variety produced more rice, we didn't need to be told to grow it anymore, we went along with it ourselves … after it was proven, many wanted it because it was really good … the new rice could produce up to eight tons per hectare, before we got a maximum of three tons.

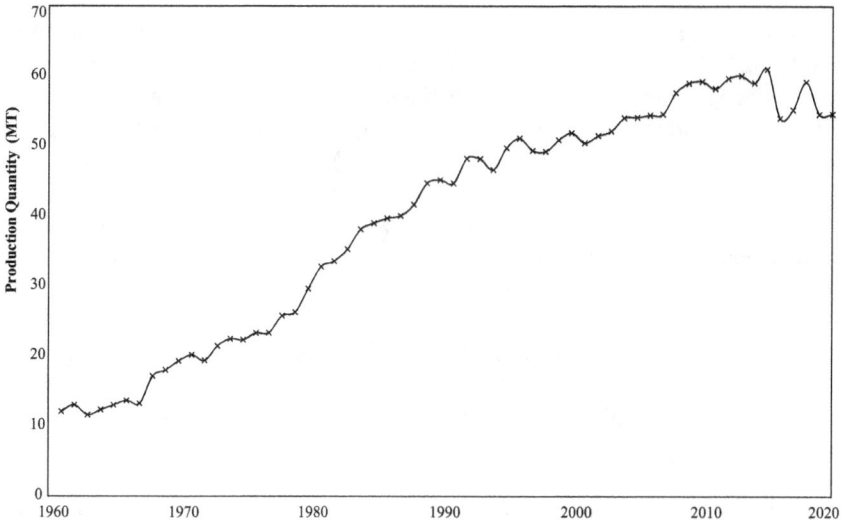

*Figure 4.3* Rice production in Indonesia over time
Source: Data from FAO 2023.

To demonstrate the yields of the new rice and associated technologies, the government ran a competition in which farmers were encouraged to plant the new rice variety, and the farmer who achieved the highest yield would receive a prize: the funds to take a pilgrimage trip to Mecca. The competition demonstrated the higher yields that could be achieved using the Green Revolution technologies and led to rapid and widespread uptake of the new rice variety and use of fertiliser and pesticides by farmers as part of their regular routine. Farmers reported that with these new technologies, starchy food sufficiency was achieved by many for the first time. As one farmer described, 'I remember when my grandmother planted the new varieties of rice, when it was harvested, some [rice] was sold and some consumed without mixing it with other foodstuffs'.

Over time, however, yields from the rice fields reportedly declined and farmers began to face issues with brown planthopper pests and fertiliser supply. This decline in rice field productivity was a push factor which led many to convert their rice fields to shrimp ponds. According to village residents, Green Revolution technologies made possible rice production on a scale large enough to support a year-round population in the village and greatly improved access to rice. However, the use of fertilisers in the fields meant that rice fields could no longer be used to catch fish, and that farmers relied on supply of affordable fertilisers and insecticides. As outlined below, rice fields were also undergoing salination. Gradually, yields reportedly declined, and by the time that shrimp farming became a viable option for farmers, many were already dissatisfied with the yields from their rice fields and open to new livelihood activities.

### Shrimp and fish farming (1980s - present)

Today, Indonesian aquaculture and fisheries products are a US$12 billion industry (FAO, 2023). Most of the value in the industry stems from a small number of products: whiteleg shrimp (24 per cent), nile tilapia (16 per cent), carrageenan seaweeds (14 per cent), catfish (13 per cent), common carp (9 per cent), milkfish (9 per cent), and giant tiger prawns (6 per cent) (Figure 4.3). The Pitu Sunggu landscape is dominated by brackishwater ponds (Figures 4.5 and 4.6) in which polyculturing of whiteleg shrimp with milkfish is common. Common carp are also grown with tilapia in some ponds, and tiger prawns are grown alongside whiteleg shrimps in others. These are some of the most widely cultivated aquaculture species in the world – whiteleg shrimps (also known as vannamei prawns or king prawns) contribute about 12 per cent of the total value of aquaculture globally and are common on supermarket shelves. Nile tilapia, common carp, and tiger prawns are also produced in large volumes, together contributing a further 8 per cent of global aquacultural production value. Indonesian currently produces about 11 per cent of the global market value for these species.

At the start of the 1980s, however, these species were little cultivated in Pitu Sunggu. Their production at scale was encouraged by the Indonesian government during the 1980s to bolster export earnings after the banning of shrimp trawling (Muluk and Bailey, 1996). This saw significant government support for, and donor

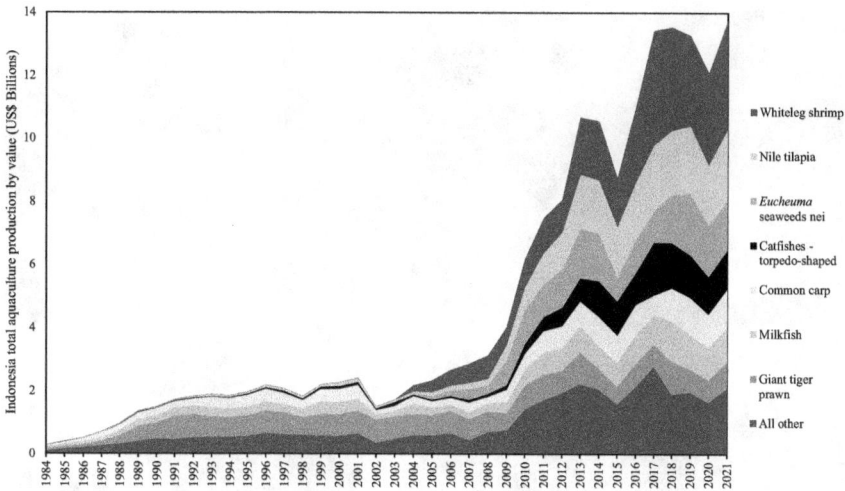

*Figure 4.4* Indonesian aquaculture production by value
Source: Data from FAO 2023.

investment in, the industry through the 1980s (Muluk and Bailey, 1996). This support provided the necessary infrastructure and market access for Pitu Sunggu farmers to begin undertaking this style of farming themselves from the 1980s (Muluk and Bailey, 1996; Hall, 2004).

### Rice field salination and early pond construction (1980s and 1990s)

Until the 1980s, Pitu Sunggu agricultural land was used almost entirely for rice fields. A few brackishwater ponds existed, but these were limited to the area between the houses lining the coastal road and the coastline, and were probably focused on fish farming with shrimp produced as bycatch (Hall, 2004). In the 1981 map shown in Figure 4.2, brackishwater ponds (*tambak*) along the coastline can already be observed, and these remain today (Figures 4.5 and 4.6).

An examination of surrounding areas on the same historical map (1981) reveals the presence of brackishwater ponds along much of the Pangkep coastline, particularly in areas that on earlier maps were described as 'swamps'. Ponds were also present along many adjacent low-lying river areas in the region, although not in Pitu Sunggu at that time. These ponds were relatively low productivity and involved polyculturing of shrimp and milkfish. This is consistent with the memory of one village resident, who said:

> In the past, ponds were only on the banks of rivers and on the seafront … ponds were only managed in a very traditional way, farmers only relied on the moss that grew in the ponds as fish food. The ponds were only harvested once a year.

*Figure 4.5* Drone image taken along the Pitu Sunggu coastline showing housing along the coastal road surrounded by brackishwater ponds

Source: Image by Nur Ihsan recorded May 2022 offshore of Pitu Sunggu village.

*Figure 4.6* Drone image taken along the Pitu Sunggu coastline showing mangroves along the coastline separating the ocean from the coastal ponds

Source: Image by Nur Ihsan recorded May 2022 offshore of Pitu Sunggu village.

The transformation of Pitu Sunggu's rice fields to brackishwater ponds began in the early 1980s, driven by a combination of push and pull factors. Seepage of salt water from the rivers and coastline and flooding at high tides led to salination of rice fields and low yields of rice. This pushed owners to convert the fields to brackishwater ponds for the cultivation of shrimp and fish. Residents reported that over time, as rice fields were converted to fishponds, neighbouring fields experienced declining yields of rice as salt water seeped into their ponds, so were pushed into converting their fields into brackishwater ponds. As one resident recalled:

> It quickly became all fishponds here. Why? Because if we dig [a pond] here, well, automatically the [neighbouring] area will also be salty. On the left, right, front and back, you will also get salty [water], and therefore you can't grow rice. That's how it started. So the process went from rice fields to ponds quickly.

This salination led to the conversion of rice fields to shrimp ponds through the 1980s and 90s, particularly in areas affected by salination, which moved from the coastal and riverbank areas inland. This created issues for those who had worked as labourers on rice farms, because pond farming requires much less employed labour and is often undertaken by the pond owners themselves. Residents report that as a result, crime increased in some parts of the village and many were driven to seek work outside the village as a result of the declining labour needs of former rice farmers.

### The Asian Financial Crisis and the tiger prawn boom (1997–1999)

The conversion of most of the remaining rice fields to shrimp ponds occurred during the Asian Financial Crisis (*Krisis Moneter* or *Krismon* in Indonesian) which began in 1997. Between 1997 and 1999, the value of the Indonesian currency rapidly depreciated such that the effective price Indonesian farmers received for export products paid in US dollars rapidly increased (Erdmann and Pet, 1999; Rimmer et al., 2013). The effect of this was an intensification in the Indonesian export-orientated fisheries sector (Erdmann and Pet, 1999). In Pitu Sunggu, the sudden increase in the prices for tiger prawns (*Penaeus monodon*) had a dramatic effect on village life and Pitu Sunggu residents remember this period as one of prosperity. It is linked in people's memories to the presidency of President Habibie, as one respondent described:

> During President Habibie's era [1998–1999] all rice fields were dug up and used as ponds. The conversion of rice fields into ponds was driven by the skyrocketing price of tiger prawns, so that the whole community was provoked to convert their rice fields into ponds. At that time, shrimp production and shrimp prices were very high. It was very easy to earn a lot of money at that time.

Residents observed that the widescale conversion during this period was no longer linked to the 'push' factors of low yields and salination of rice fields – rather, 'when Mr. Habibie was president, the price of export goods increased tremendously, so there were many rice fields which were turned into ponds, even though they were not contaminated [with salt water]'.

Residents who owned ponds remember this as a period of great wealth rather than one of crisis, as one resident recalled, '[People] really enjoyed [that time] ... Many people started to buy motorbikes ... they had a lot of money, and goods at that time were very cheap so people bought a lot of things, mainly household furniture.' Another noted the rapid transformation of the village with this influx of wealth, much of which was directed at the construction of improved housing:

> In this village, significant changes occurred when the shrimp prices were high. At that time, many residents built stone houses, and every day there were trucks bringing building materials to the village. There were so many of them, my family visiting from Makassar thought that there was a government construction project in this village, because they saw so many trucks.

The effect of the crisis on residents who did not own ponds varied. Some found work labouring on the ponds of others or received support from their newly wealthy neighbours and kin. Others left the village in search of work elsewhere – as one resident described, 'because of the conversion, farm laborers were no longer required for their services, because the landowners worked on their own ponds. [Rice] farm workers eventually migrated to find work.' Some residents found this a difficult period, as one reported, 'during the crisis, there were lots of thieves running around because there was nothing to eat. Cows and chickens were prone to being stolen. It was hard to keep livestock safe, because they were all being stolen.' The effects of the monetary crisis are still visible today in the housing in the village, with many people who owned ponds at the time of the crisis living in large houses containing ornate furniture that they report buying during the period of the tiger prawn boom.

### White spot disease and the conversion to whiteleg shrimp

However, this period of wealth was short-lived. The Indonesian Rupiah recovered within a few years, reducing the price paid to tiger prawn farmers. In addition, pond farmers were increasingly affected by 'white spot disease' which causes rapid mortality in prawn stocks (Dey et al., 2020). To combat the decline in production due to disease, the new, smaller, lower value 'whiteleg shrimp' (*Litopenaeus vannamei*) was introduced (see Flegel, 2009; Hall, 2004; Muluk and Bailey, 1996; Rimmer et al., 2013). This species dominates Pitu Sunggu ponds today, typically cultivated alongside milkfish and sometimes tilapia. Whiteleg shrimp is the largest aquacultural product in the world by value, and the most widely cultivated shrimp globally. It is widely found on supermarket shelves and known variously as 'vannemei prawns', 'king prawns', or 'Pacific white shrimp'. These shrimps have replaced tiger prawns across much of Asia since 2002 (Flegel, 2009). However, they have a lower unit value than tiger prawns and are polycultured with milkfish, which require higher applications of fertiliser. Farmers recall that after the tiger prawn boom they experienced lower prices and higher fertiliser requirements, and the resulting poor quality of the shrimp saw many flex back into their other livelihood options. As one farmer recalls, 'the price [of shrimp] went down and the quality of the shrimp also started to go down, so in the end the people were no longer interested. So they switched back to looking for crabs.'

Ponds are still a viable, but not highly profitable, livelihood activity. Pond production intensified during the 1980s and 1990s with government support, including support for fertiliser subsidies to encourage farmers to intensify their ponds. However, these subsidies are no longer available and farmers report that ponds are now less profitable than rice fields:

> Now the farmers want to return from the ponds to [rice] fields but they can't ... currently pond production is declining because there is no fertilizer [subsidy for pond farmers] ... They want to return to being rice farmers because ... currently, there is more income from rice fields than income from ponds.

The resurgence in interest in rice farming is partially a result in the shift in fertiliser subsidies available for this activity. It is also at least in part a result of a project run by Oxfam from 2010–2015 (see Muchtar, 2017) which introduced organic farming methods that have since seen favourable vegetable harvests achieved by a small number of farmers, and rice farming reinstated in a few areas of the inland hamlet Bonto Sunggu, where farmers benefit from the diversified cultivation options. However, these farmers are a small minority in the village and are all located close to the freshwater source in Bonto Sunggu. Most farmers continue to work on low productivity ponds which they harvest for their own consumption and for sale. The modest returns of the ponds following the initial boom saw farmers returning to crab fishing. Some began to experiment with a new commodity, which would grow to become the most significant income earner for coastal and riverbank residents: seaweed farming.

## Seaweed farming

### *Claiming of sea space*

Seaweed farming was first trialled in Pitu Sunggu in the early 2000s through several small-scale attempts to gather and farm seaweed which were not maintained. The first sustained farming of seaweed probably occurred around 2005, though reports vary. At first seaweed farmers faced considerable community resistance to the installation of permanent anchors in the sea, which they used to delineate an area of private use, in contrast to the traditional use of the sea as a communal resource. This process will be described in detail in Chapter 5. As more and more residents installed anchors and claimed plots for themselves their right to do so gradually became more widely recognised by their kin and neighbours. Through this process, early adopters of seaweed farming were able to claim large areas of sea space in ideal locations, while those who started farming seaweed later had to make their plots in more marginal areas which were further from shore or had unfavourable growing conditions. Importantly, the structure of sea ownership recognised by residents today was settled at a time when seaweed prices were much lower and the activity was much less profitable than it is today. It was taken up primarily by those who already earned their livelihoods by fishing in the sea, by those who had limited alternative livelihood activities, and by those who had the ability to invest in the new production techniques.

Initially, new seaweed farmers received capital for seaweed farming from interested parties and investors from outside the village, as well as from the local government and from a non-governmental organisation. Government assistance was received in the form of training programmes, seaweed seeds, and bamboo drying platforms. Village residents also developed low-cost ways of building anchors in the sea to overcome capital shortages in the establishment of seaweed plots. Pungkalawaki residents – who live along the coastline of Pitu Sunggu – were the first to take up seaweed farming in large numbers. Results from our survey (outlined in the Introduction, and with a full account provided in Langford et al. 2024) suggest that they currently represent about 58 per cent of the Pitu Sunggu seaweed farming cohort and produce 67 per cent of the village's seaweed (Figure 4.7). Large numbers of Pungkalawaki residents took up seaweed farming between 2005 and 2011, and our survey indicated that they are the largest seaweed farmers in the village, producing on average of 1.9 tonnes dry seaweed each in 2021. Kampung Baru residents adopted seaweed farming later and more gradually, with most farmers taking up seaweed farming between 2010 and 2019. Today, they represent 38 per cent of seaweed farmers and 27 per cent of production, on average producing 1.1 tonnes in 2021 – 40 per cent less seaweed than the average Pungkalawaki seaweed farmer. There are only a few Bonto Sunggu seaweed farmers – during 16 months of fieldwork we only found 7 Bonto Sunggu residents who had seaweed farms – but these few farmers on average produce higher volumes than those from other areas. In recent years, the price of seaweed has increased dramatically, including, roughly, doubling from Rp. 10,000/kg in early 2017 to over Rp. 20,000/kg by 2018, and spiking to nearly Rp. 50,000/kg from June 2021 to October 2022, before dropping again to around Rp. 15,000/kg by September 2023 (Langford et al. 2023 and refer to Chapter 2). As a result, seaweed farming has created considerable wealth for some who have access to the sea in which to undertake it.

**Employment creation**

The most labour-intensive job associated with seaweed farming is tying seaweed propagules to ropes prior to planting. This work is time-sensitive, such that a group of five people might be required to achieve the tying required to replant an area within a day. As a result, farmers do not normally do all of this work themselves but pay others to do it for them. The creation of employment through this practice is widely seen as one of the important positive social impacts of seaweed farming. This is felt particularly strongly in Pitu Sunggu, where, as discussed above, many people have traditionally had to migrate to find work. This seaweed binding work can be undertaken by women while caring for small children, as well as by elderly people and those with physical disabilities such as blindness or limited mobility

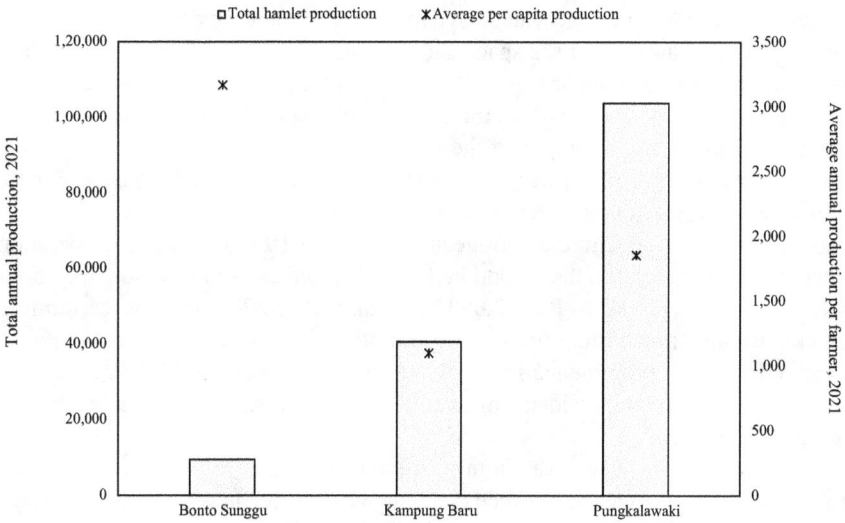

*Figure 4.7* Total and average seaweed production by hamlet

Source: Data from authors' survey.

who may not be in a position to undertake other forms of work. It is considered to be an adequate but not highly paid job compared to cultivation by the farmers themselves. It also involves sitting in the same position for long periods of time and binders who do not wear gloves may experience severe skin irritation from prolonged contact with the seaweed. Nonetheless, many people prefer it to working in the sun on the ponds. One resident observed how seaweed farming had changed labour relations between people in Kampung Baru and Bonto Sunggu:

> I used to inspire the children around my house that the people of Kampung Baru were creative and hardworking people. In the past, the people of Bonto Sunggu Hamlet were the employers of the Kampung Baru community, who were agricultural labourers, they worked on our ancestral fields. But now, the people of Bonto Sunggu Hamlet are slowly becoming labourers for the Kampung Baru community, and the hamlet's economy is growing. The growth comes from seaweed. This is proof that if you work hard and diligently, it will give you a good change in life. Please look at the community of Kampung Baru. If you look closely, the number of people who own cars is higher in Kampung Baru Hamlet [than here in Bonto Sunggu].

Chapter 8 explores the work of seaweed binders in more detail.

### *Booms and busts*

The Pitu Sunggu seaweed industry has experienced several periods of boom and bust driven by price changes. Throughout 2014 and 2015, seaweed prices declined steadily to half their previous value. As a result of this ongoing decline, many farmers sold or gave away their sea space and farming equipment and migrated to find work, often in Kalimantan or Papua. One farmer lamented that he had made several seaweed plots in the early days of the industry, but had moved away for work and given them away, not anticipating the future profitability of the industry: 'I gave them all away … and now [the new owners] don't want to give them back'. These transactions occurred before sea space was in short supply.

In 2017, the large Chinese carrageenan processor BLG commenced operation in nearby Pinrang, and in the second half of 2017, Makassar prices more than doubled from Rp. 9,400/kg to Rp. 22,000/kg (Zhang et al., 2023). Seaweed farming quickly became much more profitable, prompting many of the residents to return to the village and recommence seaweed farming. The agricultural and aquacultural livelihoods available to residents of the village are therefore closely linked to patterns of migration.

In the early years of seaweed farming, farmers struggled to maintain year-round cultivation of the species of seaweed that they were cultivating: *Kappaphycus alvarezzi,* known colloquially as 'cottonii'. The year 2017 marked an important point in the development of the industry in Pitu Sunggu because it was the year that another species was introduced to the village: *Kappaphycus striatus,* known colloquially as 'sacol'. Sacol contains the same type of carrageenan as cottonii so was easily incorporated into supply chains by traders, who often mix the species together. Sacol is more tolerant of oceanic conditions during the Pitu Sunggu dry season, so the introduction of this species meant that farmers could alternate between sacol and cottonii to achieve year-round production. Since the introduction of this species, seaweed farming activities have intensified. Households that previously undertook both fishing and seaweed farming have spent increasing time on the latter. This has also increased the demand for seaweed binders to the extent that labour supply within the village is no longer sufficient, and some farmers have to transport their seaweed to groups of binders located in other, inland villages, in order to access workers.

### Crab fishing

Over time, there have also been changes in the crab fishing activities undertaken alongside seaweed farms. Initially, village residents caught mostly mangrove crabs, which they sold at market, to local traders, or to hotels in the city. Gradually, crab trading became more common and a crab peeling business (for blue swimmer crabs) was established in the village, which bought small crabs and sold the meat in the city. At the same time, crab fishermen shifted from using nets that they sewed by hand themselves to purchasing nets from stores. Today, crab fishing is highly seasonal and prices fluctuate significantly. Over the year of our observation, prices changed from Rp. 50,000/kg in January, to a peak of Rp. 90,000/kg in April, before falling to Rp. 20,000/kg over the following months. Fishermen rarely catch large crabs but instead catch large numbers of small crabs (between 5–25 kg/day, depending on the season and the price). Crab fishers typically do not differentiate their catch based on crab size, gender, or pregnancy, but sell the entirety of their load directly to traders. There is a trader based at the pier throughout the crabbing season to buy crabs directly off the boats. The crab peeling business in the village also creates jobs for 10–20 residents, although this is far fewer than the number of people employed in seaweed binding.

While the seaweed industry has created positive benefits for many village residents, the effects on crab fishermen have been less positive. Crab fishermen, who previously had access to the entire coastal area for crab fishing, are now excluded from a large part of the coastal space and their activities have been squeezed into the gaps between seaweed farms. Many crab fishermen struggle to make an adequate living, and as the seaweed farming area is virtually all taken, it is difficult for crab fishermen to switch to the more lucrative activity. These dynamics will be explored in more detail in Chapter 5.

### Livelihood specialisation

Today, Pitu Sunggu seaweed farmers are highly specialised in seaweed farming. Most of them (60 per cent) also undertake marine fishing (Figure 4.8). Pond farming is also common, it is undertaken by 38 per cent of farmers, while some also undertake off-farm work (11 per cent), crab fishing (10 per cent), and collecting ocean shells (7 per cent). Notably, just 1 per cent of farmers surveyed reported producing rice, and no respondents reported growing fruit or vegetables or rearing livestock. This is because the land area of Pitu Sunggu is almost entirely taken up by brackish-water ponds, with some coconuts growing on the banks between them. Pitu Sunggu farmers are therefore highly export orientated in their production, and rely on the sales of seaweed and/or shrimp and fish for the money they need to purchase food.

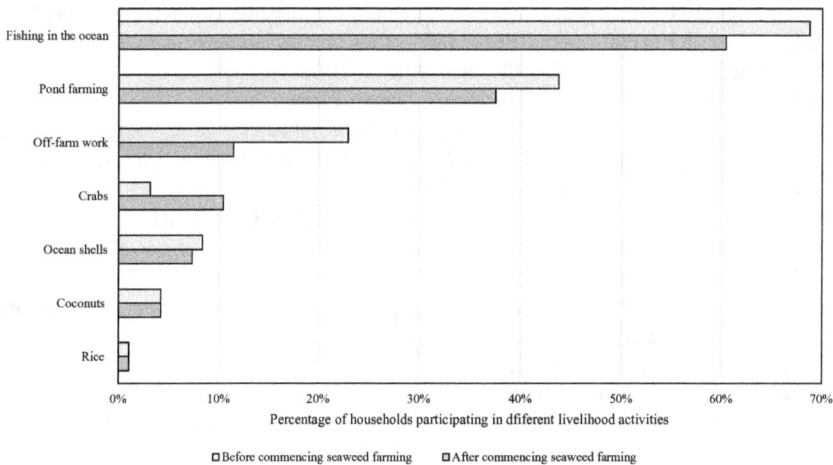

*Figure 4.8* Livelihood activities of Pitu Sunggu seaweed farmers

Source: Data from authors' survey.

This is unlike the situation in the nearby village of Laikang, where 27 per cent of seaweed farmers also grow rice and 18 per cent also grow corn. Pitu Sunggu seaweed farmers are therefore quite specialised in export-orientated products – 77 per cent of them earn more than half of their household income from seaweed farming and 83 per cent employ more than half of their household labour in seaweed farming.

### Export-driven change in Pitu Sunggu

Pitu Sunggu is a place that has seen a number of transformations of the landscape with the introduction of new crops and technologies, occurring alongside environmental changes which further push them into new productive domains. Seaweed farming was not the first and is unlikely to be the last export crop to transform village livelihoods in Pitu Sunggu. The long-term patterns in land ownership, intensification, land degradation, and infrastructure development created a period in village development where certain residents found themselves searching for a new livelihood activity. They took it up with enthusiasm, claiming rights to the sea which paid off when seaweed prices rose. This chapter has explored how historical context created a cohort of farmers poised to adopt a new crop. The next chapter shows how this process occurred and was contested, negotiated, and accepted in different contexts.

### Note

1 In the months that followed, farm-gate prices in Pitu Sunggu rose to a peak of Rp. 48,000/ kg (and Rp. 50,000/kg in nearby Makassar), before falling again through 2022 and into 2023.

### References

Bakosurtanal. 1991. "Lanskap wilayah Desa Pitu Sunggu berdasarkan fotogrametri foto udara skala 1:100.000 tahun 1981–1982." https://tanahair.indonesia.go.id/portal-web/ downloadpetacetak/Zip?skala=50K&namaFile=2011-33.zip

Cribb, R. 2000. *Historical Atlas of Indonesia*. Surrey: Curzon Press.

Dey, Bipul K., Girsha H. Dugassa, Sheban M. Hinzano, and Peter Bossier. 2020. "Causative Agent, Diagnosis and Management of White Spot Disease in Shrimp: A Review." *Aquaculture* 12 (2): 822–865. https://doi.org/10.1111/raq.12352

Drakeley, S. 2005. *The History of Indonesia*. Connecticut, US: Greenwood Press.

Druce, Stephen C. 2009. *The Lands West of the Lakes: A History of the Ajattappareng Kingdoms of South Sulawesi 1200 to 1600 CE*. Leiden, The Netherlands: KITLV Press.

Erdmann, Mark V. and Jos S. Pet. 1999. "Krismon & DFP: Some Observations on the Effects of the Asian Financial Crisis on Destructive Fishing Practices in Indonesia." *SPC Live Reef Fish Information Bulletin 5*: 22–26.

FAO (Food and Agriculture Organization). 2022. "FAOSTAT: Data." Food and Agriculture Organization of the United Nations, Rome. https://www.fao.org/faostat/en/#data

FAO (Food and Agriculture Organization). 2023. "FISHSTAT Plus – Universal Software for Fishery Statistical Time Series. Aquaculture Production 1950–2009." FAO Fisheries and Aquaculture Information and Statistics Service. Food and Agriculture Organization of the United Nations, Rome. https://www.fao.org/fishery/en/fishstat

Flegel, T. W. 2009. "Current Status of Viral Diseases in Asian Shrimp Aquaculture." *Israeli Journal of Aquaculture, Bamidgeh* 61 (3): 229–239. https://doi.org/10.46989/001c.20556

Hadrawi, Muhlis. 2018. "Sea Voyages and Occupancies of Malayan Peoples at the West Coast of South Sulawesi." *International Journal of Malay-Nusantara Studies* 1 (1): 80–95.

Hadrawi, M., N. Agus, T. Isao, and Basiah. 2019. *Jejak Kerajaan Siang-Barasa Berdasarkan Sumber Lontara.* Semantic Scholar.org. Accessed 26 February 2023. https://www.semantic-scholar.org/paper/JEJAK-KERAJAAN-SIANG-BARASA-BERDASARKAN-SUMBER-OF-Hadrawi-Agus/99b3b276fc2df41506f1d4d9aa1d137fadf7d06a

Hall, D. 2004. "Explaining the Diversity of Southeast Asian Shrimp Aquaculture." *Journal of Agrarian Change* 4: 315–335.

Hansen, Gary E. 1972. "Indonesia's Green Revolution: The Abandonment of a Non-Market Strategy toward Change." *Asian Survey* 12 (11): 932–946. https://doi.org/10.2307/2643114

Heong, Kong Luen, Larry Wong, and Joy Hasmin Delos Reyes. 2014. "Addressing Planthopper Threats to Asian Rice Farming and Food Security: Fixing Insecticide Misuse." In *Rice Planthoppers edited by* Kong Luen Heong, Jiaan Cheng and Monica M. Escalada, pp. 65–76. Dordrecht: Springer Netherlands. https://doi.org/10.1007/978-94-017-9535-7_3

Langford, A., S. Waldron, N. Nuryartono, S. Pasaribu, B. Julianto, I. Siradjuddin, R. Ruhon, Z. Zach, I. Lapong, R. Armis. 2023. *Sustainable Upgrading of the South Sulawesi Seaweed Industry.* Melbourne, Australia: Australia-Indonesia Centre. https://pair.australiaindonesiacentre.org/research/sustainable-upgrading-of-the-south-sulawesi-seaweed-industry-2

Langford, Alexandra, Scott Waldron, Jing Zhang, Radhiyah Ruhon, Zulung Zach Walyandra, Risya Arsyi Armis, Imran Lapong, Boedi Julianto, Irsyadi Siradjuddin, Syamsul Pasaribu, and Nunung Nuryartono. (2024). "Diverse Seaweed Farming Livelihoods in Two Indonesian Villages." In *Tropical Seaweed Cultivation – Phyconomy: Proceedings of the Tropical Phyconomy Coalition Development, TPCD 1, Held at UNHAS University, Makassar, Indonesia. July 7th and 8th* edited by A. Hurtado, A. Critchley, and I. Neish. 00–00. New York: Cham, Switzerland: Springer Nature.

Langford, A., J. Zhang, S. Waldron, B. Julianto, I. Siradjuddin, I. Neish, and N. Nuryartono. 2022. "Price Analysis of the Indonesian Carrageenan Seaweed Industry." *Aquaculture* 550 (737828). doi.org/10.1016/j.aquaculture.2021.737828

Makkulau, F. 2008. *Sejarah Kekaraengan di Pangkep.* Pustaka Refleksi. https://www.koleksilokal.com/2022/02/sejarah-kekaraengan-di-pangkep.html

Makkulau, F. 2017. *Tradisi Lisan: Berlomba dengan Kematian (1).* Palontaraq. https://palontaraq.id/2017/04/09/tradisi-lisan-berlomba-dengan-kematian

Muchtar, Adinda Tenriangke. 2017. "Understanding the Influence of Development Interventions of Women Beneficiaries' Perceptions of Empowerment: A Case Study in South Sulawesi, Indonesia." PhD Thesis, Victoria University of Wellington, New Zealand.

Muhaeminah, and Makmur. 2016. Masa Awal Hingga Berkembangnya Kerajaan Ajatappareng (Abad 14–18). *Pubawidya*, 4 (2), 125–135.

Muluk, Chairul and Conner Bailey. 1996. "Social and Environmental Impacts of Coastal Aquaculture in Indonesia." In *Aquacultural Development*, 1st ed., 193–209. New York: Routledge. https://doi.org/10.4324/9780429046773-11

Pelras, C. 1996. *The Bugis.* Cambridge, MA: Wiley-Blackwell.

Hidayat, Rahmi Aulia, Johan Iskandar, Budhi Gunawan, and Ruhyat Partasasmita. 2020. "Impact of Green Revolution on Rice Cultivation Practices and Production System: A Case Study in Sindang Hamlet, Rancakalong Village, Sumedang District, West Java, Indonesia." *Biodiversitas* (Surakarta) 21 (3): 1258–1265. https://doi.org/10.13057/biodiv/d210354

Rimmer, M. A., Ketut Sugama, Diana Rakhmawati, Rokhmad Rofiq and Richard H. Habgood. 2013. "A Review and SWOT Analysis of Aquaculture Development in Indonesia." *Reviews in Aquaculture* 5 (4): 255–279. https://doi.org/10.1111/raq.12017

Settle, W. H., H. Ariawan, E. Astuti, W. Cahyana, A. L. Hakim, D. Hindayana, and A. S. Lestari. 1996. "Managing Tropical Rice Pests through Conservation of Generalist Natural Enemies and Alternative Prey." *Ecology* (Durham) 77 (7): 1975–1988. https://doi.org/10.2307/2265694

Topografische Inrichting and Topografische Dienst. 1927. "Gouvt. Selébès en Onderh: zuidwest/Topographische Inrichting." Jakarta: Topografische Inrichting, cartographer. and Netherlands. Topografische Dienst. http://nla.gov.au/nla.obj-234654080

United States Army Map Service. 1943. Southern Celebes 1:125,000 / prepared under the direction of the Chief of Engineers, U.S. Army, by the Army Map Service, Indianapolis Unit. Indianapolis Unit, cartographer. and Allied Land Forces. South-East Asia. Directorate of Surveys. and Great Britain. Army. Royal Engineers. Map Production Company, 110. http://nla.gov.au/nla.obj-234438708

Van Gorsel, J. T. 2022. "Geological Investigations of Sulawesi (Celebes) before 1930." *Berita Sedimentologi*, 48 (1), 79–114. https://doi.org/10.51835/bsed.2022.48.1.391

Zainal, A., and A. Aprasing. 2014. The Emergence of Early Kingdoms in South Sulawesi. *Journal of Humanity*, 2 (1), 23–48. https://doi.org/10.14724/02.02

Zhang, Jing, Scott Waldron, Zannie Langford, Boedi Julianto, and Adam Martin Komarek. 2023. "China's Growing Influence in the Global Carrageenan Industry and Implications for Indonesia." *Journal of Applied Phycology* June: 1–22. https://doi.org/10.1007/s10811-023-03004-0

# 5 From communal access to private ownership

## Negotiating rights to the sea

*Zannie Langford, Radhiyah Ruhon,*
*Zulung Zach Walyandra, Risya Arsyi Armis, and*
*Imran Lapong*

### Enclosure

Looking out across the waters of Pitu Sunggu, the ocean is visibly full of seaweed farms, tightly packed together with only small gaps between them for boats to pass through. Just twenty years ago, seaweed farming was virtually unknown in this area, and fishermen moved around the space freely, catching fish, netting crabs, and trading with offshore islands. The sea was a communal resource which could be used by anybody, and nobody could be prevented from using it. However, today it is divided into individually owned plots of sea space (Figure 5.1) which can be bought, sold, rented, and inherited.

Seaweed farming has caused a dramatic change in the way that the sea is used and managed. Unlike fishers and crab netters, who typically move around across large

(a)                                                (b)

*Figure 5.1* a) A seaweed farmer working on his farm; b) Pitu Sunggu seaweed farms from above

Image from drone footage by Nur Ihsan recorded May 2022 offshore of Pitu Sunggu village.

DOI: 10.4324/9781003183860-8

areas in search of their catch, seaweed farming requires that farmers be able to claim use of an area of sea space for the full farming cycle – around six weeks (Figure 5.1). During the time that the seaweed is growing, they must be able to exclude other ocean users – such as crab netters and passing boats – from their area of the sea, as boats passing through a seaweed plot can damage the seaweed or become tangled in the ropes supporting the farm. In addition, in order to set up their farming apparatus, farmers must install anchors to attach their farm to the sea floor, known as seaweed farm 'foundations'.[1] These foundations delineate the area of the farm and are expensive to set up. To justify the investment, farmers need to be able to guarantee that they will be able to continue to use them for longer than just one cycle – normally for several years. This requires a different type of right to use the sea: one that is both exclusive and permanent. This is a significant change from the customary use of sea space, in which people have traditionally enjoyed the non-exclusive right to temporarily access the sea: they were free to use it, but so was everybody else.

The success of the seaweed farming transition has therefore depended on the conversion of the ocean from a public access area to space divided into discrete, individually owned farming plots, where farmers can work the same area of sea space over many months or years. It relies on excluding other ocean users from seaweed farming areas, and, in Pitu Sunggu, the excluded users are primarily crab netters, who with the seaweed farming transition have been pushed out of shoreline areas, and now must set their crab nets further offshore or in the gaps between seaweed farms. The concept of land 'enclosure' has been widely used to interpret the drivers and effects of 'the division or consolidation of communal ... lands' into 'carefully delineated and individually owned and managed farm plots' (*Britannica* 2013, n.p.). Early writing on enclosure in Europe explored the multiple interacting forces that led to enclosure of common land from the 12th to the 19th century, driven not only by landholding elite, but also by farmers themselves:

> Contrary to the popular idea that enclosure was wholly a landlord's movement ... there was a distinct effort on the part of the peasantry ... to abandon the open-field system and escape compulsory co-operation with the lazy and shiftless ... there was no compulsion on the customary tenants ... to make them enclose; theirs was a purely spontaneous movement prompted by a desire to escape obsolete restrictions. There was also another motive – the need of self-protection. The growth of large grazing farms, and the consequent over-stocking of the commons, led the small men to enclose as the only way to keep some of the pasture for their own use.
>
> (Curtler 1920, pp. 65–66)

As Curtler notes, enclosure resulted from multiple divergent efforts on the part of different groups of people operating under interacting pressures over a long period of time, resulting in uneven geographies of enclosure. He emphasises the agency of farmers within these processes, as actors in the process driven by both entrepreneurialism (desiring to 'escape compulsory co-operation' with others), and the increasing pressures they faced as a result of broader structural changes.

More recent work on enclosure has similarly emphasised both 'top-down' and 'bottom-up' processes of enclosure. In Indonesia, enclosure has often been driven by large-scale land acquisitions for agriculture (such as palm oil plantations), mining or conservation (for example, Goldstein 2016; Hall et al. 2015; Ito et al. 2014; McCarthy et al. 2012; Pichler 2015; Schoenberger et al. 2017; Semedi and Bakker 2014). However, enclosure is also a process undertaken 'from below', by smallholder farmers themselves (see, for example, Alkhalili 2015; Castellanos-Navarrete and Jansen 2015; Curry and Koczberski 2009). Investor- and farmer-driven processes often interact as smallholders both respond to external pressures (for example, Giorgio et al. 2022) and entrepreneurially pursue new agricultural opportunities, such as those offered by export commodities (for example, Olofsson 2021).

To date, the Indonesian seaweed industry has been resistant to large-scale agribusiness as a result of the practicalities of farming seaweed, which is seasonally variable, requires high labour inputs, and is subject to high levels of risk due to environmental changes (as will be described in detail in Chapters 6 and 7). Consequently, processes of enclosure have occurred within villages largely in the absence of claims on sea space from outside the village. In this context, enclosure has been driven by coastal residents themselves. Hall et al. (2011, p. 145) describe smallholder driven processes of 'intimate exclusion', in which 'neighbours and kin who share common histories and social interaction ... exclude one another from access to land as part of a strategy to accumulate capital'. They note that 'these are "everyday" processes, mundane and piecemeal, that do not grab headlines. But cumulatively, they have the effect of producing agrarian classes with differential access to means of production' (Hall et al. 2011, p. 145). The role of export commodities in triggering the reconfiguration of land access systems has been studied in several contexts. Rights to use land are often gained through the application of labour. In subsistence agricultural systems, constant labour application on land is typically required to maintain its productivity, which reinforces the ownership claim of the user. There are differences in labour use between different types of land-based agricultural activities – annual crops and perennial crops – that impact on the establishment of property rights. Li (2014) describes how the production of cocoa by Sulawesi highlanders generated enclosure of previously communal land. She emphasises the materiality of cocoa as central to this process, since

> [t]ree crops like cacao do something by their permanence. When highlanders planted cacao in their fields, the presence of the trees disrupted the cycle in which they cleared a patch of forest, used it for a few seasons, then left it to fallow. The trees also changed the ownership status of the land, transforming it into individual property, since no one else could use the land thereafter. Excluding other users, and other uses, wasn't new: planting a field of corn also required exclusion, at least until after the harvest. The new element was permanence.
>
> ( pp. 84–85)

Similar processes occur in Vanuatu where production of kava in large quantities for export, which requires lower labour inputs than traditionally produced food crops, enables the use of much larger areas of customary land than was possible under traditional cultivation systems (Langford 2019, p. 26; 2022). In such cases, changing agricultural livelihoods shifted the way that communal land was used and claimed, in some cases laying the groundwork for the creation of what is now recognisable as more formal land markets, in which it is possible to buy, sell, rent, inherit, and lend against plots of previously communal land. This chapter explores how Pitu Sunggu village transitioned from a system of shared communal rights to access the sea to a system comprised primarily of private ownership rights – in less than two decades. It outlines how this system was negotiated by seaweed farmers and other ocean users, how it is (tentatively) maintained and stabilised today, and the potentially conflicting agendas of provincial and national government marine zoning programmes and local marine space governance. We begin with the first seaweed farmer.

## The first seaweed farmer

At the turn of the century, seaweed farming was not yet established as a livelihood option in Pitu Sunggu. On the land, coastal residents cultivated shrimp and fish in brackishwater ponds, and in the sea they caught fish and crabs which they sold, traded, or consumed. As discussed in Chapter 4, marine fish were an important dietary supplement. The sea was viewed as a communal resource: anyone could use the sea to find sustenance, and no one had the right to exclude others. As one village resident put it, '[I]n the sea, no one may forbid you from searching … just like when we came to the village, there was no rule to forbid us from searching in the sea to find sustenance'. Crab fishing was a major livelihood strategy in this region. Crabs are caught by installing temporary gill nets in the ocean which are left for hours or overnight (depending on the net size and other factors) and are checked regularly for crabs caught in the net. They are more permanent than net fishing because they occupy an area of the sea for up to a few days, but are still relatively transitory, since crab fishermen move around to optimise their harvests of crabs with different tidal and seasonal fluctuations in ocean conditions.

From our interviews and oral histories, it appears that seaweed farming was first successfully introduced to the village sometime around 2007,[2] most likely by a man we will call Dodi. Dodi recalled that a family member from Bone Regency had come to see him to tell him about seaweed cultivation, reporting that the results were very good. He decided to try it, and travelled to Makassar to buy ropes, which were not available in the village, and then rented a car and drove to the regency of Bone, some 170km away, to buy seeds to start the cultivation. Dodi describes the community resistance to the new activity:

> When I first started farming seaweed, there were many challenges … the crab fishermen did not agree with my seaweed farming, because my farm displaced their crab nets … They went to the house of the village head and protested against my farming … [they said] 'why put a rope in the sea?' Many people were angry, because the seaweed ropes prevented them from looking for crabs … but I persisted. Even at night, I went out to sea to guard it. I brought a light and a flashlight to guard the seaweed and tell the crab

fishermen who were looking for crabs at night not to cross my seaweed longlines. I did it with the help of my neighbor. I went out to sea after dinner, and guarded it until 9pm, when the crab fishermen were no longer active ... Often [people would damage my seaweed farm]. They broke my rope, longline and anchor. Some used boat propellors, some used machetes ... For a year, every day there was an argument.

Early adopters of seaweed cultivation report that their activities were not welcomed by other village residents. As one early farmer described,

In the beginning, many people hated this seaweed. They said, "what kind of job is that? After being lowered into the sea, [the seaweed] is raised again?" They really hated it. But I did it anyway, [although] I felt embarrassed ... people laughed at us at that time.

As others began to experiment with seaweed farming, conflict over the use of sea space intensified, beginning a period of conflict that lasted for several years and involved direct confrontation between those who sought to claim areas of the sea for seaweed farming and those who wished to maintain the system of communal access. The conflict was concentrated around crab netters and seaweed farmers, as one farmer described:

People here argued because there were those who want to catch crabs and those who wanted to plant seaweed. How can you put a crab net if there is a seaweed plot there? The seaweed [ropes] will get stuck in the net. That's the reason they fought.

Prior to seaweed farming, crab fishers were responsive to changing ocean conditions and would move their nets around and change the type of net they used depending on prevailing conditions. They would test locations in several areas by installing a net there, and if the catch in that location was good, would install additional nets. They resisted the construction of seaweed farms as it prevented them from moving around in this way and reduced the area available to them. The imposition of seaweed farms excluded them from areas of the sea, and deprived them of the access rights they had previously enjoyed. In particular, residents of the coastal hamlet of Pungkalawaki (who had traditionally specialised in marine capture livelihoods) opposed the construction of new plots. As one farmer described:

In the past, none of the Pungkalawaki residents wanted to plant seaweed ... if anyone wanted to build a new foundation, it was prohibited by the Pungkalawaki people, because they only liked to net crabs and the area for crab netting was large when it was not planted with seaweed ... [The situation lasted for] almost five years. People would stand guard at the wharf and if someone wanted to build a foundation, the Pungkalawaki people would take away the ropes they were using ... It used to be difficult, almost every day people wanted to fight.

Fierce resistance to seaweed farming persisted for a number of years. Residents recalled conflict between the new seaweed farmers and the crab netters, who were angry about the decreasing area available for them to catch their crabs. As one early seaweed farmer described, '[when they] passed our houses, they always provoked fights ... they were annoyed and asked "why did you make the foundation where we usually hang our crab nets?"'. The residents of Pungkalawaki, including the hamlet Head, unilaterally forbade seaweed farming along the coast of Pungkalawaki (the borders of which were marked by the mouths of the Bawapitu and Langoting Rivers). At that time, every Friday at the mosque it was announced that seaweed farmers should not make plots along the coastline of the village. Despite this, seaweed farming gradually began to encroach on the village coastline, often established by residents of neighbouring villages. Residents of the coastal hamlet grew increasingly frustrated with their inability to exclude people from the area. One of the residents expressed regret that even though these seaweed farmers had violated the rules, they could not be punished for their actions. One influential village member, Pak Vino, who had vehemently opposed the construction of seaweed farms in the region, described to us how over time he became increasingly anxious about the situation. Seeing the expansion of seaweed farms along the coastline of the hamlet, he realised that soon there would be no space left. Accepting this, he relented and began to install his own farms along the coastline in order to avoid missing out on claiming an area of sea space for himself. His actions were observed by other village residents and were followed by a rush for sea space and a rapid dissipation of resistance to the practice. Seaweed farmers quickly became a majority within the village.

**Claiming the sea: first come, first served**

From 2007 to 2011 there was a rapid increase in the number of seaweed farmers, from 16 to 50, and these were mostly from Pungkalawaki, the coastal hamlet that had initially resisted the construction of farms (Figure 5.2). This was followed by another period of uptake and consolidation from 2011 to 2019, when the total number of seaweed farmers in the village rose from 50 to 93 (largely due to residents of Kampung Baru hamlet taking up the practice), while existing farmers continued to accumulate plots of sea space.

Crab fishermen who established their own seaweed farms were no longer in a position to protest against the farms of others, and as more and more people took up this activity, resistance dissipated. Specialised crab fishers no longer actively protested against seaweed farming, as they found themselves surrounded by family and friends who had plots. As one crab fishermen said,

I was bothered by the seaweed farms because when I went to the island [to net crabs] there would be people there installing seaweed [in the crab netting locations] ... [but I] didn't say anything, I just felt it in my heart because [seaweed farming] cannot be forbidden. Because here we are all family, so you can't say anything.

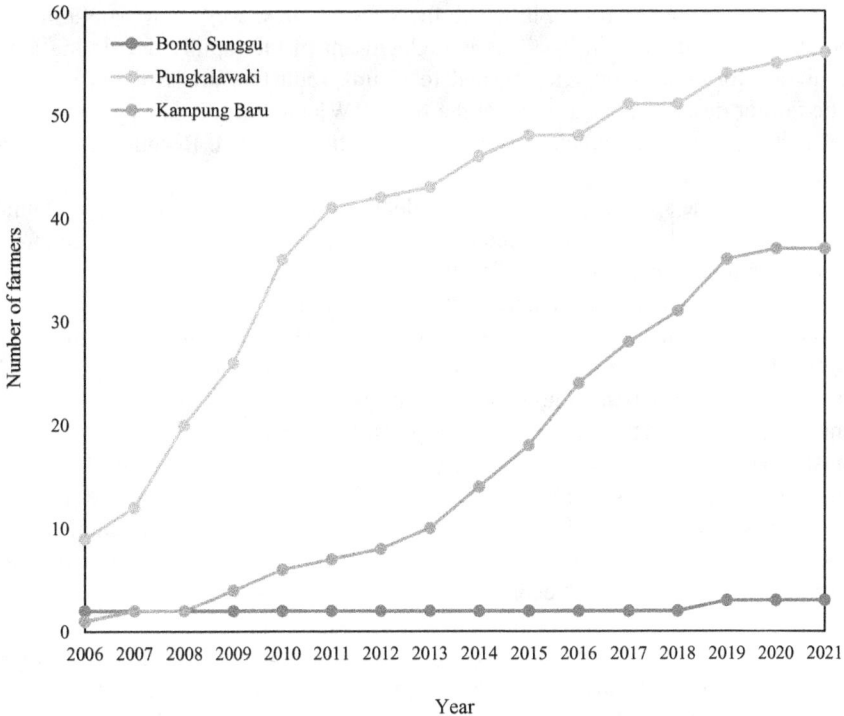

*Figure 5.2* Increase in number of seaweed farmers over time by hamlet

Source: Data from authors' survey.

During fieldwork we spoke with several crab netters who complained about the challenges of crab fishing in the seaweed dominated area. However, the situation has become less clear over time as many coastal residents undertake a combination of activities. Sometimes crab netters have felt unfairly marginalised by the encroachment of seaweed farming, and at other times they have been the beneficiaries of it. As one farm extension officer put it, '[T]here is no conflict between seaweed farmers and crab fishermen because the two professions are carried out by the same person'.

As seaweed farmers rushed to claim areas of the sea for their farm locations, their selection of sea areas was complicated by the variable suitability of the coastal space for seaweed farming. Oceanic conditions can change rapidly and often vary across small areas. Farmers could only assess if seaweed would grow well in a particular location by trial and error, such that they would 'test' an area by constructing a foundation, planting seaweed in it, and observing its growth. As one farmer said,

> In the beginning, there were still not many people working on seaweed. So we spread the cultivation area. We checked the condition of the waters – "Oh, so it's like this". Then we moved to another location and then checked again the condition there. And so on until we got to the waters that were quite far away.

When farmers found an area in which the seaweed grew well, they would install additional foundations in that location. Cognisant of the rapidly decreasing area available for cultivation, they rushed to claim productive areas of sea space – one farmer described the attitude at the time, '[W]hen you see that your seaweed [growth] is good, immediately expand your cultivation area. Because later there will be no space left.'

Although the spatial variability in productivity contributed to the rapid claiming of space, it also led to abandonment. Many farmers would install foundations in plots and then abandon them if they were unproductive, susceptible to disease, too muddy, too vulnerable to fish grazing, or too far away to be easily used. Unlike some horticultural and forestry crops that entail substantial initial and ongoing investments, it is relatively easy to establish and maintain ownership for seaweed plots, simply by leaving foundations in the water. Thus, even if unproductive, many households retained claims to the plots. Furthermore, many of these previously abandoned plots have been brought back into operation as seaweed farming methods have changed – through, for example, the introduction of the double-line method which made farms more resilient to waves, the introduction of the sacol seaweed species which was suitable for cultivation during the dry season, and higher prices which make it economic to farm areas of lower productivity.

In addition, as seaweed is highly vulnerable to disease caused by a sudden change in ocean conditions (brought about by factors such as high rainfall), some farmers have pursued strategies of geographic diversification, in which they build plots in a range of different areas to minimise the risk of losing all their seaweed after a significant weather event (as will be discussed further in Chapters 6 and 7). As a result, many farmers have claimed large areas of sea space, and sea space ownership is unequally distributed in the village: the top 25 per cent of seaweed farmers produce 51 per cent of the village's seaweed, while the bottom 25 per cent produce just 6 per cent. In addition, there are a number of households that want to farm seaweed but are excluded from the industry due to a lack of sea space. The largest producers of seaweed in the village are those who laid claim to plots early in the development of the industry (Figure 5.3).

Some of these seaweed farmers have rights to very large areas of sea space. One farmer described how he had worked proactively to establish his foundations. Every time he heard someone say they had found a good growing area, he would go to the location and build a foundation. In the end, he had so many foundations that he was unable to work all of them. As he boasted to us, '[M]y foundations have never been full. I have planted more than thirty thousand seeds but the site is not full yet … [my location is very] wide.' Another farmer told us that during the rush for sea space he had built so many plots that he was unable to find them all again later, because he had left them idle for a while and the foundation anchors had been buried in sand.

Other farmers – typically those who commenced later – only have access to small areas, sometimes sharing a single plot between multiple family members, or entering into sharecropping arrangements with other residents. One couple described how, after returning to the village from working on another island for

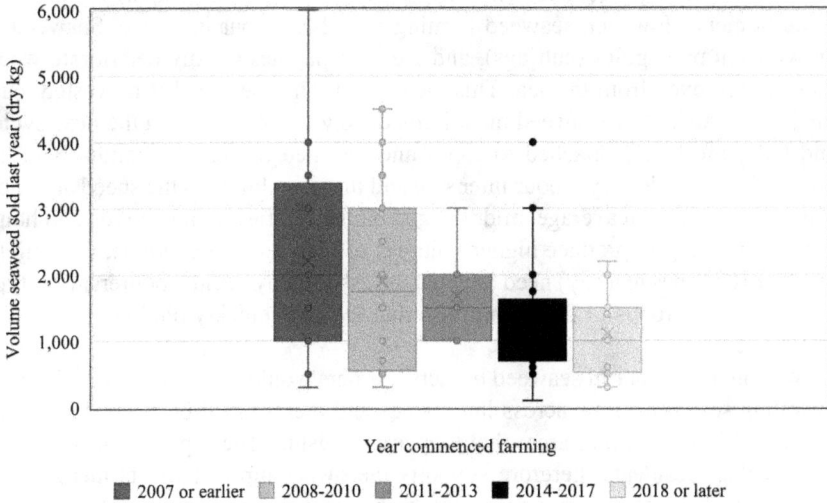

*Figure 5.3* Scale of production of farmers commencing seaweed farming at different times
Source: Data from authors' survey.

several years, they had found that most of the sea had already been claimed and had had to work hard to establish their livelihood in the village. They had first slowly accumulated money by catching crabs, and eventually saved enough to build a small seaweed foundation in a 'boat lane' – an area of the sea proximate to the river kept empty to allow boats moored upstream to travel out to sea. They reported that they kept the foundation as small as possible to try to avoid conflict and allow boats to still pass by. They gradually accumulated money by farming this small plot, hoping to eventually rent or buy a larger plot. As they explained, their inability to access sea space meant that for them, '[I]t's been a long process. Not like people who have a lot of money [to rent a good space].' Despite the unequal division of sea space amongst village residents, owners of seaweed farming areas feel justified in their ownership, which they see as grounded in their own labour and entrepreneurialism. As one seaweed farmer noted when discussing the distribution of sea space in the village, '[E]veryone already has a location, except for those who showed up late'. However, this system of ownership is maintained not only by seaweed farmer labour, but by the ability of seaweed farmers to access the labour of others.

## The development of markets for land, labour, and capital

### Labour and capital markets

The rush for sea space was driven by the pressure villagers felt to secure areas for themselves and was compounded by the uncertain production conditions for seaweed, since the yield from different locations varied considerably and in ways that varied seasonally, such that farmers could not be sure that the plots they had secured would be productive. This led to the claiming of large areas of space by some farmers. However, seaweed farming is highly labour intensive. Seaweed is grown from propagules (cuttings), and these propagules rapidly deteriorate when they are removed from the sea. This means that when seaweed is harvested, cuttings from the mature seaweed must immediately be taken to seed the next cycle, and these must be reattached to ropes and returned to the sea within about 24 hours. This work is very labour intensive and there are limits to the speed at which ropes can be tied. An average, middle-aged binder can tie around one rope an hour. Farmers wishing to produce higher volumes of seaweed (for example, wanting to re-plant 100 ropes in a day) need to enlist the assistance of paid labourers, who typically work in groups of five or more so that they can quickly bind large amounts of seaweed.

Without the labour of seaweed binders, farmers would not be able to profit from operating seaweed farms across large areas and there would be fewer incentives to maintain ownership of areas that they are not using. The supply of labour from other village residents therefore supports the ownership of large farmers. More recently, as farming has increased with high prices, farmers have begun sourcing labour from nearby inland villages because wages are lower and there is higher availability.

In addition to labour, farmers need access to capital to benefit from ownership of sea space, to pay for ropes and the purchase new seeds when changing seasonally between seaweed species or when their crop is affected by disease (as happens regularly). Sea space ownership is therefore founded not only on claims to the sea, but also the ability to access the labour and capital necessary to benefit from those claims. Labour is typically provided by groups of mostly female seaweed binders, who work together to attach seaweed to longlines (as will be discussed further in Chapter 8). Capital is provided by the farmer themselves, by banks through the People's Business Credit (*Kredit Usaha Rakyat* (KUR)) scheme, by pawnshops, and by seaweed traders, who provide financial services to farmers in addition to buying their seaweed (as will be discussed further in Chapter 9). We found that it is common for farmers to borrow from multiple sources concurrently, and to repay old debt using new debt from other sources. The system of sea ownership has therefore been enabled and sustained by the development of labour and financial markets within the village. These features have supported the ownership of large areas of sea space by some farmers, leading to a situation in which the space which is productive for seaweed farming has been fully claimed – creating a condition of scarcity required for the assetisation of land (Visser 2017; Langford, Lawrence, and Smith, 2021).

### Sea space markets

The filling of the sea space in Pitu Sunggu has meant that newcomers to the industry are unable to claim areas of the sea in productive growing areas, and instead need to rent, borrow, or buy space from other seaweed farmers. Seaweed farming is viewed as a highly profitable activity, and seaweed farmers see their farms as important assets and sources of wealth. However, some farmers have areas larger than they can manage themselves or have claimed areas which they don't farm. Some who claimed space in the sea have since moved away or grown too old to work their plots. Meanwhile, as prices paid for dried seaweed more than doubled between July 2021 and October 2022, seaweed farming came to be viewed as a much more lucrative activity. Some people from the village who moved away in order to find work began to return to try their hand at seaweed farming. As one farmer described, '[T]his year … the nomads, city people, everyone – they came home [to work on seaweed]'. As a result, there was demand for space in the sea from those without claims to it, and some people who had claimed areas of the sea but did not wish to use it, leading to the development of a market for sea space.

*Renting*

The most common type of sea space transaction is the leasing of a foundation to another farmer. Renting practices are highly seasonal, as the seaweed farming year is divided into two seasons. The 'dry' season runs roughly from May to October, in which the species cultivated is mostly 'sacol', and the region of cultivation is mostly in areas close to the coast (100–500m from the shoreline). This gives way to the 'wet' season, roughly from November to April, in which the 'cottonii' species of seaweed is grown in areas further from the coast (>500m from the coast). These locations can be a long way from the coast (up to 5km) as in this area there are several offshore islands and a large expanse of shallow waters that farmers make use of. The seasonality of seaweed farming is clearly visible on satellite imagery (Figure 5.4). Farmers report that the areas suitable for cultivation in the dry season are full, and this area is closely packed with seaweed farms during these months.

In the wet season, farmers can achieve good seaweed growth in outer areas but also experience a high level of risk associated with crop losses due to sudden changes in ocean conditions (as will be discussed further in Chapter 6). Consequently, the dry season is widely considered to be the most profitable season for seaweed farming in Pitu Sunggu, but many farmers do not have foundations in locations suitable for cultivation during this season. In addition, variable spatial productivity means that some plots are not productive, while others yield good seaweed harvests. As one farmer explained,

> In that sea, all fertility is not the same. There are indeed points in the sea that are considered rather fertile, there are also those that seem to be barren land. If only all locations in the sea were fertile, no one would fight over space, no one would need to rent space. That's why people rent the sea space, because in the right position, which is good, which is fertile, they don't have a place.

Due to the shortage of space suitable for dry season cultivation, the leasing of space during this season is common. One respondent, who owns around 900 stretch ropes, mentioned that although he owns several foundations in suitable dry season locations, this area is still not big enough to fit all of his 900 ropes, so he rents three

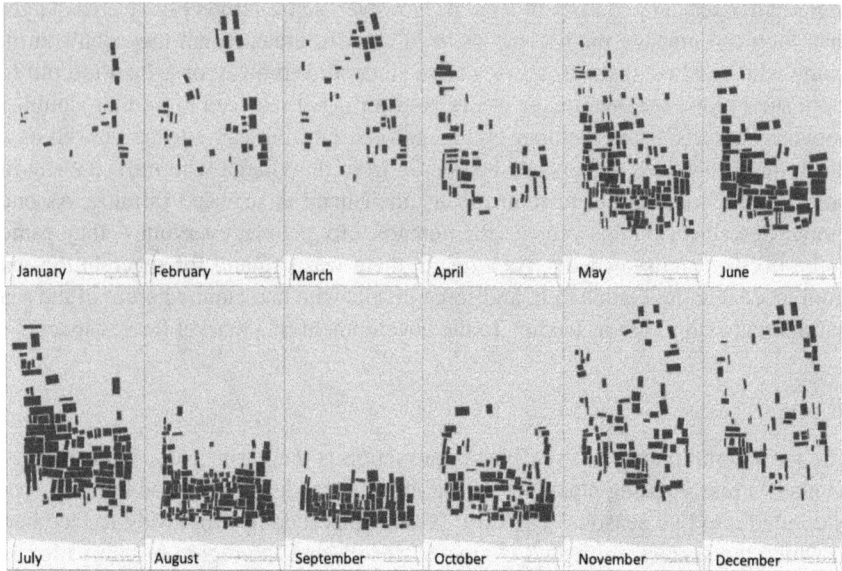

*Figure 5.4* Seasonal change in sea space use for seaweed farming in Pitu Sunggu in 2022
Source: Image created by Zannie Langford using data from Planet Labs for 2022.

additional foundations during the dry season. However, during the rainy season leasing happens infrequently because many farmers own foundations in waters that are quite a long way from the coast. In fact, according to one of the farmers, there is still enough area in the outer waters for new foundations to be laid. Indeed, there seems to be a class of 'senior farmers' who own plots close to the shore, while many newer entrants to the industry only have space further out to sea. Typically, farmers from Pungalawaki hamlet who started earlier (Figure 5.2 above) dominate this class of senior farmers, while farmers who started later (many of whom are residents of Kampung Baru) are less likely to own foundations close to the coast.

As a result, foundations are commonly leased during the dry season, where a fee is paid to the foundation owner for its use. There is no set price for plot rental, rather the price depends on the growth rate of the seaweed in that area (referred to as the 'fertility' of the plot), the size, location, and duration of lease (often a year or 'season', but flexible arrangements are also common). With the doubling of seaweed prices that occurred during the 2022 financial year (as described in Chapter 2), demand for sea space also grew, putting upward pressure on rental prices. As one renter described:

> The owner will set the price. Many people want to plant seaweed, but because we don't have any space, those who do have space will raise the price for us to rent ... Previously it was Rp. 500,000 (~$50) [for the season], then it increased to Rp. 1 million (~$100). My cousin increased the price like that too, he told me that 'seaweed is already valuable. If you don't want to rent the

space [at that price], someone else will take it' … Many people are looking now, even though the rent is already now up to Rp. 3 million ($300).

Lease prices in the village vary widely between Rp. 500,000 and Rp. 5,000,000 per foundation ($50–$500), with a duration that typically spans a planting season but in some cases lasts until it is reclaimed by the owner. Farmers differentiate between the 'quality' of different plots in assessing a foundation rental price. For example, one farmer referred to a location close to the mouth of the river as a 'first class' location. The renting of plots provides an avenue for young people and new farmers to commence farming, but the high price of leases make this a somewhat risky venture and excludes people who do not have the capital required. Many new farmers took up seaweed farming by using plots belonging to a family member, or by establishing small plots in unfavourable locations. Barriers to entry of seaweed farming have therefore increased significantly since the industry was initiated about two decades ago.

*Buying*

Foundations can also be sold if a farmer is willing to permanently part with their right to their foundation. We found only a few reports of sales of foundations, and these had occurred several years earlier at varying prices. One sale occurred a few years ago for around Rp. 1 million (~$100) for a 500 stretch plot (capable of holding 500 'double' ropes), and another in 2021 for Rp. 5 million ($500) for a 1,000 stretch plot. These prices are currently considered very low relative to rental prices and seaweed farm profits after the recent price increase. As a result, sales of seaweed farming foundations are uncommon, and farmers report that it is currently very difficult to find a foundation for sale, as they only become available when farmers are quitting the business permanently. When asked whether he would ever consider selling his foundation, one farmer responded, 'Never. I will bequeath this foundation to my children … nothing can interfere [with that].' Most sales that occur tend to be between relatives and are therefore at prices below that which would be expected on an open market. Foundation owners not using their plots prefer to rent them out rather than sell them, as this provides a regular source of income and retains their right to use the plot in the future if production conditions change or prices remain high.

*Borrowing*

Although the market for sea space appears to be maturing, there are still some farmers who prefer to lend their land without charging a fee. These farmers feel that since anyone may find themselves in need of help at any time, it is important to cultivate a supportive relationship between farmers. As one such farmer said,

I have never bought [a foundation], nor have I rented [one]. At most, if people want to borrow [my foundation], I just lend it to them, no charge …

I have a foundation ... that can fit around 1,000 ropes, it hasn't been filled in 7 years ... [It is] not [for rent], people there just use it ... [if I want to use it] I just come, I say that I want to fill it. People here [are like that] ...we just exchange information with friends, that such and such [belongs to] such and such, [but you can] just use it for now.

Some farmers appear to have close working relationships with each other to the extent that they do not even need to ask each other's permission before using a plot. Farmers who benefit from this sharing system report that they have been greatly assisted by it when they were just starting in the industry. The practice is common among family members and friends and requires a level of trust between lender and borrower because of the potential for conflicts to arise.

## Conflict over sea space

Conflicts occur between different users of the ocean as a result of the sea space ownership system that emerged through the development of the seaweed industry. Rising prices of seaweed through 2021–2022 intensified conflicts between sea users, including old conflicts that had faded away and new conflicts that have arisen. These conflicts have mostly been resolved quickly and informally, although some parties appeared to continue to harbour negative sentiments. As one farmer involved in a conflict put it, conflicts may seem to be 'quickly resolved ... but are like a "thorn in the flesh" ... They must arise again. Not all of them [can be considered finished]. The people's mindsets are different. There is social jealousy.' We observed three main types of conflicts: between seaweed farmers and crab netters, between seaweed farmers and boat users, and between seaweed farmers and other farmers.

### With fishers

The first common type of conflict over sea space occurs between seaweed farmers and fishers, including crab netters. As a result of the shrinking space available to crab fishers to hang their nets, they often work in close proximity to seaweed farms. Gill nets for catching crabs are installed on the bottom of the ocean in 'sets' of 12 stretches of 100m, and 2–3 stretches are often lined up sequentially such that they may reach several kilometres in length. The location and direction of the nets on the sea floor are difficult to spot from the surface such that crab nets often overlap each other on the seabed, and may become tangled. One crab netter complained that during the crabbing season, there were so many crab netters that 'the nets on the seabed are like cobwebs'. Conflicts can occur because crab netters working in close proximity to seaweed farm plots may inadvertently damage the plots if their boat collides with the seaweed farm. In addition, crab nets often become tangled in the seaweed farm ropes. A tacit agreement has been reached that when this occurs, the crab fisher must cut their own net. Crab nets are easier to repair than seaweed ropes, and crab fishers do not want to be liable for the potentially high costs of damaging a seaweed farm. Similarly, crab fishermen and fishers are both liable for any damage they may cause to seaweed farms whilst moving between them. Most fishermen reluctantly accepted this rule. As one fisher complained,

> Why is it that we fishermen are just catching fish, but seaweed farmers are angry if we cut the ropes or [damage] the foundation, when we only do that because there are no more roads [for boats] that we can use ... [there are] unlimited [seaweed] foundations.

Another fisherman complained about the negative impact that they perceived that seaweed farming had had on their livelihood:

> Since I was small, I have been catching fish ... [but] since seaweed has been here, I rarely catch big fish ... there have been many changes ... in the past it was easy ... [but] now [the fishing location] is far away ... [because] there are [seaweed] foundations. If the current is fast, [the boat] can float [with the waves] into people's foundations ...

Some seaweed farmers expressed anger that crab fishers damage their foundations. However, the overlap in livelihood activities between crab fishers and seaweed farmers curbed these conflicts. As one seaweed farmer explained,

> Crab netters are not completely to blame, because the distance between seaweed foundations is indeed narrow. What can he do but install the crab net right on the sideline of the foundation? So when the boat is hit by strong waves, they cannot control the rudder. It automatically goes up into the seaweed. This is bad for the crab netter, but he really can't control the rudder too much.... [he has to] keep throwing [the net into the sea and] keep one hand on the wheel [for steering].

Seaweed plots are often left empty for days (between cycles) or months (between seasons) and crab fishers can use an empty foundation for laying crab nets. They report that they vacate the plots when the farmer who owns it indicates that he will imminently be planting it with seaweed. As one crab fisher said,

> we understand each other ... if ... the next day he wants to fit the plot with seaweed, he makes a sign, and from the sign we know that tomorrow the foundation will be fitted with seaweed so we can no longer put nets there ... [we give in] because we who are crab netters are not bound to one location, it is different from seaweed farmers because they have to make a foundation beforehand.

Crab netters therefore appear to have workable, but lesser rights to use the sea than seaweed farmers. They can net crabs in the gaps between seaweed foundations, but must cut their own net if they become entangled with a seaweed farm. They can use empty spaces marked by seaweed foundations, but must vacate them when the owner moves to use them. They can still find empty spaces not marked by foundations, but often have to travel further out to sea, using valuable petrol to do so. Crab fishers therefore feel that on the whole, their rights have diminished as a result of the introduction of seaweed farming. They also report that while crab prices are higher today than they were fifteen years ago, this is offset by smaller catches in both the size and number of crabs.

The introduction of seaweed farming has not, however, had a uniformly negative impact on crab fishers. Seaweed often breaks off seaweed longlines as a result of wind, waves, and changes to seaweed health. This phenomenon is particularly common in March, when Pitu Sunggu experiences high winds and waves. Broken seaweed becomes entangled in crab nets, and when this occurs in large quantities, crab netters can obtain significant quantities of seaweed (Figure 5.5). During 2022, when the majority of this research was undertaken, the crab season ran roughly from April to June with crab prices ranging from Rp. 20,000 to 75,000/kg. Although the prices are highly variable, they were much higher than a few years ago when farmers reported prices of only around Rp. 15,000/kg. During the same period (April–June 2022), seaweed prices increased from Rp. 33,000 to 37,000 Rp/kg dry – the equivalent of around Rp. 4,400 to 5,000/kg wet. As seaweed collected in the crab nets in much greater quantities than crabs did, crab fishermen found this to be a significant extra income stream. The crabbing season runs roughly from April to June, during which time crab fishermen typically catch 7–10 kg.

Outside of this season, catches of crabs are limited, but many crab fishers still install crab nets in order to catch the broken seaweed – going 'fishing' for seaweed, using old, damaged crab nets to catch falling seaweed. As one crab netter commented, 'It is good if there is (strong) wind because we can catch the (broken) seaweed'. This is especially common in March, when high winds mean that there is significant seaweed breakage. Some seaweed farmers expressed annoyance at the crab netters benefitting from the broken seaweed since they were simultaneously experiencing losses of their own seaweed, but as the seaweed could no longer be assigned to any individual owner, they were not able to prevent this and did not seek to collect broken seaweed themselves, leaving this to other village residents. Indeed, some residents with no other livelihood made a habit of searching for broken seaweed along the shore amongst the mangroves, which they then dried and sold for a little income. In these ways, income from seaweed spreads beyond the farmers themselves. Fishermen also appear to benefit indirectly from seaweed farming in some ways: one of the most common fish species we observed in fishermen's baskets were rabbitfish (*Siganus sp.*), a species commonly observed around seaweed farms because it grazes on the seaweed.

Today, fishers and seaweed farmers seem to have reached an uneasy truce, although conflicts occasionally break out between them. Many residents of Pitu Sunggu even insisted that such conflicts no longer exist, because crab netters and seaweed farming are overlapping livelihoods often undertaken by kin or neighbours, as well as frequently 'by the same person'.

*Figure 5.5* Seaweed caught in gill nets used by crab fishers
Source: Image by Zannie Langford, Pitu Sunggu, May 2022.

### With other seaweed farmers

Conflicts also occurred between seaweed farmers, particularly as prices rose through the financial year 2022. In general, people expressed a sense of resentment over the unequal division of sea space in the village. As one large farmer put it, '[P] eople fight over the sea, because people who didn't get space for cultivation end up feeling angry, whereas before [in the early days of the industry], they themselves didn't want to!'. Small and beginning farmers were indeed dissatisfied with the system, but simultaneously felt that they were not really able to contest the situation since they themselves participated in the process. As one small farmer commented, 'We think it's unfair (that some have lots of space), but what can we do? We want to make the location full too.' Some of this resentment centred on the rights of farmers to maintain ownership of plots they were not using, as happened when farmers had too many plots to operate at full capacity or when they retired from farming seaweed but still refused to sell their plots, preferring to lease them for an ongoing source of income. Complaining about a retired seaweed farmer who continued to benefit from his 'landlord' status, one resident said, '[I]in our opinion it is unfair ... he has lots of locations, but he just rents out the places ... so our feelings must hurt, but what can we do because he feels that he completely owns that place'.

In addition to these generalised tensions, conflicts occur between seaweed farmers over specific farming foundations. This occurs in two main ways. Firstly, conflict arises between lessors and lessees when lessees damage the foundation - that is, the anchor ropes which demarcate the location of the farm. Several cases

were cited where ropes were damaged and the lessee did not repair them before va-
cating the foundation. Some farmers reported that they chose not to lease out empty
plots because they were worried about such damage. Conversely, one farmer, who
commonly rents space from others, said that he sought to avoid potential conflicts
with lessors by always taking action to improve the foundation that he had rented.

The second type of conflict that arises is due to a lack of clarity around who
owns different foundations. In recent years there have been several instances where
people have dishonestly claimed ownership of a foundation and have rented out or
sold the plot. This often occurs when farmers have left plots empty for an extended
period of time, particularly if the plots are located offshore from other villages. In
one such instance, a farmer who had established several plots in a nearby village
returned to check on these plots only to find them in use by someone else, who had
paid a sale price to a third party claiming to be the owner. In the end, the original
farmer assented to the loss of his space, as he foresaw an extended dispute if he
took the matter further. However, the farmer learnt from that experience that sea
space which is left unoccupied, not leased, and not regularly checked, may be
claimed by others. In this 'use it or lose it' situation, leasing a foundation to others may be a way of retaining an ownership claim on it. These conflicts are not
always resolved in favour of the occupant however, and there have been a number
of cases where people have been forced to pay rent or purchase prices twice when
the original owner returns to claim the foundation. In her work amongst Sulawesi
highlanders, Li similarly noted issues raised in land markets by uncertain ownership status, finding that 'Land purchasers … preferred to buy land that had already
been sold at least once, so they could be sure it was an unencumbered *lokasi*, fully
detached from its origins in someone else's work' (Li 2014, p. 93). This reflects the
ambiguity in ownership rights established through informal mechanisms.

### *With other boat users*

Finally, seaweed farmers can come into conflict with other boat users when they try
to establish plots in new areas encroaching on boat lanes. In Pitu Sunggu, boats are
moored overnight either in an upstream bend of the river or at the pier on the coast.
In the morning, boat users travel from the mooring location out to sea using boat
lanes. These are unoccupied stretches of sea which are allocated by the community
to be used for transport and should therefore not be filled with seaweed farms or
obtrusive fishing gear. Satellite imagery along the South Sulawesi coastline reveals
that at most river mouths, there is an area of space from the mouth of the river to
beyond the seaweed farming area which is clear of farms (see Langford, Waldron
et al. 2021). These passageways are presumably boat paths (Figure 5.6).

These gaps left for boat transit are appealing to new farmers who face the prospects of high rental prices to access space for farming. In one case, a crab netter
named Adil sought to establish his own seaweed farm and established a foundation in part of an area used as a boat lane. Other seaweed farmers using the boat
lane became annoyed and complained to the head of the hamlet, who reprimanded
Adil. Adil felt this reprimand was unfair, since he himself was disturbed by the
presence of seaweed farms, which meant that he had to slowly manoeuvre around

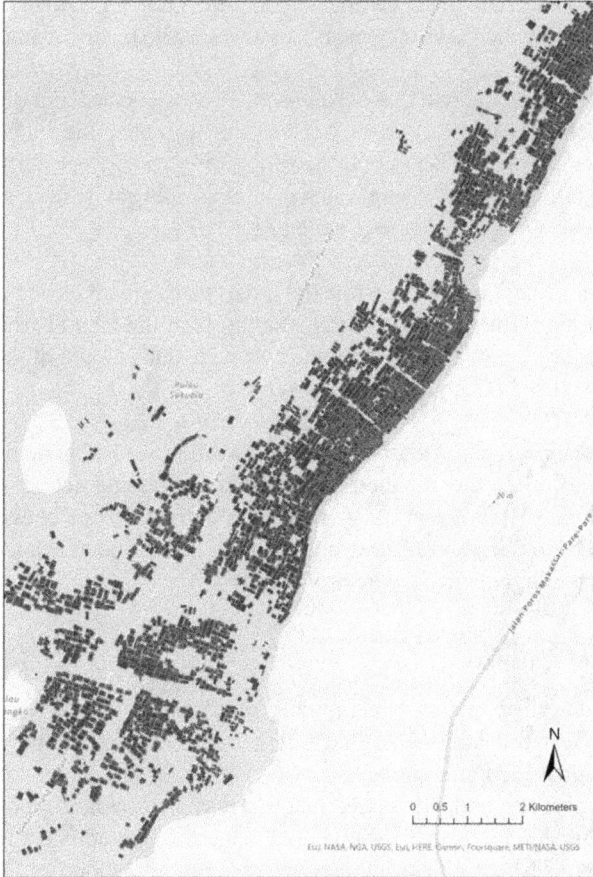

*Figure 5.6* Distribution of seaweed farms observed at any time from 2018–2020

Source: Image created by Zannie Langford using data from Planet Labs for 2018–2020.

them when catching crabs. He felt that given the inconvenience he suffered, he was also entitled to inconvenience others by establishing a foundation. Consequently he complained, '[A]ren't these *all* [boat] lanes? We [crab netters] also want to go out [to sea], but [seaweed farms] block the road.' Despite the warning, Adil maintained his plot in the boat lane. When farmers threatened that his longlines would be broken by seaweed propellers, he reportedly responded that '[I]f the problem is that the rope will break, it can be reconnected' and continued to farm the area regardless.

Some farmers also reported that they established new plots in controversial areas through subterfuge, believing that once installed, their right to farm in the foundation location would be respected. As one farmer admitted,

I asked about the location [I wanted to claim] first, but there was no owner for it yet. I asked all the neighbors about the location but they all said no one had that place yet. So I secretly made the foundation. I thought, if someone asks about it tomorrow, only then I will tell them that it was *I* who made the foundation.

Although it is understandable that ocean users inconvenienced and disadvantaged by the presence of seaweed farmers feel that they too have the right to establish plots, the blocking of boat lanes and the filling of gaps between farms has made transport difficult during the dry season, when plots are heavily used and crab netting is at its peak. As one ocean user lamented,

> There are no [rules], and meanwhile the [boat] paths are all closed ... it used to be good here, [the boat path] was straight, [and the area in front of] the river was wide ... now they have let the boat path at the mouth of the river be completely closed off [by the seaweed farms].

Similarly, farmers complained that the gaps between farms had in many cases been filled. As one retired farmer recalled, '[B]ack when I was still active [as a farmer], all the cultivation plots were in order. There was a dividing line between the tiles. But now, it is all out of order.' These conflicts raise the issue of sea space governance in an increasingly contested environment.

### Governance of sea space

As has been described, informal community rules and norms have formed property rights in seaweed farming in Pitu Sunggu. When asked about whether they had sought formal recognition of legal ownership from the regency or provincial government, most respondents replied that this was unnecessary, since the current system of ownership recognition within the village was operational and mostly uncontested. The main area in which they saw a role for regulation from higher levels of government (i.e. beyond the hamlet and village government) as desirable was in preventing the construction of new plots in boat lanes. Regulation of ownership of individual plots was not viewed by farmers as the role of the provincial government, but a matter for the community themselves. Village residents felt secure in their ownership of individual plots, since ownership was widely recognised within the village. As one farmer said, 'if you want [a land ownership certificate], we don't have it... [but] we all already know [who owns it] ... as long as there is a rope there, it means that the others can't [build a new foundation]'. The system of ownership was enforced by the community itself via the village government structure, which occasionally plays a role resolving conflicts. Residents have confidence in the authority of the village government and community members in protecting ownership rights, even if the owner is no longer living in the village. As one farmer said,

> [I]f someone tries to occupy a location left empty by a community member, they will ban it because the community believes it is still the location of the person who occupied it first ... even though the person is not there. Because here kinship bonds are still strong.

The method for testing if an apparently empty area of sea is used is simple, as one farmer told us: '[I]f someone doesn't believe that someone owns a location, we order him to drop the anchor. When the anchor is pulled and there is a rope going up, it means someone already owns that location.'

The rules for claiming sea space and access to it – namely that the establishment of a foundation gives the maker the exclusive right to use that area for seaweed farming, and to exclude other ocean users from the area while the plot is in use – were defended by community members as their own. As one resident put it, 'Only we, the people of this area, have made such rules'. This was true even when residents noted the unequal distribution of space in the village – as one farmer said, 'I think it is fair, because those who came first own the location. Those who were slow to claim a place made an error – but such are the rules we have made for the sea.' The right to build foundations in boat lanes was contested, but still ultimately respected, as occurred in the case above. Residents sought to avoid doing so if at all possible due to the negative reactions this was likely to provoke from other boat users, but as one farmer put it, it remains the case that '[Y]ou can't [forbid other people from building foundations in boat lanes]. Who knows, we may get scolded too. Unless the government regulates it … [But] if we manage it ourselves, it will be difficult.' In this way, patterns of acceptable behaviour that emerged informally over time through extended conflict have become accepted as community formed rules.

These rules have not been formally recognised by village or provincial government however, and the relationship between possession and recognition is complex (see Lund 2021). One attempt at such formalisation was made a number of years ago when the Pitu Sunggu village government participated in an initiative with several neighbouring villages to develop regulations relating to the use of coastal space in the region. As one person involved with the initiative told us, they sought to determine a minimum distance from the shore at which seaweed could be planted, a task which was challenging due to the many small islands in the region and the mobility of seaweed farmers and fishermen from the islands and coastal villages, who often worked in the sea several kilometres from their place of residence and in sea areas closer to other villages than their own. As one person involved with the initiative said, '[W]e wanted to make an agreement between Village A and Village B and also with the island … [but] that's difficult because we also don't understand things like that'. In the end, such regulations did not eventuate and further efforts have not been pursued, in the opinion of one fisheries official, '[B]ecause [it seemed that] the conflict had subsided … [the community] already found a solution so we didn't follow up on it … we saw it to be safe'.

It appears that as demand for sea space intensified with rising prices, previously observed customs are no longer followed. As one farmer said,

There used to be limits … not a rule from the government but the result of community agreement … it was not permissible to build foundations at the top of the estuary… [and where] a foundation had been erected, no one may place another foundation within [that area] or closer to the coast. Community members could build foundations on the outer coast. The same goes for road [boat lane] rules. There were rules regarding the distance between one foundation to another ... Because it's not just seaweed farmers who want to use the location. In addition to them there are the crab netters, and those who look for rebon shrimp (*acetes*) in the interior of the coast close to the mangroves ... in the past the arrangement of foundations in the sea was neat ... Very neat. Roads, boundaries, foundations were very visible ... [But] when the price of seaweed soared ... society got greedy and kept building the foundations until the rules were broken ... So now it is hard. Only certain locations can be accessed because there are already many obstructing foundations, boat lanes are also closed.

Despite this, there was understandable hesitation about a role for formal government intervention, since farmers cannot be sure whether they would benefit from additional protection, or lose access to their own plots. This is by far the most common stance of farmers interviewed in our study. In a recent village forum with government officials, no farmers dared to raise issues related to seaweed farm space. The village government feels that they are capable of regulating the situation themselves but to date have not taken action on this – as one of them explained, 'there's no need to make a PERDES [formal government regulation] ... we had planned it, but then we wondered what the procedural problems were, how to establish a legal basis. It seemed like a long [process] ... [So] now [there are] only the rules of the game [between] fishermen.'

Some efforts to set rules for sea space allocation and use have been set by other regions in South Sulawesi. In Wajo Regency, for example, a 2012 regulation outlines limits on sea space use for seaweed farming. The regulation specifies that shipping lanes must be at least 200m wide, that the distance between seaweed farming plots should be at least 10m, the distance from any given seaweed farm plot to the shore should be at lease 50m, and that any given farmer should not own an area exceeding 2 hectares for longline cultivation (Pemerintah Kabupaten Wajo 2012). However, we were informed that these regulations are not enforced and that compliance is limited. In nearby Laikang village, rules have been set specifying processes for sea space sales, among other things. In Daiama village in Rote Ndao, rules for seaweed farming specify that farmers must seek a licence from a customary leader before establishing a new plot, and must report to this leader again at harvest time – although in practice monitoring of compliance is limited and violations are not strictly punished (Satria et al. 2017). These initiatives from other locations may offer insights that would support regulation in Pitu Sunggu, although farmers emphasised that they felt that limits on individual ownership areas were not practical as this would interfere with their strategy of geographic diversification (owning many plots in different areas to allow them to move seaweed around to avoid disease and maximise growth). In addition, although village residents stressed their ability to resolve conflicts internally, in practice this was complicated by the fact that people from other villages also farm and fish in Pitu Sunggu waters, and vice versa. Indeed, across Indonesia 10 per cent of farmers report that their farms are located outside their village (BPS 2022). Finally, it is worth noticing that prices crashed at the end of 2022 and in 2023 sea space limitations have not been a major issue due to drastic declines in production. It is clear that any effort to formally regulate and plan sea space use should carefully consider the context and unintended negative impact of such regulation and consult carefully with the communities involved.

## Future developments

### Intensification

The needs of village residents for sea space governance in the future will probably depend on the developments in the broader seaweed industry and in other marine industries. In Pitu Sunggu, prices increased dramatically following the construction of a major Chinese processing plant, BLG, in 2017, and this happened again in the 2022 financial year as prices doubled post COVID-19 (Langford, Zhang et al. 2022a; Zhang et al. 2023). As seaweed farming became more lucrative, contestation over sea space intensified, and governance systems for sea space have been put under additional stress. If prices continue to increase, it may be necessary to adapt these systems, although to date periods of intensive use of sea space have been temporary due to the effect of price volatility on demand for space. In addition, there is also the potential for seaweed farmers to transition out of seaweed farming and into higher-value products. In nearby Laikang, for example, some seaweed farmers have installed lobster grow out systems in areas of the sea previously claimed for seaweed farming. Lobsters are a very high-value product, and they use much smaller areas of sea space but they require high ongoing inputs. In Laikang, the activity is supported by a nearby fish processing facility which makes low-cost fish off-cuts available as lobster food – such inputs are essential to support lobster farming because lobsters are carnivorous and if not fed regularly and sufficiently will devour each other. Nonetheless, lobsters and other high-value marine aquaculture offer a pathway for further intensification of marine space use, and it is likely that sea space ownership developed for seaweed farming will form the basis of this use. As the use of marine space changes and intensifies, village residents, village governments, and, potentially, wider regional governments, may face challenges regulating an increasingly competitive sector.

### Expansion

Initiatives to develop the seaweed industry often emphasise the large areas of space available to expand cultivation across more of Indonesia's coastline, and indeed globally. Pitu Sunggu is in South Sulawesi, reportedly Indonesia's largest seaweed producing province. As such they experience sea space shortages which are still emerging in other areas. In our field sites on Timor, Rote, and Semau islands in Nusa Tenggara Timor (NTT), surveyed farmers did not see access to sea space as a major issue. No farmers in these locations viewed the issue as 'very problematic', and only 30 per cent of farmers saw it as 'somewhat problematic' (Langford, Turupadang et al. 2022; 2023). This is compared to farmers in Pitu Sunggu, 23 per cent of whom found the issue 'very problematic' and a further 24 per cent found it 'somewhat problematic' (Langford, Waldron et al. 2023; Langford et al. 2024). These figures are for existing seaweed farmers, indicating that 47 per cent of Pitu Sunggu farmers who currently farm seaweed consider access to sea space to be an issue – the figures would probably be higher if the survey reflected the views of the community as a whole. In addition, these figures are from our 2021 survey, which was undertaken prior to the 2021–2022 price increases which intensified competition for sea space.

As efforts to expand seaweed cultivation into new areas proceed, it would be wise to develop some level of spatial planning in relation to sea space use prior to the establishment of the industry in new areas. While detailed monitoring of intra-village space distribution may not be feasible or desirable, some level of planning to allocate boat lanes and environmental protected areas (covering sensitive ecosystems such as coral reefs and seagrass meadows, as well as locations free of fishing and seaweed farming to support ecosystem sustainability) may prove important.

## Conclusion

This chapter has described how seaweed has transformed patterns of ownership of sea space in an Indonesian village. Ultimately, the inability of crab netters to exclude seaweed farmers from the communal area used to catch crabs led to the enclosure of previously communal space by seaweed farmers. These seaweed farmers were often also crab netters, who participated in these processes of enclosure driven by entrepreneurialism as well as by the need to secure an area of space for their own use in a context of rapidly reducing public space. Construction of a foundation is recognised by the community to give the foundation owner a series of rights, including: the right to use the area and its foundation to farm seaweed at any time; the right to exclude others from farming seaweed in the area; the right to exclude other ocean users from operating in the area while it is in use for seaweed farming; and the right to lease, sell, or gift the foundation to another person. These rights are not formally recognised by higher levels of government and are recognised to varying extents by different types of actors (for example, government, banks, traders, family members, other seaweed farmers, other fishers).

Community understanding of the conditions of sea space use and access underpin the patterns of development of the industry. In Pitu Sunggu, the unequal distribution of sea space has led to the creation of a large labour trade for casual binding work. In addition, the high density of seaweed farms has impinged on the activities of other ocean users such as fishers and crab netters by reducing the space available for these activities and by impeding the passage of boats. Spatial planning could help new seaweed farming communities elsewhere in Indonesia to avoid some of the issues that have arisen in Pitu Sunggu and to support more environmentally sustainable outcomes. Understanding the pattern of sea space use for seaweed farming is critical to planning for a sustainable industry. These patterns of sea space ownership also affect the way that seaweed is farmed. Farmer decision-making about how to manage their farms is constrained by their access to different areas of sea space, as the next two chapters discuss.

## Disclaimer

Our best guess is that seaweed farming was introduced to Pitu Sunggu around 2007, probably between 2005 and 2007. However, a number of farmers report farming before this time. This could be because they experimented with seaweed without adopting it permanently. Alternatively, it could be because they farmed in neighbouring villages.

## Notes

1 Translated from the Indonesian word *pondasi* used to describe the seaweed farm foundations.
2 There is some uncertainty as to the date that farming was successfully achieved in this village as farmers in nearby villages were also beginning to experiment with seaweed farming around the same time, and prospective farmers often farmed outside the bounds of the village they resided in.

## References

Alkhalili, Noura. 2017. "Enclosures from Below: The Mushaa' in Contemporary Palestine." *Antipode* 49 (5): 1103–1124. https://doi.org/10.1111/anti.12322

*Britannica* (2013). Enclosure. *Encyclopedia* Britannica. https://www.britannica.com/topic/enclosure

BPS (*Badan Pusat Statistik*). 2022. "Hasil Survei Komoditas Perikanan Potensi Rumput Laut 2021 Seri 2." *Badan Pusat Statistik*. https://www.bps.go.id/publication/2022/08/29/269de33babc6e3d52bbae5b6/hasil-survei-komoditas-perikanan-potensi-rumput-laut-2021-seri-2.html

*Britannica.* 2013. "Enclosure." *Encyclopedia Britannica.* https://www.britannica.com/topic/enclosure

Castellanos-Navarrete, Antonio and Kees Jansen. 2015. "Oil Palm Expansion Without Enclosure: Smallholders and Environmental Narratives." *The Journal of Peasant Studies* 42 (3–4): 791–816. https://doi.org/10.1080/03066150.2015.1016920

Corson, Catherine and Kenneth Iain MacDonald. 2012. "Enclosing the Global Commons: The Convention on Biological Diversity and Green Grabbing." *The Journal of Peasant Studies* 39 (2): 263–283. https://doi.org/10.1080/03066150.2012.664138

Curry, George N. and Gina Koczberski. 2009. "Finding Common Ground: Relational Concepts of Land Tenure and Economy in the Oil Palm Frontier of Papua New Guinea." *The Geographical Journal* 175 (2): 98–111. https://doi.org/10.1111/j.1475-4959.2008.00319.x

Curtler, William Henry Ricketts. 1920. *The Enclosure and Redistribution of Our Land.* Oxford: Clarendon Press.

Goldstein, Jenny E. 2016. "Knowing the Subterranean: Land Grabbing, Oil Palm, and Divergent Expertise in Indonesia's Peat Soil." *Environment and Planning* A 48 (4): 754–770. https://doi.org/10.1177/0308518X15599787

Giorgio, Olivia del, Brian E. Robinson, and Yann le Polain de Waroux. 2022. "Impacts of Agricultural Commodity Frontier Expansion on Smallholder Livelihoods: An Assessment through the Lens of Access to Land and Resources in the Argentine Chaco." *Journal of Rural Studies* 93: 67–80. https://doi.org/10.1016/j.jrurstud.2022.05.014

Hall, Derek, Tania Li, and Philip Hirsch. 2011. *Powers of Exclusion: Land Dilemmas in Southeast Asia.* Hawai'i: University of Hawai'i Press.

Hall, Ruth, Marc Edelman, Saturnino M. Borras, Ian Scoones, Ben White, and Wendy Wolford. 2015. "Resistance, Acquiescence or Incorporation? An Introduction to Land Grabbing and Political Reactions 'from Below'." *The Journal of Peasant Studies* 42 (3–4): 467–488. https://doi.org/10.1080/03066150.2015.1036746

Ito, Takeshi, Noer Fauzi Rachman, and Laksmi A. Savitri. 2014. "Power to Make Land Dispossession Acceptable: A Policy Discourse Analysis of the Merauke Integrated Food and Energy Estate (MIFEE), Papua, Indonesia." *The Journal of Peasant Studies* 41 (1): 29–50. https://doi.org/10.1080/03066150.2013.873029

Langford, A. 2019. "Vanuatu Custom Land Management Policy Outlook." *Custom Land Management Policy Outlook.* Vanuatu: Port Vila, 1–38.

Langford, A. 2022. "Developing Food Markets in Vanuatu: Re-examining Remote Island Geographies of Food Production and Trade." *World Development Perspectives* 28 (100463): 1–10. doi.org/10.1016/j.wdp.2022.100463

Langford, A., G. Lawrence, and K. Smith.(2021. "Financialisation *for* Development? Asset-making on Indigenous Land in Remote Northern Australia." *Development and Change* 52 (3): 574–597. doi.org/10.1111/dech.12648

Langford, A., W. Turupadang, M. D. R. Oedjoe, C. Liufeto, B. Bire, M. Yohanis, Y. Suni, and S. Waldron. 2022. *Seaweed Farmer Resilience in Eastern Indonesia after COVID-19.* Canberra, Australia: Australian National University Indonesia Project.

Langford, Alexandra, Welem Turupadang, and Scott Waldron. 2023. "Intterventionist Industry Policy to Support Local Value Adding: Evidence from the Eastern Indonesia Seaweed Industry." *Marine Policy* 151 (105561). doi.org/10.1016/j.marpol.2023.105561

Langford, A., S. Waldron, N. Nuryartono, S. Pasaribu, B. Julianto, I. Siradjuddin, R. Ruhon, Z. Zach, I. Lapong, and R. Armis. 2023. *Sustainable Upgrading of the South Sulawesi Seaweed Industry.* Melbourne, Australia: Australia-Indonesia Centre. https://pair.australiaindonesiacentre.org/research/sustainable-upgrading-of-the-south-sulawesi-seaweed-industry-2

Langford, A., S. Waldron, Sulfahri, and H. Saleh. 2021. Monitoring the COVID-19-Affected Indonesian Seaweed Industry Using Remote Sensing Data. *Marine Policy* 127(104431): 1–10. doi.org/10.1016/j.marpol.2021.104431

Langford, A., J. Zhang, S. Waldron, B. Julianto, I. Siradjuddin, I. Neish, and N. Nuryartono. 2022. "Price Analysis of the Indonesian Carrageenan Seaweed Industry." *Aquaculture* 550 (737828): 1–13. doi.org/10.1016/j.aquaculture.2021.737828

Leynseele, Yves Van and Malin Olofsson. 2023. "Unpacking Land-Associated Assemblages 'from Below': Smallholders' Land Access Strategies at the Commercial Tree-Crop Frontier." *Political Geography* 100: 102792. https://doi.org/10.1016/j.polgeo.2022.102792

Li, Tania Murray. 2014. *Land's End.* Durham, NC: Duke University Press. https://doi.org/10.1515/9780822376460

Lund, Christian. 2021. *Nine-Tenths of the Law.* New Haven, CT: Yale University Press. https://doi.org/10.2307/j.ctv1b0fw9d

McCarthy, John F., Jacqueline A. C. Vel, and Suraya Afiff. 2012. "Trajectories of Land Acquisition and Enclosure: Development Schemes, Virtual Land Grabs, and Green Acquisitions in Indonesia's Outer Islands." *The Journal of Peasant Studies* 39 (2): 521–549. https://doi.org/10.1080/03066150.2012.671768

Olofsson, Malin. 2021. "Expanding Commodity Frontiers and the Emergence of Customary Land Markets: A Case Study of Tree-Crop Farming in Venda, South Africa." *Land Use Policy* 101: 105203. https://doi.org/10.1016/j.landusepol.2020.105203

Pemerintah Kabupaten Wajo. 2012. *Peraturan Daerah Kabupaten Wajo Nomor 4 Tahun 2012 Tentang Pengelolaan Sumberdaya Perikanan Kabupaten Wajo.* Pemerintah Kabupaten Wajo.

Pichler, Melanie. 2015. "Legal Dispossession: State Strategies and Selectivities in the Expansion of Indonesian Palm Oil and Agrofuel Production." *Development and Change* 46 (3): 508–533. https://doi.org/10.1111/dech.12162

Satria, A., N. H. Muthohharoh, R. A. Suncoko, and I. Muflikhati. 2017. "Seaweed Farming, Property Rights, and Inclusive Development in Coastal Areas." *Ocean and Coastal Management* 150: 12–23. https://doi.org/10.1016/j.ocecoaman.2017.09.009

Schoenberger, Laura, Derek Hall, and Peter Vandergeest. 2017. "What Happened When the Land Grab Came to Southeast Asia?" *The Journal of Peasant Studies* 44 (4): 697–725. https://doi.org/10.1080/03066150.2017.1331433

Semedi, Pujo and Laurens Bakker. 2014. "Between Land Grabbing and Farmers' Benefits: Land Transfers in West Kalimantan, Indonesia." *The Asia Pacific Journal of Anthropology* 15 (4): 376–390. https://doi.org/10.1080/14442213.2014.928741

Visser, Oane. 2017. "Running Out of Farmland? Investment Discourses, Unstable Land Values and the Sluggishness of Asset Making." *Agriculture and Human Values* 34 (1): 185–198. https://doi.org/10.1007/s10460-015-9679-7

Zhang, Jing, Scott Waldron, Zannie Langford, Boedi Julianto, and Adam Martin Komarek. 2023. "China's Growing Influence in the Global Carrageenan Industry and Implications for Indonesia." *Journal of Applied Phycology* June: 1–22. https://doi.org/10.1007/s10811-023-03004-0

# 6 Environmental and socio-economic constraints to marine seaweed farming

*Zannie Langford, Radhiyah Ruhon,
Zulung Zach Walyandra, Imran Lapong,
and Risya Arsyi Armis*

### Farming the sea

Farming the sea is a relatively new phenomenon. While land-based plants have a history of thousands of years of domestication (Wikfors and Ohno 2001), there are only a few documented cases of seaweed domestication prior to the 1900s, including that of *porphyra* in Japan, best known today for its use in making nori for sushi rolls. More recent life-cycle control has been possible for *laminaria* species of seaweed in China since the 1950s, and *undaria* species (*wakami* – widely used in 'seaweed salads') in Japan since the 1970s (Wikfors and Ohno 2001). Nevertheless, seaweed farming is a relatively new activity and the development of farming systems and the selection of appropriate species is still unfolding. Seventy years ago, the global supply of seaweed was primarily from harvests of wild sources, with very little of it being farmed. Since the development of methods to farm commercially important species of seaweed, farmed seaweed has rapidly overtaken wild harvests as the largest contributor to the supply of global seaweed (Figure 6.1).

Farming seaweed allows much larger quantities to be produced than through wild harvest. However, farming techniques have only been developed for a small range of species. In the Philippines, it took years to develop methods to successfully farm carrageenan seaweeds and another decade to replicate that success in Indonesia under what would appear to be very similar conditions. Farming techniques developed in the 1970s and 80s are still used today in only slightly modified forms. They involve a basic process of propagating seaweed using cuttings, which are attached to ropes and suspended in the sea for around six weeks before being harvested and divided into two parts – one for sale and one for propagation. The replanting of propagules does not involve control over the reproduction cycle of the seaweeds, as occurs in domestication, which would enable selection for higher performing variants. Rather, propagules are selected for signs of vigour such as colour, thallus thickness, and evidence of new branches. Technologies for mechanised planting of seaweed are being trailed by a few companies, but the industry remains labour-intensive and dominated by smallholder farmers.

Farming seaweed is not an easy task, and seaweed farmers have had to solve a range of biophysical challenges in order to develop viable seaweed businesses. These techniques are both an art and a science, as they rely on farmer innovation

DOI: 10.4324/9781003183860-9

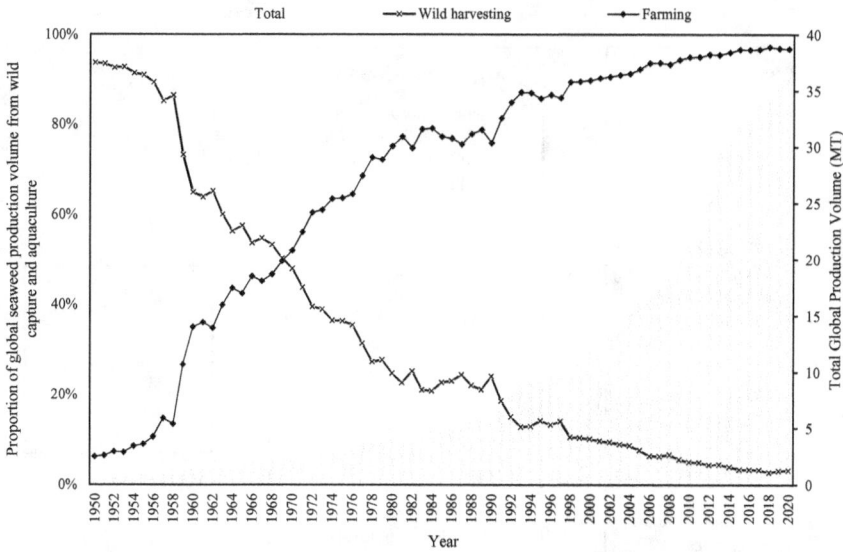

*Figure 6.1* Global seaweed production from farming and wild harvesting 1950–2020
Source: Data from FAO (2023).

and experimentation, sharing of anecdotal evidence, subjective judgement making, and development of the different farming styles, assets, and risk appetites of individual farmers. Scientific studies have yet to resolve deterministically how a range of interacting ocean conditions affect seaweed growth. Observations of farm performance are complicated by the fact that farmer decisions are not based solely on maximising productivity or income, but also on socio-economic factors that affect production choices. These factors – such as access to marine space in areas with different oceanic conditions, farmer risk tolerance, conditions of access to capital and labour, relationships with kin, traders, and other farmers, and other livelihood options – all affect the choices farmers make when farming seaweed (Figure 6.2). The next two chapters will explore how farmers in Pitu Sunggu grow seaweed, focusing on the two species of tropical seaweed grown in the village: *Kappaphycus alvarezzi* and *Kappaphycus striatus*. This chapter explores the biophysical and socio-economic factors that constrain farm productivity and farmer decision-making. The next chapter (Chapter 7) describes farmer decision-making under these constraints, and explores how Pitu Sunggu farmers have adapted farming approaches to local conditions and how they manage risk under increasingly intensified and uncertain growing conditions. These two chapters build on a wide body of research examining the impact of ocean conditions on seaweed growth in laboratory and field trials, as well as key works combining scientific and practical knowledge to describe key seaweed farm management decisions, particularly the work of Neish (2008).

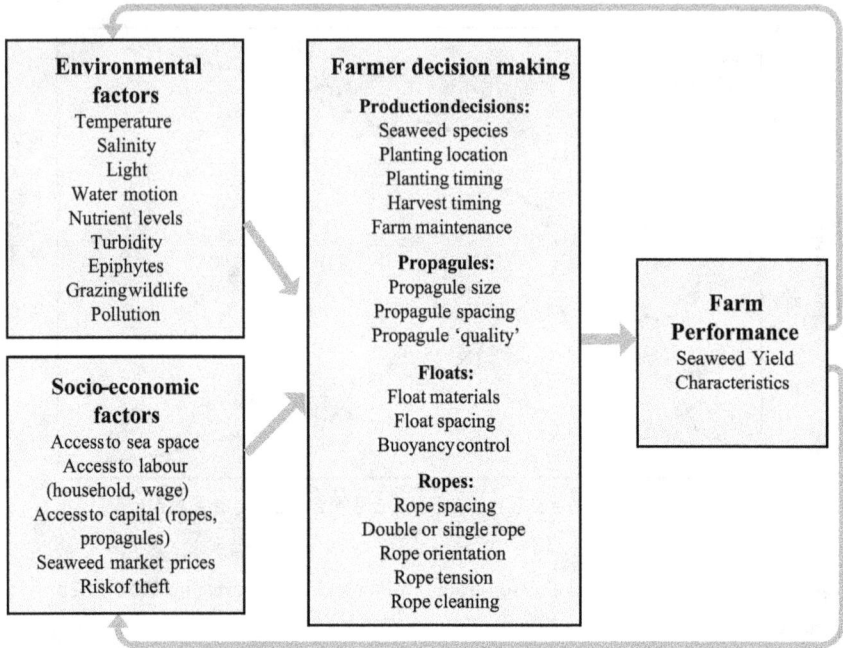

*Figure 6.2* Factors affecting seaweed farmer decision-making

Source: Authors' schematic.

### The longline farming system

There are two species of eucheumatoid seaweed grown in Pitu Sunggu: *Kappaphycus alvarezii* (known colloquially as 'cottonii') and *Kappaphycus striatus* (known colloquially as 'sacol'). These two species have slightly different growth patterns. Cottonii is faster growing, while sacol appears to be more tolerant of heat stress. Cottonii has been better studied in the literature as it was the species of seaweed originally grown widely across Indonesia. Sacol was introduced to Pitu Sunggu just a few years ago as a 'dry season' species more suitable for cultivation at this time of year. Both species are grown using the same farming apparatus. Pitu Sunggu seaweed farmers use a longline farming method suitable for deeper waters (Figures 6.3 and 6.4). In this method, seaweed propagules are attached to ropes which are typically 25m in length in Pitu Sunggu. These ropes are suspended at the surface using floats (normally empty plastic bottles) and are secured in place by attaching them to a set of ropes which are fixed to the ocean floor using anchors.

A seaweed farming cycle consists of the set-up ('planting'), maintenance, and harvest stages. In the set-up stage, farmers acquire seaweed propagules and attach them to ropes with plastic bottles attached at regular intervals to act as floats. These ropes are then taken to sea and attached to the plot foundation, which consists of anchors on the sea floor with ropes connecting them. The ropes with the seaweed attached are then connected to these anchoring ropes. The maintenance phase

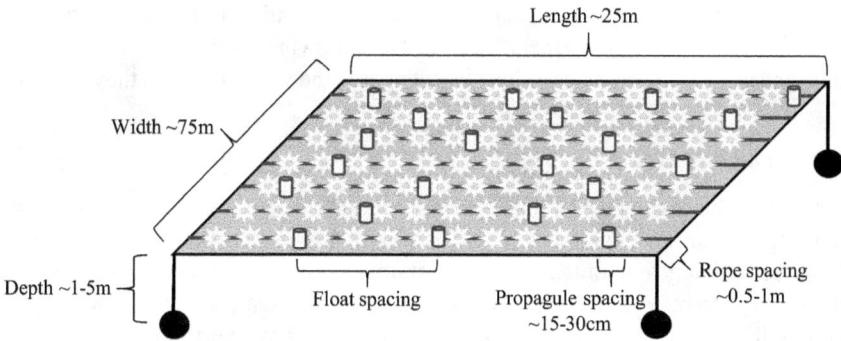

*Figure 6.3* Longline farm set up and approximate dimensions of seaweed farm plots in Pitu Sunggu (not to scale)

Source: Authors' schematic.

*Figure 6.4* A long line seaweed farm in Pitu Sunggu

Source: Image from drone footage by Nur Ihsan recorded May 2022 offshore of Pitu Sunggu village.

involves regular checking on the seaweed farm to monitor growth, and potentially making the decision to harvest the seaweed or move it to a different location. This phase may also involve removing epiphytic algae or barnacles from the seaweed (Figure 6.5). The harvest phase involves removing the seaweed-filled ropes from the ocean and bringing them to shore. The seaweed is removed from the ropes by running the line between two pieces of wood to break off the seaweed. If required, some of the seaweed is put aside to be used as propagules and the remainder is dried and sold. The ropes and bottles are cleaned in preparation for re-use. A number of farming decisions must be made at the various stages of seaweed farming, which are influenced by a large number of socio-economic and environmental variables that are outside the farmers' control. The following sections outline key variables faced by farmers, as described by them, and the ways in which they respond.

**Environmental factors**

Farming in the ocean takes place in an environment where seawater moves around and changes on an hourly basis, and differs to farming on land where the medium (soil) changes on much longer timeframes. This means that ocean conditions change regularly, and also vary significantly across small distances. Within the same village, plots only a hundred metres apart can have very different environmental conditions. As cited in Chapter 5, one farmer said, 'In that sea, all fertility is not the same. There are indeed points in the sea that are considered rather fertile, there are also those that seem to be barren land.' Farmers have developed a working understanding of the impacts of some of these environmental factors, although this understanding is partial and diligent farmers continuously observe the activities of others in order to interpret changes affecting their own farms and develop strategies to resolve them. *In situ* experiments exploring the connection between environmental parameters and seaweed growth rates have yielded inconclusive results due to the high number of parameters and their frequent variability. For example, van Oort et al. (2022) did not find a clear relationship between growth rate and measured environmental variables and observed unexpected decreases in biomass and certain points in seaweed growth, noting that the correlation between oceanic conditions and growth rates was difficult to determine. Simatupang et al. (2021) similarly found no correlation between *in situ* production and any of the environmental factors that they measured, positing that measuring a greater range of environmental variables may be necessary to reveal correlations. A range of variables have an impact on seaweed growth rates, and these interact in complex ways that have prevented determination of the relative impact of different variables on growth rates to date, although work in this area is ongoing.

Further complicating this analysis is the adaptability and morphological plasticity of seaweed. Although cultivated carrageenan seaweeds typically are genetically similar, seaweed grown in different conditions can have very different physical appearances and characteristics. One study of seaweed cultivated across Indonesia noted that the same species grown in different locations differed in colour, shape, growth rates, and carrageenan content (Simatupang et al. 2021). Another noted that wild samples of the same species of seaweed collected at different times of the year (in the wet and dry seasons) had very different optimum growth conditions (Araujo et al. 2014). The observable traits of seaweed cultivars are affected by both genetics and environment (Ratnawati et al. 2020). Recent research has shown that seaweed cultivars sourced from different sites, when grown together in a new location, converge in their morphology and colour (Simatupang et al. 2021). Farmers in our study contended that seaweed propagules moved between locations needed time to acclimatise to new conditions. On one occasion seeds brought from another region were sold to farmers in Pitu Sunggu, where they had poor survival rates. One farmer complained that 'The seeds that were bought from a different location are not suitable for the water conditions in this village: here is muddy, but at there it was sandy'. Similarly, in Pitu Sunggu farmers observe that there are two 'types' of sacol – a 'round' shaped one and a 'branched' type, with the former appearing in plots closer to the shore and the latter in plots further away. However, they observe that if these two types swap locations, they will eventually change their morphology to the opposite 'type'. The responsiveness of seaweed to environmental conditions – changing observable characteristics such as colour, shape, and growth rate (alongside unobserved characteristics such as carrageenan yield and characteristics) – complicate the development of a deterministic model of the impact of oceanic conditions on seaweed cultivation in the absence of a large-scale, long-term, multi-location field trial. Nevertheless, in Pitu Sunggu, some patterns can be observed and located within existing research.

### Ocean conditions

#### Temperature

Surface water temperature has a significant effect on eucheumatoid seaweeds both through direct effects on physiological processes, and through indirect effects that alter the surrounding environment (such as by affecting water motion and growth of other organisms such as epiphytes) (Neish 2008). The Indonesian Ministry of Marine Affairs and Fisheries (*Kementerian Kelautan dan Perikanan* (KKP)) recommends growth in waters from 26–32°C (MKPRI 2019, p. 18), although other studies have suggested a wider range of optimal growth temperatures of 20–30°C for both cottonii and sacol (Araujo et al. 2014; Lideman et al. 2013, Mairh et al. 1986; Ohno and Orosco 1987; Preisig and Hans 2005). Tolerance of high temperatures is an important consideration in understanding the impact of climate change on seaweed growth. In a laboratory, under otherwise ideal conditions, cottonii can grow (slowly) at temperatures up to 36°C (Kumar et al. 2020), although in another study mortality was induced at this temperature (Terada et al. 2016). Another laboratory study noted that temperatures of 33–35°C caused extreme whitening which impairs thalli growth (Largo et al. 1995), while in field trials where temperatures ranged from 29.5–35.5°C ice-ice developed and epiphyte infestation was triggered (Largo et al. 2020). These studies indicate that 36°C represents an upper limit on the growth of cottonii, while temperatures above 32°C stress the seaweed and in combination with other factors may lead to mortality. Although studies on sacol are limited, this species appears to be somewhat more tolerant of exposure to high temperatures than cottonii (Critchley et al. 2004; Hurtado et al. 2006), but is slower growing. Optimum growth rates have been reported for sacol at a wide temperature range of 23–31.6°C (Borlongan et al. 2017; Lideman et al. 2013; Ohno and Orosco 1987). Notably, seaweed propagules from the same species may exhibit adaptation to different ocean conditions. One study found that wild specimens collected in the dry season had a higher tolerance of both temperature and salinity variation than those collected in the wet season (Araujo et al. 2014). This suggests that seaweeds have some ability to adapt to different ocean conditions, although the temperature range in which seaweeds grow well appears to be narrow.

We recorded water temperatures in Pitu Sunggu on 113 days from March to September (roughly covering the dry season in Pitu Sunggu) and found that temperatures during the day (irradiance >100 lux) ranged from 28.7–35.7°C (excluding outliers), with outlying high temperatures of up to 37.7°C (Q1 = 31.3°C, median = 32.3°C, Q3 = 33.1°C). This indicates that the normal temperature range of Pitu Sunggu surface water during the dry season is at the upper limit of seaweed survival rates and regularly exceeds temperatures at which ice-ice and mortality have been induced in other studies. Farmers in our study were highly conscious of the problems caused by high temperatures for seaweed growth. They noted that high temperatures occurred at the surface when the heating of the sun was not relieved by rainfall or water movement and perceived that these conditions lead to seaweed stress and often death (Figure 6.5). During the dry season, they reported that surface water temperatures are highest in shallower areas (which have less water motion) which, in Pitu Sunggu, are located further offshore. As a result, during this season they cultivate plots closer to shore rather than the hotter outer plots and most grow sacol rather than cottonii, which they believe is more resilient to temperature stress. In Kupang and Semau in Nusa Tenggara Timor (NTT), where rainfall is even lower, most farmers grow sacol only, and note the difficulty of growing it in the high ocean temperatures of the dry season when wind speeds and rainfall are low (Langford, Turupadang et al. 2022).

In addition to the direct effects on physiological processes, temperature also indirectly affects seaweed growth by affecting the activity of other organisms

*Figure 6.5* Seaweed affected by high temperature
Source: Image by Radhiyah Ruhon, April 2022, Pitu Sunggu.

associated with epiphytic growth (Msuya and Porter 2014). In Pitu Sunggu, farmers complain that 'the hotter it is, the more the molluscs stick' and consequently during the dry season, when it is hottest, they face considerable issues with small molluscs attaching to the seaweed. These molluscs create multiple problems since they both directly impede seaweed growth (by restricting access to light and nutrients) and if not removed regularly also make the seaweed and ropes heavier, sinking them deeper and further reducing light exposure. During the dry season farmers also experience issues with epiphytic algae in parts of the sea area, similarly linked to the temperature, salinity and water motion conditions prevalent at the time. Temperature therefore has a considerable impact on several aspects of seaweed production which, if not managed, can be fatal to seaweed.

*Salinity*

Surface water salinity is one of the key variables affecting seaweed growth. Ocean surface water salinity undergoes long-term, seasonal, and daily changes, driven by patterns of ocean currents which vary with seasonal rainfall (Iskander and Suga 2022; Linsley et al. 2017; Purba et al. 2021). Ocean salinity in Indonesia at the equator ranges from around 27–37psu (Purba et al. 2021) but may be significantly lower in surface water close to shore as a result of rainfall and surface water runoff. Several studies have examined the tolerance of cottonii to different salinity conditions, although sacol has been less well studied and further research is needed to assess differences in salinity tolerance between the two species. In one study, cottonii tolerated salinity in the range of 25–45psu, but at 15psu died within three days (Hayashi 2011). Another study found good growth rates of cottonii in salinity levels of 25–35psu, with lower (but still positive) growth at 20 and 40psu (Yong et al. 2014). Salinity tolerance also varies between specimens. One study found that salinity tolerance of wild specimens collected during the rainy and dry season varied considerably. Samples collected during the dry season had salinity optima at 25 and 35psu while salinity levels of <15psu and >45psu were detrimental to the seaweed (Arujo et al. 2014). Samples of the same species collected during the rainy season had optimal salinity levels at 15psu and did not grow at 35psu or above (Arujo et al. 2014). This suggests the seaweed has considerable ability to adapt to different ocean salinity. These studies indicate that *Kappaphycus* are tolerant of salinities within the normal range of ocean salinity in Indonesia (27–37psu) but become stressed at salinities below 20psu and begin to die at around 15psu (though some specimens have adapted to grow at lower salinities). The Indonesian government advises seaweed be grown in conditions of 28–34psu (MKPRI 2019, p. 18).

Our measurements of surface water ocean salinity in Pitu Sunggu at regular intervals from March to September in seaweed growing locations recorded salinities of up to 31.8psu, below the maximum recommended growing salinity. However, after rainfall, surface water salinity can fall drastically as a freshwater layer forms on the surface of the water, creating a hyposaline environment which can cause widespread occurrence of ice-ice disease and seaweed death. To avoid exposure to freshwater, farmers immediately travel to their seaweed farms after a period of prolonged or intense rainfall and fill the plastic bottles used as floats with water to sink their seaweed below the layer of hyposaline water near the surface. Seaweed farms can remain like this for several days, but should be returned to the surface as quickly as possible after the freshwater disperses to avoid excessive grazing by fish, attachment of barnacles, and poor seaweed health due to lack of sunlight. Farmers who fail to protect their seaweed from hyposaline environments are likely to experience widespread losses, which are common after rainfall events. One farmer lamented after he suffered severe losses after failing to sink much of his seaweed: 'I didn't have time to open all the bottles, because there were so many. Those that were opened, those are the survivors.'

Rainfall affects not only salinity but also temperature. Farmers in Pitu Sunggu report that both species grow best with some rainfall, but will die if this is excessive. As one farmer told us, 'The cottonii cannot stand a lot of fresh water, but if there is no fresh water it's not good either. In the middle is just right.' Farmers do not cultivate cottonii during the dry season (except for small amounts intended for use as propagules at the start of the next season), and typically cease its cultivation by May at the latest unless the rain persists, in which case they will continue growing it. As one farmer said, 'It all depends on the rain. If there is still some rain then cottonii will grow.' Since rainfall affects both temperature and salinity, it is not clear whether this the growth of cotonii is related to the higher temperature of water without rainfall, or the higher salinity. However, given that surface water in Pitu Sunggu does not appear to reach hypersaline environments, cottonii stress under low rainfall environments is probably linked to high temperatures.

Apart from the effect of extreme salinity drops, salinity also indirectly impacts seaweed growth by altering the growth of other organisms. In particular, farmers report that fresh water increases the growth of epiphytic algae (moss) on the seaweed. Moss can be a major issue for farmers as it is difficult to remove and stunts seaweed growth by reducing its exposure to light intensity (Jalil et al. 2020), as discussed further below.

*Light*

Like most plants, eucheumatoid seaweeds require optimal amounts and types of light for growth. Eucheumatoid seaweeds are grown close to the surface of the water (<1m depth). Growth is impaired under conditions of low light, while excessive light can damage seaweed growth (Neish 2008, p. 33). Farmers in Pitu Sunggu observe that their seaweed is highly sensitive to light across even very small differences in depth (Figure 6.6). Depth can vary along the line depending on proximity to floats, so conscientious farmers adjust their buoyancy mechanisms regularly to maintain optimal light exposure, as will be discussed in Chapter 7 under the heading 'Float spacing'. Farmers contend that there are differences between cottonii and sacol in this respect. They say that sacol is slow-growing and therefore requires maximum light exposure at all times to maintain growth. Cottonii is said to be more resilient to lower light levels in rainy season with high cloud cover or when seaweed is submerged further below the surface. Light exposure is also affected by other factors such as water turbidity and the growth of epiphytic algae on the surface of the seaweed.

*Water motion*

Water motion has a significant effect on seaweed growth. Some water movement is required for growth. In still water, an unmixed boundary layer can cause seaweed to consume all nutrients in the adjacent water and fill it with its waste products, causing simultaneous starvation and poisoning effects (Neish 2008). Water motion is essential to transport nutrients to the seaweed and remove waste products. It can also clean the seaweed of mud and some epiphytic algae, which can build up during periods of low water motion. Water motion also mixes high temperature surface water with cooler water below, reducing exposure to temperature extremes. The extent of water motion, including wave action, varies seasonally and locationally in Pitu Sunggu and has a significant impact on seaweed farms. While farmers report benefits from moving waters, excessively large waves can damage seaweed and break seaweed propagules from the ropes, especially if the propagules have reached a large size, or are of poor health and have become brittle. The large size of the seaweed at the end of a growing cycle increases the risk of losses due to waves. Farmers therefore want to grow seaweed in a 'Goldilocks' zone of water motion – as one farmer said, 'It's difficult if there are no waves, but too many waves are difficult too'. Farmers adapt to differences in water motion through several mechanisms, including changing rope orientation, using small propagules to reduce the risk of breakage, harvesting early to avoid losses, and ceasing farming during periods of high waves, as will be discussed in Chapter 7.

*Nutrient levels*

Seaweeds require nutrients to grow. This includes macronutrients such as nitrogen, phosphorous, and ammonia, as well as micronutrients. Nitrogen has been widely recognised as a limiting factor for seaweed growth and, as a result, seaweeds are thought to have the potential to remove excess nitrogen from coastal waters. In Australia, researchers are working on using seaweeds as 'biofilters', which would be grown at estuaries proximate to the Great Barrier Reef to absorb nitrogen from rivers with high levels of fertiliser runoff to prevent it from reaching the reef (Australian Seaweed Institute 2023). In some locations in Indonesia where seaweed is grown offshore from intensively fertilised agricultural regions, seaweed growth has been observed to be higher (personal communication May 2020). There have been suggestions that marine seaweed farms could be fertilised to increase growth. This practice is not widespread due to the high cost of fertiliser and its rapid dissipation in coastal waters, nor is it advised as it would probably have unintended environmental consequences for marine ecosystems. However, it does highlight the ways that land-based and marine activities are inter-related. For example, in Pitu Sunggu, as seaweed farming has expanded and become more profitable, many farmers have given up or reduced their work on terrestrial fish and shrimp ponds, including ceasing or reducing the use of nitrogen fertiliser in these brackishwater ponds, many of which are in coastal or riverbank areas. It is possible that any reduction in fertiliser runoff to coastal marine areas is partially responsible for the declines in seaweed growth rates that farmers have observed over time.

Micronutrients are also required for seaweed growth. In the early days of the seaweed industry, farming apparatuses were fixed to the sea floor in shallow coastal areas, and seaweed grown in deeper waters performed poorly, leading to the assumption that seaweed requires some combination of micro and macro nutrients not found in the open sea (Neish 2008). Longline seaweed farming is common in shallow offshore deeper waters close to the coastline, and the amounts and types of nutrients required by seaweed is still not well understood. Further research is required to reveal these interactions, as seaweed growth can be limited by the absence of one or more necessary nutrients. Seaweed farmers in Pitu Sunggu perceive some areas of the ocean seem to be 'fertile' where others are 'barren'. In the dry season in which flooding is not a significant risk, farmers in Pitu Sunggu prefer to cultivate plots close to the river mouth where they have observed growth rates are higher. In addition, some farmers report that areas of the sea that experience a 'fallow' period of low cultivation may regain their productivity. It is not clear whether this anecdotal evidence reflects random factors or systematic effects, although they have also been observed by Neish (2008) who notes that over-farmed areas may experience productivity declines, which recover after several fallow years. Steenbergen (2017) described the environmental collapse of seaweed farming in an over-farmed area, and Blankenhorn (2007) recommended observing maximum planting densities to stay within sustainable production limits. Further research is needed to explore the environmental impact of seaweed farming and the ways that these could be ameliorated (such as through maintaining proximate marine protected areas (Le Gouvello et al. 2017) or observing maximum planting densities (Blankenhorn 2007; Neish 2008).

*Turbidity*

One of the main issues faced by seaweed farmers in Pitu Sunggu during the dry season is the high turbidity of the area close to the shoreline, which causes mud to stick to the seaweed. In some cases, sedimentation can lead to a rapid decline in productivity of seaweed farming (Limi et al. 2018). Neish (2008) recommends that farmers regularly shake seaweed lines to remove sediment. Turbidity affects seaweed growth because it coats the seaweed, blocking the light and the water motion it needs to grow. Farmers in Pitu Sunggu report that in areas close to the shoreline during the dry season, mud sticks to the seaweed and is difficult to remove. However, mud from rivers also introduces nutrients to the water. One farmer reported that notwithstanding the coating issue, 'if its muddy, the seaweed will grow fast. Even the main stalks also produce some new branches.' Turbidity therefore has both negative and positive impacts on the seaweed growth but must be managed as excessive mud can be detrimental. It further introduces issues in drying the

*Figure 6.6* Farmers routinely shake the longline to remove sediment
Source: Image by Zulung Zach Walyandra, August 2022, Pitu Sunggu.

seaweed, as mud must be washed off the seaweed before being dried and sold. One Pitu Sunggu farmer who regularly failed to wash his muddy seaweed was widely criticised by other farmers, who felt that traders would penalise them all for the low-quality seaweed that he introduced. Seaweed marking and the interaction with production practices will be discussed further in Chapter 9.

*Ephiphytes*

Farmers in Pitu Sunggu experience considerable issues with epiphytes, including a range of epiphytic microalgae, macroalgae, and molluscs. Small molluscs – known locally by a generic name of *picing-picing* and probably including various members of the *Dreissenidae* family – grow on the surface of the seaweed itself. They cause significant problems for farmers because at certain times of the year and in certain locations (particularly in the outer locations during the dry season from June onwards) they become numerous and fast-growing. They attach to the surface of the seaweed and stunt its growth (Figure 6.8). These molluscs are heavy and weigh down the seaweed, causing it to sink further below the surface and reduce access to light and surface water motion. As one farmer said, 'If the molluscs stick, the seaweed will be stunted and sunk. Meanwhile, sinking seaweed is not good since seaweed needs waves during growth.' In addition, a larger species of mollusc grows on the seaweed ropes, further contributing to this sinking effect. Farmers

*Figure 6.7* Molluscs growing on seaweed
Source: Image by Radhiyah Ruhon, June 2022, Pitu Sunggu.

use several strategies to prevent and remove barnacles, as will be discussed under 'Farm maintenance' in Chapter 7.

In addition, locations close to the coast experience issues with the growth of macroalgae on the seaweed. Farmers report that moss is even more of a problem than mollusc growth because it severely stunts seaweed growth, prevents new branches from forming, and causes the seaweed to become brittle so that it is then easily broken off the ropes. In addition, it is very difficult to remove – as one farmer lamented, 'if it is only molluscs stuck to the seaweed it will be easy to remove, but when its moss, we don't know what to do'. Farmers note that seaweed covered with moss will remain small even when mature. This is probably due to several effects including damaging the seaweed, blocking the light, and competing for nutrients. It has been widely observed that epiphytic algae growth on longline cultivated seaweed significantly reduces seaweed growth rates (e.g. Jalil et al. 2020). Farmers who experience severe problems from moss say that they may persist for quite a long time. In June, when moss and molluscs are a particular problem, we observed farmers harvesting their stunted seaweed and using it as propagules, complaining that although the propagules were clearly unhealthy, they were all they had. Farmers who experience ongoing issues with seaweed growth may give up production altogether. Throughout June we observed worsening sacol production and some farmers eventually dried their remaining crop and gave up on farming for the season.

Farmers also experience problems with ephiphytic microalgae, which grow on the surface of the seaweed and reduce its growth by blocking light and competing for nutrients (Neish 2008) and by damaging the seaweed tissue (Vairappan et al. 2008).

Farmers consider this a less serious problem than those discussed above as these microalgae can easily be removed by shaking the seaweed longlines. Regular farm maintenance is sufficient to prevent excessive microalgae build up, as will be discussed in Chapter 7.

*Grazing wildlife*

Grazing by fish and turtles can be a serious problem for farmers. In our survey of Pitu Sunggu farmers, 39 per cent of farmers reported this as a very significant problem, with a further 36 per cent reporting it to be somewhat significant. In our survey in NTT, these Figures were slightly higher at 43 per cent and 40 per cent respectively (Langford, Turupadang et al. 2022). The extent to which grazing is a problem varies by location and season, in some areas being a minor nuisance while in others preventing seaweed from reaching maturity entirely. Built structures in the oceans typically attract fish populations and seaweed farming may impact fish assemblages in nearby areas (Kelly et al. 2020). In particular, fish from the *Siganidae* family (rabbitfish) are well known to graze on seaweed (Kelly et al. 2020). In Pitu Sunggu, one of the most common fish species we observed in fishers' baskets was the golden rabbitfish (*Siganus guttatus*) (Figure 6.9), a species that farmers often see grazing on seaweed farms. Farmers note that grazed seaweed cannot be used for seed because fish preferentially graze on the new branches of the seaweed which are needed for ongoing growth. Farmers report that fish grazing is normally a serious issue in April. Although it is detrimental to seaweed farms, catches of

*Figure 6.8* Golden rabbitfish (*Siganus guttatus*). Farmers refer to most varieties of rabbitfish as *baronang*

Source: Image by Radhiyah Ruhon, March 2022, Pitu Sunggu.

species of herbivorous reef fish often increase after the commencement of seaweed farming (Hehre and Meeuwig 2016) suggesting that Pitu Sunggu fishers may benefit from larger populations of these market fish. The interactions between seaweed and associated non-seaweed activities feature strongly in the coastal systems and livelihoods.

*Pollution*

Seaweed farms can also be affected by pollution levels, including visible plastic pollution and unobserved micropollutants. Larger pieces of plastic are less of an issue in Pitu Sunggu than in nearby Laikang, where some farmers adjust the depth of their seaweed to avoid high rates of plastic pollution floating around seaweed farms during the wet season (Langford et al. 2023). Seaweed farming can also introduce pollutants to the ocean, particularly through the use of plastic in seaweed farming. Our study identified high rates of microplastics in Pitu Sunggu waters, as well as microplastics on the surface of seaweed and in marine life, although most were from sources not associated with seaweed farming (Hovey et al. 2023). The extent to which ocean pollutants affect seaweed growth and final product purity are not well studied and require further research.

**Major resulting production issues**

The environmental parameters described above combine to create different ocean conditions in different locations at different times of the year which have to be carefully managed. A major issue for seaweed farmers is the occurrence of ice-ice disease following sudden environmental changes such as a period of high surface water temperature or prolonged rainfall.

*Figure 6.9* Intensive use of plastic bottles as floats in the dry season
Source: Image by Zulung Zach Walyandra, June 2022, Pitu Sunggu.

*Managing seasonal change*

As a result of shifting interacting ocean variables with seasonal weather patterns, Pitu Sunggu experiences two distinct seaweed farming seasons, the wet season and the dry season. During the wet season from November to April, cottonii is culti- vated in offshore locations (typically >500m from the shoreline). During the dry season, which reaches its peak from August to October, sacol species is typically cultivated in areas <500m from the shoreline. There is a transition period between the two seasons from May–July in which both species can be cultivated. Cottonii typically declines in productivity in this period and sacol comes into season, but these are broad trends rather than strict rules: some farmers succeed in cultivating outside these seasonal patterns. Figure 6.11 shows the locations of active seaweed farms during the 2022 dry and wet seasons in Pitu Sunggu (see Langford et al. 2021 for details on mapping process).

The seasonal switch between seaweed species has important implications for farmers' livelihoods and the broader seaweed industry. As discussed in Chapter 5, the different suitability of inner and outer waters to cultivation of different spe- cies in different seasons means that not all farmers in Pitu Sunggu can cultivate seaweed all year round. There is a class of 'senior' farmers who established plots early in the industry development stage and have access to good near-shore dry season locations as well as offshore wet season locations. With fewer options, jun- ior farmers often struggle during the dry season when confined to unproductive offshore sites. Seaweed farming is therefore not a year-round livelihood activity for all farmers. In our farmer survey, we noted that farmers who grew only cottonii ceased farming during the dry season, as shown in Figure 6.12.

Seasonality causes other disruptions. Even farmers with suitable spaces for both wet and dry season cultivation will need to completely change species twice each year. This means that instead of using their own seaweed harvest to propagate a new cycle, they must purchase propagules from other farmers to seed a new season, which typically means buying seeds from farmers in other regions with different seasonal patterns. Farmers in Pitu Sunggu often buy seeds from Wajo Re- gency, which sits on the western coast of South Sulawesi so experiences different seasonal changes in ocean conditions. The need to purchase seeds twice yearly is a significant capital outlay for the farmers, as discussed in Chapter 8. In addition, the cultivation of two different species means that farmers have had to develop production strategies for both species, which react differently to changing ocean conditions. For example, sacol reportedly grows slightly more slowly than cottonii and needs more light, but is more resistant to temperature stress.

Further, the cultivation of two different species in the same area means that both species typically enter the same value chain mixed together. This is particu- larly the case in the transition season (May–July) and when fishermen 'catch' both types of seaweed in their nets and dry it for sale. Both seaweed farmers and fisher- men then sell both species mixed together. This behaviour is widely practised as sacol and cottonii contain the same type of carrageenan – kappa-carrageenan – unlike *Eucheum denticulatum* (known colloquially as 'spinosum') which contains

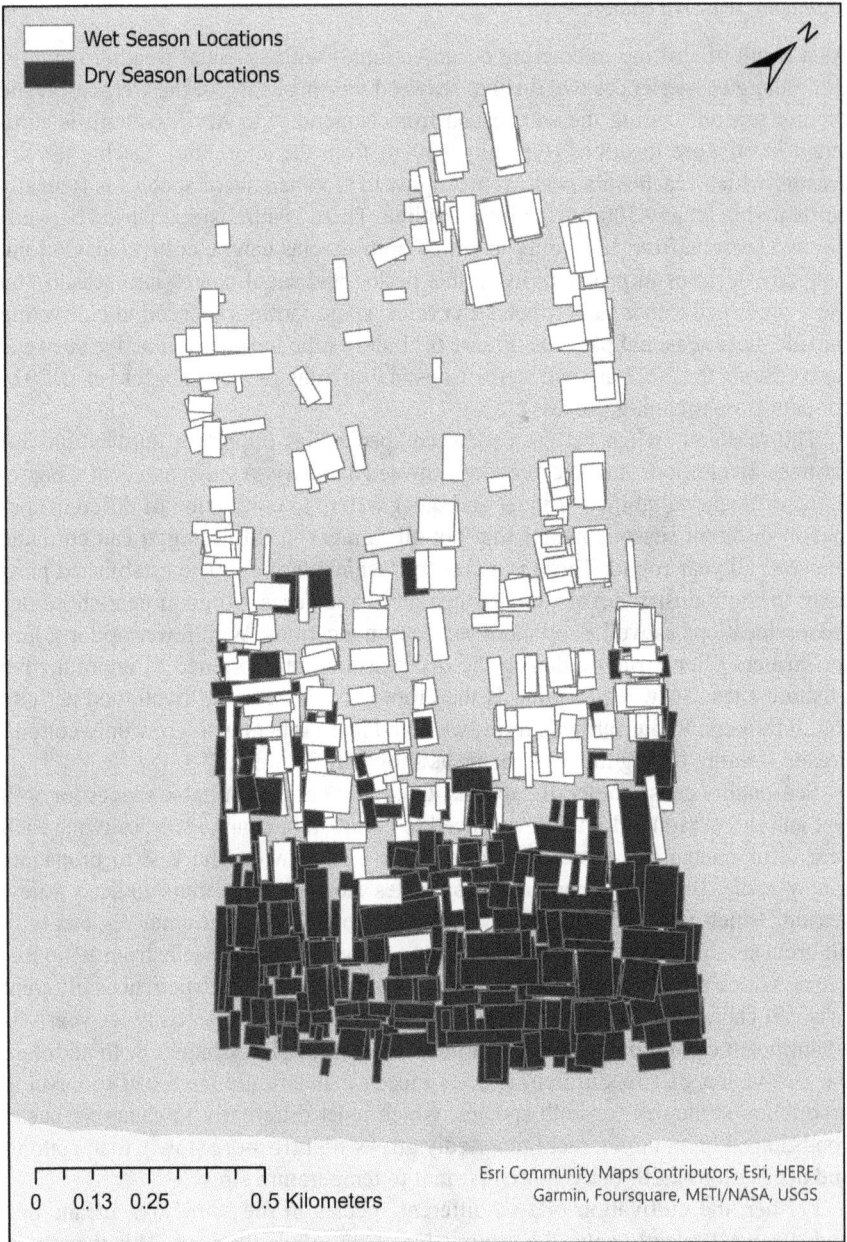

*Figure 6.10* Seasonal variation in area planted with seaweed in the wet season (Nov–April) and the dry season (Aug–Oct)

Source: Authors' analysis of satellite imagery in 2022. See Langford et al. 2021 for methodology.

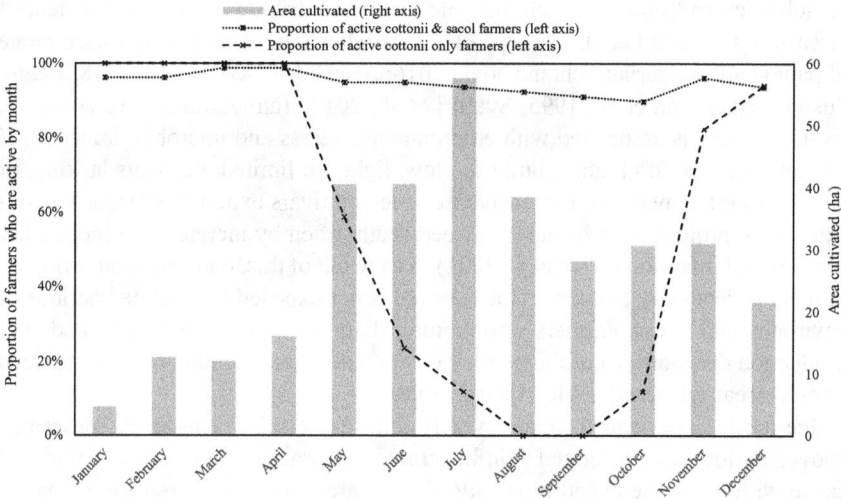

*Figure 6.11* Seasonal variation in area planted with seaweed and proportion of farmers who are active

Source: Data for area under cultivation calculated from analysis of PlanetLabs imagery for 2022 (see Langford et al. 2021 for methodology). Data for number of cultivating farmers from Authors' survey (see Langford et al. 2024).

iota-carrageenan. However, sacol has slightly different physical characteristics to cottonii. It has thicker thallus and branching, which means that it dries more slowly and does not normally reach as low a moisture content as cottonii. Traders take advantage of these characteristics by mixing the two species together to achieve an optimal moisture content that is at the higher end of the tolerated moisture content range to maximise the weight the seaweed mix is sold for. This practice has implications for seaweed marketing, as discussed in Chapter 9.

Seasonal shifts in farming locations and species are common in many parts of Indonesia. During our fieldwork in North East Rote Ndao in NTT, farmers similarly report transitioning between species and locations (Langford, Turupadang et al. 2022; 2023). The distinctiveness of seasonal change varies a lot. Some seaweed farming areas experience very distinct change and periods of inactivity in seaweed farming, while others enjoy year-round cultivation. This is partly influenced by coastline morphology including bays and peninsulas. In our Laikang village research, for example, farmers positioned outside the bay grew spinosum which was not grown within the bay (Langford et al. 2024).

*Managing ice-ice disease*

A second key production management skill for farmers is the need to regularly and consistently manage the occurrence of ice-ice disease. Ice-ice is a condition of seaweed whereby it turns white (like ice) and dies (Figure 6.13). Ice-ice is a 'complex pathobiotic syndrome in which multiple factors contribute to the development of disease signs' (Ward et al. 2021, p. 14) and has been associated with a wide range of pathogenic bacteria (Achmad et al., 2016; Aris, 2011; Azizi et al., 2018; Largo, Fukami, Nishijima et al., 1995; Syafitri et al., 2017; Tahiluddin and Terzi, 2021). Ice-ice disease is associated with environmental stress and microbial infection. If temperatures are too high, salinity too low, light too limited, nutrients lacking, or damage to the seaweed too extensive, seaweed cultivars experience stress which if extreme or prolonged will cause seaweed death, often by increasing susceptibility to microbial infection (Ward et al. 2021). As a result of this, constant monitoring by farmers is necessary to ensure that seaweed is not exposed to extended periods of environmental stress. Farmers who monitor their seaweed farms closely and alter production decisions according to their observations can maintain healthy seaweed plots in areas where others fail (Neish 2008).

Ice-ice is experienced most severely during the rainy season. As described above, periods of prolonged rainfall create hyposaline environments that can cause widespread seaweed mortality if steps are not taken to reduce exposure, for example by sinking the seaweed below the surface. Many farmers cease cultivation in the rainy season due to the high risk of loss. One farmer described the situation as follows, 'working on seaweed in the rainy season is like balancing an egg on a horn tip – there are many things that go wrong'. Farmers report that ice-ice disease caused by hyposaline surface water layers will rapidly cause death of

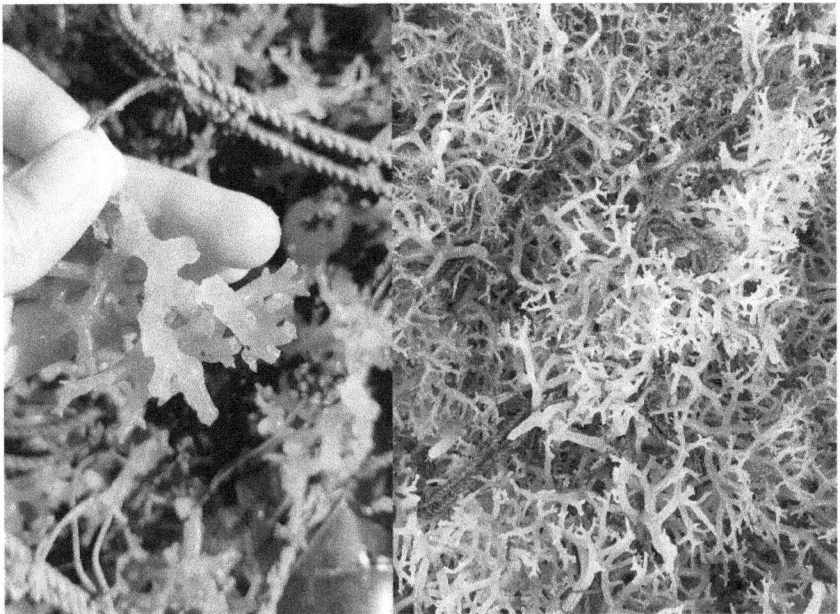

*Figure 6.12* Ice-ice disease on sacol

Source: Photo credit: Radhiyah Ruhon, Pitu Sunggu January 2022.

the entire propagule. Significant losses after rainfall are common and expected. As one farmer explained, 'well, that is how we all work here. Sometimes we get results, sometimes we do not. What matters is that we don't complain, we keep the spirit up.' Losses from ice-ice disease can be complete or partial. Farmers affected by ice-ice typically examine the seaweed to determine if there are portions which can be recovered for use as propagules. As one farmer told us, 'sometimes when the seaweed is damaged, there are still some we can save and use as seedlings'.

Ice-ice malaise also occurs in a less severe form in association with other environmental stressors. One farmer noted that careful observation of the patterns of occurrence of ice-ice is necessary to diagnose the problem:

> If the ice-ice affects the seaweed variably, you can tell it is not from rain. If it is caused by rain, the ice-ice affects the whole cultivar evenly, and affects the seaweed close to the floats [at the water surface] more. But if you see that it is [only] the submerged seaweed that has turned white, you know it is something else.

The parts of a seaweed propagule affected by ice-ice disease do not recover, but if only small areas of the cultivar have been damaged, the cultivar may recover if moved to a more favourable environment. In such cases the part affected by ice-ice will fall off the seaweed and the remaining portion will continue to grow. As one farmer explained, when the damage is minor 'if the seaweed is relocated, it usually recovers … if the damage is still at the tip of the branch, it can come off later, that's where a kind of new branch will grow… [But] if the thallus is already white, it's not good.' As this farmer notes, if the thallus is affected by ice-ice disease, the propagule will not recover. Farmers who observe ice-ice on their seaweed must therefore decide whether to move the seaweed or harvest it before damage becomes more severe. This requires careful and consistent monitoring of the seaweed to avoid and manage ice-ice outbreaks.

## Socio-economic factors

In addition to the environmental factors described above, farmers also work within a series of socio-economic constraints that affect their production decisions. These include access to sea space (as discussed in Chapter 5), labour (Chapter 8), capital (Chapter 9), and market prices. This section provides an overview of how these factors affect production decisions.

### *Access to sea space*

The most important constraint faced by farmers is access to sea space, which leads farmers to farm in different locations, with different strategies and at different intensities. Farmers who have large areas of sea space are able to plant their seaweed less intensively (i.e. at a lower rope spacing). Using a low rope spacing can sometimes lead to higher seaweed growth rates than on very closely spaced ropes. As a result, farmers with large areas of sea space sometimes enjoy higher returns on their labour as a result of the large areas of 'land' at their disposal. In contrast, farmers with smaller plots in less favourable areas are often forced to over-apply labour on their farms (through closely spacing ropes which are labour-intensive to prepare) and undertake more frequent monitoring and replanting activities. Some farmers with very limited access to sea space maintain small plots of closely spaced seaweed longlines – as little as 50cm apart, around half the recommended spacing. Seaweed growth rates generally decline with increasing planting density (Neish 2008) but overall yields per area may increase. For example, if a farmer with lim-ited space installs twice as many longlines in an area, but these longlines yield only 70 per cent of what they would have in a less densely spaced area, the farmer is experiencing a 30 per cent lower yield per labour (because the amount of labour used per rope is constant) but a 40 per cent higher yield per land (because twice as many ropes at 70 per cent of the yield is applied). Farmers therefore manage their production parameters – rope spacing, maintenance frequency and practices – in response to their sea space access, as will be discussed below.

### Access to labour

Seaweed binding and planting are labour intensive. Very small seaweed farmers may be able to manage replanting using only household labour, but larger farmers re-quire access to teams of casual wage labourers to bind seaweed. This is especially the case as seaweed binding should occur in a time period of less than six hours between harvest and replanting (MKPRI 2019). We observed that on average, one binder normally completes one longline per hour (although younger and agile people may be able to work at up to double this rate). A farmer wishing to replant 60 ropes would therefore require a team of ten binders working a six-hour day to complete the task. During our research, binding labour was in short supply in Pitu Sunggu so it was becoming increasingly common for farmers to send their seaweed to inland villages for binding. Several middlemen had started offering binding services where, for a fee, they would collect seaweed from farmers, transport it to binding groups in other villages, and return the completed longlines at the end of the day. As one farmer lamented, 'The binders here are overwhelmed with work. Lots of farmers want them to tie their seeds. Usually, we look for available labour in the village, but none are available.' As a result of labour shortages and high seaweed prices, binding wages increased several times during the course of our study, from Rp. 4,000 in June 2022 to Rp. 5,000–5,500 in October (not including transportation fees and twisting of longlines together to make double-ropes). Farmers must also time their harvest around the availability of binding labour and often 'take turns' with each other in scheduling the services of binding groups. Shortages of labour sometimes result in somewhat longer production cycles than would otherwise be the case, particularly during the period of transition from cottonii to sacol production in May, and in Janu-ary, when high rainfall events mean that farmers must frequently harvest seaweed early to avoid losses from ice-ice disease, increasing the demand for binding services.

### Access to capital (ropes, seeds, boats)

Capital is required for several key seaweed farming activities, including purchasing boats, ropes, and propagules. Farmers also require funds for ongoing gasoline costs, and to pay binding labourers.

### Boats

In our survey of 96 seaweed farmers in Pitu Sunggu, we found that 80 per cent had a motorised boat, 10 per cent an unmotorised boat, and 10 per cent had no boat (presumably requiring them to borrow one). Boat size and motor power affect the distance from shore at which farmers can cultivate, but distant plots are expensive to maintain due to fuel costs. Thus, access to a boat and capital for fuel have a bearing on which plots are farmed. Farmers also factor in the costs associated with distance when making decisions about frequency of plot maintenance and their farming cycle.

### Ropes

Ropes are required for farming seaweed and a lack of capital to buy ropes is a barrier to entry for some farmers. Farmers with an insufficient number of ropes might choose to cultivate less seaweed than they otherwise would and need to clean ropes quickly after use and return them to production. Farmers with a surplus of ropes have more flexibility in the cleaning process. Some farmers even leave their ropes in fishponds to allow the fish to graze on epiphytic molluscs.

### Propagules

Seaweed farmers require propagules to commence a farming cycle. Farmers try to use their own seaweed wherever possible to avoid the high cost of purchasing propagules. However, this is often not possible if they have been affected by an event such as high rainfall that leads to high seaweed loss or during the transition period between species. In this instance farmers buy seeds, and many farmers borrow money from traders to finance this cost (farmer-trader relations are discussed in detail in Chapter 9). Many farmers try to minimise the quantity of purchased propagules to reduce loans, preferring instead to multiply the propagules themselves (as Chapter 7 will discuss). This means that they may farm several short cycles (of around 25 days each) to produce propagules before commencing production for sale. As a result, risk averse farmers in particular operate significantly under capacity for the initial periods of a new planting season or after a rainfall event. Shorter cycles at the beginning of the seasons may also be a response to perceived higher seasonal risk when ocean conditions may not yet be suitable for the new species. Smallholder farmers therefore may produce lower volumes than would be considered 'optimum' as a result of capital constraints and risk minimisation. Farmers' decisions about production are based largely on their appetite for risk, which stems from both their personal choices and from their household circumstances. Farmers with accumulated wealth may be more willing to take more risk by investing heavily in propagules at the start of a new season, while most farmers prefer to purchase only a small quantity of seeds.

### Market prices

An increasing number of people have taken up seaweed farming to capitalise on the rise in seaweed prices in recent years (see Langford, Zhang et al. 2022 for price analysis). This includes many people who had previously migrated to other islands or abroad to find work, and recently returned to Pitu Sunggu to try their hand at seaweed farming. This has intensified the use of sea space even in suboptimal growing conditions. Where previously low yields and high labour and capital input may have deterred farmers from cultivating in low-performing areas, high prices incentivise farmers to utilise the areas. Although the environmental impacts of seaweed farming are not well studied, it is likely that this intensification puts additional pressure on marine ecosystems. Farmers in Pitu Sunggu were observed growing seaweed in areas in which it became covered in mud, molluscs, and epiphytes, was frequently affected by ice-ice disease, often lost to high waves, and grazed by fish. High seaweed prices driving increased cultivation can mean re-allocation of household resources away from other household activities. For example, many farmers with brackishwater ponds report neglecting or abandoning this work during periods of high seaweed prices.

### Risk of theft

As a result of high market prices, farmers' production decisions are also increasingly influenced by the risk of theft. There were several cases of seaweed theft from longlines in the ocean over the course of our study. These were thought to occur at night, mainly suspected to be by residents of offshore islands. One farmer reported that he harvested his sacol early at 32 days (rather than the typical 45–60 days) because he was afraid of losing seaweed if he let it grow too big. Such behaviour has implications for carrageenan seaweed processors, as Chapters 2 and 3 discussed. In some parts of Indonesia, villages have implemented farming curfews which restrict activity in the sea at night in order to reduce the incidence of theft. In our study in Laikang village, residents have been urged not to stay at sea past 6pm to avoid being suspected of theft. Meanwhile, in Daima village in Nusa Tenggara Timor (NTT), village rules were made prohibiting farming after 6pm to avoid theft (Satria et al. 2017). In both Pitu Sunggu and Laikang, during the course of our study there were even cases of seaweed being stolen from the pier or drying platform where it was being dried. Considering the price of seaweed reached nearly Rp. 50,000/kg (dry) in 2022, this is not entirely surprising. A binding wage in the village is about Rp. 5,000/hr, so stealing just one kilogram of dried seaweed is equivalent to about two days' work (at five hours per day). Farmers' decision-making is therefore linked to market prices in important ways.

# Conclusion

This chapter outlined key environmental and socio-economic factors affecting seaweed farming in Pitu Sunggu. Changing ocean conditions have a major impact on seaweed growth and survival, including temperature, salinity, light, water motion, nutrient levels, turbidity, epiphytes, grazing wildlife, and pollution. In addition, socio-economic factors such as access to sea space, labour and capital, market prices, and socio-economic risks restrict the options available to farmers. Although these factors limit the opportunities available to farmers, in isolation they do not determine farm performance. The next chapter will explore farmers' decision-making in response to these constraints, including how farmers adjust activities to manage environmental conditions within their socio-economic constraints.

# References

Achmad, Marlina, Alimuddin Alimuddin, Utut Widyastuti, Sukenda Sukenda, Emma Suryanti, and Enang Harris. 2016. "Molecular Identification of New Bacterial Causative Agent of Ice-Ice Disease on Seaweed *Kappaphycus alvarezii*." *PeerJ Preprints*. https://doi.org/10.7287/peerj.preprints.2016v1

Araujo, P. G., A. L. N. L. Ribeiro, N. S. Yokoya, and M. T. Fujii. 2014. "Temperature and Salinity Responses of Drifting Specimens of *Kappaphycus alvarezii* (*Gigartinales, Rhodophyta*) Farmed on the Brazilian Tropical Coast." *Journal of Applied Phycology* 26: 1979–1988.

Aris, Muh. 2011. "Identifikasi, patogenisitas bakteri dan pemanfaatan gen 16s-rrna untuk deteksi penyakit ice-ice pada budidaya rumput laut (*Kappaphycus alvarezii*) [Identification, Pathogenicity of Bacteria and the Use of Gene 16SrRNA for Ice-ice Detection on Seaweed Aquaculture (*Kappaphycus alvarezii*)]." Ph.D. thesis. Bogor Agricultural University, Indonesia.

Aslan, La Ode Muhammad, Wa Iba, Andi Besse Patadjai, Ruslaini, and Manat Rahim. 2021. "A Preliminary Study of the Effect of Different Seedling Sources on Growth of Seaweed *Kappaphycus alvarezii* Cultivated in Konawe Selatan and Bombana Regency, Southeast (SE) Sulawesi, Indonesia." *IOP Conference Series. Earth and Environmental Science* 763 (1): 12018. https://doi.org/10.1088/1755-1315/763/1/012018

Australian Seaweed Institute. 2023. "Seaweed Biofilters." Accessed 11 May 2023. https://www.australianseaweedinstitute.com.au/seaweed-biofilters

Azizi, Azhani, Nursyuhaida Mohd Hanafi, Mohd Nazir Basiran, and Chee How Teo. 2018. "Evaluation of Disease Resistance and Tolerance to Elevated Temperature Stress of the Selected Tissue-Cultured *Kappaphycus alvarezii* Doty 1985 Under Optimized Laboratory Conditions." *3 Biotech* 8 (321): 1–10. https://doi.org/10.1007/s13205-018-1354-4

Blankenhorn, S. U. 2007."Seaweed Farming and Artisanal Fisheries in an Indonesian Seagrass Bed – Complementary or Competitive Usages?" PhD Thesis. University of Bremen, Germany.

Borlongan, Iris Ann, Grevo S. Gerung, Shigeo Kawaguchi, Gregory N. Nishihara, and Ryuta Terada. 2017. "Thermal and PAR Effects on the Photosynthesis of *Eucheuma denticulatum* and *Kappaphycus striatus* (So-called Sacol Strain) Cultivated in Shallow Bottom of Bali, Indonesia." *Journal of Applied Phycology* 29 (1): 395–404. https://doi.org/10.1007/s10811-016-0956-7

Critchley, Alan. T., Danilo Largo, W. Wee, G. Bleicher L'honneur, Anicia Q. Hurtado, and J. Schubert. 2004. "A Preliminary Summary on *Kappaphycus* Farming and the Impact of Epiphytes." *Japanese Journal of Phycology*, 52 (Supplement): 231–232.

Hayashi, Leila, Gabriel S. M. Faria, Beatriz G. Nunes, Carmen S. Zitta, Lidiane A. Scariot, Ticiane Rover, Marthiellen R. L. Felix, and Zenilda L. Bouzon. 2011. "Effects of Salinity on the Growth Rate, Carrageenan Yield, and Cellular Structure of *Kappaphycus alvarezii* (*Rhodophyta, Gigartinales*) Cultured in Vitro." *Journal of Applied Phycology* 23 (3): 439–447. https://doi.org/10.1007/s10811-010-9595-6

Hehre, E. James and Jessica J. Meeuwig. 2016. "A Global Analysis of the Relationship Between Farmed Seaweed Production and Herbivorous Fish Catch." *PloS One* 11 (2): e0148250–e0148250. https://doi.org/10.1371/journal.pone.0148250

Hovey, R., S. Werorilangi, W. Umar, H. Hasyim, N. M. R. Harusi, K. F. Dexter, R. Ruhon, Z. Zach, and A. Langford. 2023. "End of Life of Plastics Used in Seaweed Aquaculture in South Sulawesi." Melbourne, Australia: Australia-Indonesia Centre.

Hurtado, A. Q., A. T. Critchley, A. Trespoey, and G. Bleicher L'honneur. 2006. "Occurrence of Polysiphonia Epiphytes in Kappaphycus Farms at Calaguas Is., Camarines Norte, Phillippines." *Journal of Applied Phycology* 18 (3–5): 301–306. https://doi.org/10.1007/s10811-006-9032-z

Iskandar, Mochamad Riza and Toshio Suga. 2022. "Change in Salinity of Indonesian Upper Water in the Southeastern Indian Ocean during Argo Period." *Heliyon* e10430: 1–12. https://doi.org/10.1016/j.heliyon.2022.e10430

Jalil, Abdul R., Muhammad F. Samawi, Hasni Y. Azis, Ilham Jaya, Abdul Malik, Ibrahim Yunus, and Muhammad A. Achmad Sohopi. 2020. "Comparison of Physical-Chemical Conditions for Seaweed Cultivation in the Spermonde Archipelago, Indonesia." *Aquaculture, Aquarium, Conservation & Legislation* 13 (4): 2071–2082. www.bioflux.com.ro/docs/2020.2071-2082.pdf

Kelly, Emily L. A., Abigail L. Cannon, and Jennifer E. Smith. 2020. "Environmental Impacts and Implications of Tropical Carrageenophyte Seaweed Farming." *Conservation Biology* 34 (2): 326–337. https://doi.org/10.1111/cobi.13462

Kumar Yushanthini Nair, Sze-Wan Poong, Clair Gachon, Juliet Brodie, Ahemad Sade, Phaik-Eem Lim. 2020. Impact of Elevated Temperature on the Physiological and Biochemical Responses of *Kappaphycus alvarezii* (*Rhodophyta*). PLoS One. doi.org/10.1371/journal.pone.0239097

Langford, A., W. Turupadang, M. D. R. Oedjoe, C. Liufeto, B. Bire, M. Yohanis, Y. Suni, and S. Waldron. 2022. "Seaweed Farmer Resilience in Eastern Indonesia after COVID-19." *Australia National University Indonesia Project*.

Langford, A., S. Waldron, Sulfahri, and H. Saleh. 2021. "Monitoring the COVID-19-Affected Indonesian Seaweed Industry Using Remote Sensing Data." *Marine Policy* 127 (104431): 1–10. doi.org/10.1016/j.marpol.2021.104431

Langford, Alexandra, Scott Waldron, Jing Zhang, Radhiyah Ruhon, Zulung Zach Walyandra, Risya Arsyi Armis, Imran Lapong, Boedi Julianto, Irsyadi Siradjuddin, Syamsul Pasaribu, and Nunung Nuryartono. (2024). "Diverse Seaweed Farming Livelihoods in Two Indonesian Villages." In *Tropical Seaweed Cultivation – Phyconomy: Proceedings of the Tropical Phyconomy Coalition Development, TPCD 1, Held at UNHAS University, Makassar, Indonesia. July 7th and 8th* edited by A. Hurtado, A. Critchley, and I. Neish. 00-00. New York: Cham, Switzerland: Springer Nature.

Langford, A., J. Zhang, S. Waldron, B. Julianto, I. Siradjuddin, I. Neish, and N. Nuryartono. 2022b. Price Analysis of the Indonesian Carrageenan Seaweed Industry. *Aquaculture* 550(737828). doi.org/10.1016/j.aquaculture.2021.737828

Largo, Danilo B., Kimio Fukami, and Toshitaka Nishijima. 1995. "Occasional Pathogenic Bacteria Promoting Ice-Ice Disease in the Carrageenan-Producing Red Algae *Kappaphycus alvarezii* and *Eucheuma denticulatum* (*Solieriaceae, Gigartinales, Rhodophyta*)." *Journal of Applied Phycology* 7 (6): 545–554. https://doi.org/10.1007/BF00003941

Largo, Danilo B., Kimio Fukami, Toshitaka Nishijima, and Masao Ohno. 1995. "Laboratory-Induced Development of the Ice-Ice Disease of the Farmed Red Algae *Kappaphycus alvarezii* and *Eucheuma denticulatum* (*Solieriaceae, Gigartinales, Rhodophyta*)." *Journal of Applied Phycology* 7 (6): 539–543. https://doi.org/10.1007/BF00003940

Largo, Danilo B., Flower E. Msuya, and Ana Menezes. 2020. "Understanding Diseases and Control in Seaweed Farming in Zanzibar." *FAO Fisheries and Aquaculture Technical Paper*, 662: 1–49.

Le Gouvello, Raphaëla, Laure-Elise Hochart, Dan Laffoley, François Simard, Carlos Andrade, Dror Angel, Myriam Callier et al. 2017. "Aquaculture and Marine Protected Areas: Potential Opportunities and Synergies." *Aquatic Conservation* 27 (S)1: 138–150. https://doi.org/10.1002/aqc.2821

Lideman, G. N., T. N. Nishihara, and R. Terada. 2013. "Effect of Temperature and Light on the Photosynthesis as Measured by Chlorophyll Fluorescence of Cultured *Eucheuma denticulatum* and *Kappaphycus sp.* (Sumba Strain) from Indonesia." *Journal of Applied Phycology* 25: 399–406.

Limi, Muhammad Aswar, La Sara, Taane La Ola, Lukman Yunus, Suriana Suriana, Sitti Aida Adha Taridala, Hartina Batoa, Awaluddin Hamzah, Syamsul Alam Fyka, and Meilan Prapitasari. 2018. "The Production and Income from Seaweed Farming after the Sedimentation in Kendari Bay." *Aquaculture, Aquarium, Conservation & Legislation* 11 (6): 1927–1936.

Linsley, Braddock K., Henry C. Wu, Tim Rixen, Christopher D. Charles, Arnold L. Gordon, and Michael D. Moore. 2017. "SPCZ Zonal Events and Downstream Influence on Surface Ocean Conditions in the Indonesian Throughflow Region." *Geophysical Research Letters* 44 (1): 293–303. https://doi.org/10.1002/2016GL070985

Mairh, O. P., U. Soe-Htun, and M. Ohno. 1986. "Culture of *Eucheuma striatum* (*Rhodophyta, Solieriaceae*) in Subtropical Waters of Shikoku, Japan." *Botanica Marina* 29: 185–191.

MKPRI (Menteri Kelautan dan Perikanan Republik Indonesia). 2019. *Keputusan Menteri Kelautan dan Perikanan Republik Indonesia Nomor 1 / Kepmen-KP/2019 Tentang Pedoman umum pembudidayaan rumput laut.*

Msuya, Flower and Marilyn Porter. 2014. "Impact of Environmental Changes on Farmed Seaweed and Farmers: The Case of Songo Songo Island, Tanzania." *Journal of Applied Phycology* 26: 2135–2141. doi.org/10.1007/s10811-014-0243-4

Neish, Iain C. 2008. "Good Agronomy Practices for *Kappaphycus* and *Eucheuma*: Including an Overview of Basic Biology." *Seaplant.net* Monograph no. HB2F 1008 V3 GAP.

Ohno, M. and C. A. Orosco, 1987. "Growth Rate of Three Species of *Eucheuma*, Commercial Red Algae from the Philippines." In *Scientific Survey of Marine Algae and Their Resources in the Philippine Islands* edited by I. Umezaki, pp. 77–81. Kyoto: Laboratory of Fishery Resources, Graduate School of Agriculture, Kyoto University.

Preisig, R. and R. Hans. 2005. *Historical Review of Algal Culturing Techniques.* Zurich: Academic.

Purba, Noir P., Widodo S. Pranowo, Anthony B. Ndah, and Pieldrie Nanlohy. 2021. "Seasonal Variability of Temperature, Salinity, and Surface Currents at 0 Degrees Latitude Section of Indonesia Seas." *Regional Studies in Marine Science* 44 (101772): 1–9. https://doi.org/10.1016/j.rsma.2021.101772

Ratnawati, Pustika, Nova F. Simatupang, Petrus R. Pong-Masak, Nicholas A. Paul, and Giuseppe C. Zuccarello. 2020. "Genetic Diversity Analysis of Cultivated *Kappaphycus* in Indonesian Seaweed Farms Using Coi Gene." *Squalen* 15 (2): 65–72. https://doi. org/10.15578/squalen.v15i2.466

Satria, A., N. H. Muthohharoh, R. A. Suncoko, and I. Muflikhati. 2017. "Seaweed Farming, Property Rights, and Inclusive Development in Coastal Areas." *Ocean and Coastal Management* 150: 12–23. https://doi.org/10.1016/j.ocecoaman.2017.09.009

Simatupang, Nova Francisca, Petrus Rani Pong-Masak, Pustika Ratnawati, Agusman, Nicholas A. Paul, and Michael A. Rimmer. 2021. "Growth and Product Quality of the Seaweed *Kappaphycus alvarezii* from Different Farming Locations in Indonesia." *Aquaculture Reports* 20: 100685. https://doi.org/10.1016/j.aqrep.2021.100685.

Steenbergen, Dirk J., Cliff Marlessy, and Elisabeth Holle. 2017. "Effects of Rapid Livelihood Transitions: Examining Local Co-Developed Change Following a Seaweed Farming Boom." *Marine Policy* 82: 216–223. https://doi.org/10.1016/j.marpol.2017.03.026

Syafitri, E, S. B. Prayitno, W. F. Ma'ruf, and O. K. Radjasa. 2017. "Genetic Diversity of the Causative Agent of Ice-Ice Disease of the Seaweed *Kappaphycus alvarezii* from Karimunjawa Island, Indonesia." *IOP Conference Series. Earth and Environmental Science* 55 (1): 12044: 1–8. https://doi.org/10.1088/1755-1315/55/1/012044

Tahiluddin, Albaris and Ertuğrul Terzi. 2021. "Ice-Ice Disease in Commercially Cultivated Seaweeds *Kappaphycus* and *Eucheuma*: A Review on the Causes, Occurrence, and Control Measures." *Marine Science and Technology Bulletin* 10 (3): 234–243. https://doi. org/10.33714/masteb.917788

Terada, Ryuta, Triet Duy Vo, Gregory N. Nishihara, Keisaku Shioya, Satoshi Shimada, and Shigeo Kawaguchi. 2016. "The Effect of Irradiance and Temperature on the Photosynthesis and Growth of a Cultivated Red Alga *Kappaphycus alvarezii* (Solieriaceae) from Vietnam, Based on In Situ and In Vitro Measurements." *Journal of Applied Phycology* 28 (1): 457–467. https://doi.org/10.1007/s10811-015-0557-x

Vairappan, Charles S, Chong Sim Chung, A. Q. Hurtado, Flower E. Soya, Genevieve Bleicher L'honneur, and Alan Critchley. 2008. "Distribution and Symptoms of Epiphyte Infection in Major Carrageenophyte-Producing Farms." *Journal of Applied Phycology* 20 (5): 477–483. https://doi.org/10.1007/s10811-007-9299-8

van Oort, P. A. J., N. Rukminasari, G. Latama, A. Verhagen, and A. K. van der Werf. 2022. "The Bio Economic Seaweed Model (BESeM) for Modelling Tropical Seaweed Cultivation – Experimentation and Modelling." *Journal of Applied Phycology* 34 (5): 2627–2644. https://doi.org/10.1007/s10811-022-02799-8

Ward, G. M., C. S. B. Kambey, J. P. Faisan, P. L. Tan, C. C. Daumich, I. Matoju, G. D. Stentiford, D. Bass, P. E. Lim, J. Brodie, and S. W. Poong 2021) Ice-Ice Disease: An Environmentally and Microbiologically Driven Syndrome in Tropical Seaweed Aquaculture. *Reviews in Aquaculture* (14): 414–439. https://doi.org/10.1111/raq.12606

Wijayanto, Dian, Azis N. Bambang, Ristiawan A. Nugroho, and Faik Kurohman. 2020. "The Impact of Planting Distance on Productivity and Profit of *Eucheuma cottonii* Seaweed Cultivation in Karimunjawa Islands, Indonesia." *Aquaculture, Aquarium, Conservation & Legislation* 13 (4): 2170–2179.

Wikfors, Gary H. and Masao Ohno. 2001. "Impact of Algal Research in Aquaculture." *Journal of Phycology* 37: 968–974. https://doi.org/10.1046/j.1529-8817.2001.01136.x

Yong, Wilson Thau Lym, Siew Hoo Ting, Yoong Soon Yong, Vun Yee Thien, Siew Hie Wong, Wei Lie Chin, Kenneth Francis Rodrigues, and Ann Anton. 2014. "Optimization of Culture Conditions for the Direct Regeneration of *Kappaphycus alvarezii* (Rhodophyta, Solieriaceae)." *Journal of Applied Phycology* 26 (3): 1597–1606. https://doi.org/10.1007/ s10811-013-0191-4

# 7    Farmer decision-making in the Indonesian seaweed industry

*Zannie Langford, Radhiyah Ruhon,*
*Zulung Zach Walyandra, Imran Lapong,*
*and Risya Arsyi Armis*

## Introduction

Farmers make decisions in response to the environmental and socio-economic conditions in which they operate, which were described in Chapter 6. Farmer decision-making is a personalised process. Some farmers are very diligent in monitoring their farms. As one young farmer who kept detailed written notes of his observations and interventions said, '[T]o enable the seaweed to survive, every tool must be used'. Farmers often checked weather forecasts when making decisions about when to replant, although they did so with uncertainty as they thought that weather patterns were becoming increasingly unpredictable. As one farmer complained, '[W]e know nothing about the weather, because it keeps changing'. Farmers were generally positive about the role of farmer extension advice. One farm extension officer told us that farmers were very interested in their advice, and we noted that attendance at and interest in farming information events was high. As one farmer explained, 'I think it is a good thing to be involved in training activities so that we can develop, not stick to the same old ways. There may be better [cultivation] methods.' Farmers did not necessarily accept all the extension advice they were given but viewed it as one source of information that they critically examined and selectively incorporated. One farmer explained to us that farming advice provided by extension officers was sometimes correct, but not always, and often was too general and not tailored to the specific conditions of Pitu Sunggu. Similarly, some outreach events were rejected outright by farmers. For example, information in a session on the application of a fish-repelling technology was perceived by farmers to be much too expensive, unlikely to be effective and to address an issue that was not a major concern in the region, and so was dismissed. Diligent farmers reported that 'you have to monitor the seaweed everyday if you want good growth', and that frequent, close attention to the seaweed was key to success, rather than adherence to general rules. Other farmers suggested that monitoring farms every few days was sufficient. As one farmer explained, '[I]f we check on the seaweed every day, we cannot distinguish the growth changes. So we can get bored of it.'

During the sixteen months that two of the authors spent in Pitu Sunggu, farmers were frequently observed having in-depth discussions about growth conditions and rates. Farmers developed individual theories about how to improve growth, which

DOI: 10.4324/9781003183860-10

they tested. For example, one farmer theorised that cultivars should be grown in a more distant off-shore site for several weeks before moving them close to shore so that they would develop a thicker thallus which could be attached more securely to the ropes to avoid losses. However, the results of this approach were observed to be inconsistent, and farmers selectively incorporated this strategy into their own approach. In this way, through trial-and-error and experimentation farmers have developed strategies for growing seaweed that are highly location specific and tailored to weather patterns and ocean conditions.

Experimentation is also needed to deal with increasingly unpredictable conditions. In 2022, for example, rainfall persisted longer than usual, meaning that existing strategies for managing the seasonal transition between species were ineffective. Several studies have noted difficulty determining the relationship between environmental parameters and farm performance given the large number of interrelated variables affecting seaweed growth (e.g. Simatupang et al. 2021; van Oort et al. 2022). This section describes the decision-making processes adopted by seaweed farmers in Pitu Sunggu in response to environmental and socio-economic conditions.

## Production decisions

### Seaweed species

Farmers respond to seasonally changing ocean conditions by alternating between cultivating cottonii in the wet season and sacol in the dry season, as described in Chapter 6. These two species have slightly different production dynamics. Sacol is slower growing but with a thicker thallus so farmers report that it will be heavier when dried and is less labour intensive to cultivate due to less frequent re-seeding. Farmers enter the new year cultivating cottonii and continue until it becomes unprofitable due to declining yields or there is an increased incidence of ice-ice disease, at which point farmers cease production and dry their remaining cottonii. As the end of the wet season approached, one farmer told us that his cottonii growth 'has started to decline. There is not much time left for it, because in April or May it cannot grow well.' Another explained that 'in the open [outer] area[s], the seaweed growth has begun to slow down. It needs to be transferred here [closer to the shoreline].'

As the wet seasons ends, farmers with suitable dry season plots will begin cultivating sacol. Farmers typically wait until after March to begin sacol cultivation as March is a time of high winds and waves which lead to losses of seaweed. Normally, farmers will purchase a small amount of sacol propagules from other areas of South Sulawesi, often from Wajo Regency which sits on the opposite side of the peninsula and has different seasonal patterns. Farmers will begin growing these cultivars with the goal of multiplying propagules over a few short (~25–30 day) cycles. Under ideal conditions, after one cycle each cultivar will be used to produce four more cultivars. These will be multiplied until farmers have enough propagules to fill their available cultivation area, after which time they will be grown for sale with a small proportion of each cycle carried over to seed the next cycle. Sacol will be grown until the start of the wet season when they transition to cottonii. This seasonal transition between species has been taking place since about 2017, when sacol was introduced to the village. Prior to this, farmers only cultivated cottonii which meant that seaweed farming was more seasonal. The introduction of sacol for dry season cultivation has greatly increased the profitability of seaweed farming. Indeed, some farmers consider the dry season to be more profitable than the wet season as there is a lower risk of losses to bad weather.

### Planting location

The second key production decision farmers make is the planting location. As mentioned, location varies between the wet and the dry season, but there are also more subtle determinants. Conforming to findings that the planting location has the greatest influence on the growth rate of each species of seaweed (MKPRI 2019), respondents in Pitu Sunggu say that growth 'just depends on farming location'. An important aspect of decision-making is that in some locations different species can be grown out of season. As one farmer who successfully cultivates cottonii during the dry season explained, '[S]ometimes it just depends on the location – if the cottonii is located in a perfect location, it can also grow well [in the dry season]'.

Many large farmers own plots in different areas and use a strategy of geographic diversification to reduce the risk of losses and manage changes in ocean conditions. They often their seaweed regularly for signs of health, such as colour and shape. If they notice one of their plots is affected by ice-ice and the damage is minor, they can move the seaweed to a new location to allow the seaweed to recover. Alternatively if the damage is severe and the seaweed is not expected to recover, they can restart cultivation using seeds from one of their other locations.

Smaller farmers with only one plot do not enjoy this flexibility. They must continue to farm the same area regardless of conditions and often struggle with problems of low yields, epiphytes, mud, disease or grazing. They are also exposed to a greater degree of risk, since if their single plot is destroyed by ice-ice, they are left with no propagules from which to recommence farming. One farmer lamented that they had borrowed money for propagules in order to pay the seaweed binders, only to see their plot destroyed, leaving them with no money to repay the loan or reinvest in a new cycle. Seaweed farming can be a precarious business, and over the last twenty years many farmers have ceased production entirely after negative experiences such as these.

### Planting and harvest timing

Seaweed farmers also vary the timing of planting and harvest to minimise losses, sometimes harvesting earlier than industry guidelines recommend. Seaweed farms in Pitu Sunggu experience periods of different environmental conditions associated with different risks, including the wet and dry season, as well as a period of strong winds and large waves in March which can cause rope breakages. Losses during this season can be severe. After one period of high waves we observed a seaweed farm with no seaweed attached to the ropes at all – just a set of empty ropes suspended in the sea. On another occasion, after a period of high waves, one farmer lamented:

> I hadn't harvested my seaweed because there was no vacant space on the pier, I was actually just waiting for an empty drying space, but suddenly big waves hit the seaweed [causing losses]. My seaweed was already very big and ready to be harvested.

Planting and harvest decisions during such periods depend on farmers' risk tolerance, and many farmers choose to cease production if they perceive risks of loss to be too high. Planting during the wet season entails risk due to high rainfall events, while planting in March entails risk due to high waves. The risks have caused some farmers to opt out of seaweed production. One farmer explained that he had ceased growing seaweed in the wet season because there were so many things that could damage the crop. As he explained, '[I]t was on purpose that I was late to return to seaweed cultivation [this year]. Because the weather was unpredictable, I thought, maybe I will plant [in the wet season] only to lose.' Another told us that he persisted through the wet season but planned to cease cultivation before March: '[W]hen the waves threaten, I'll take a break [from seaweed farming]'.

Most farmers in Pitu Sunggu reduce exposure to risk by harvesting a portion of their longlines before March. As one farmer said, '[I]t is normal to have intense winds in March. Thus, people do not want to deploy all of their seaweed longlines.' Farmers that are risk averse or that foresee a risk of high waves may also harvest early. However, early harvesting has negative impacts on downstream processing because it means that seaweed may not reach the maturity required for high-quality carrageenan.

Farmers also harvest early for other reasons. Some farmers harvest early if the extra weight of molluscs means that seaweed does not float close enough to the surface to grow adequately or if it causes the seaweed to become fragile and susceptible to losses from wave action. Similarly, seaweed severely affected or unlikely to recover from moss growth or grazing may be harvested simply to avoid further losses. In addition, if seaweed appears to be fragile and farmers judge that it is likely to be easily broken off the longlines, they may harvest early to pre-empt such breakages and minimise losses. Similarly, seaweed which is regularly (partially) affected by ice-ice disease may be harvested early due to low growth rates and high risk of loss. As one farmer explained, '[S]ometimes they try to leave it there so they have propagules [for the next cycle], but if it keeps turning white [due to ice-ice disease] they will just dry it'. Harvest decisions are therefore linked to an assessment of both environmental risks and growth rate projections.

Farmers may also harvest early for socio-economic reasons. With an increasing incidence of theft in Pitu Sunggu during 2022, some farmers had harvested their seaweed earlier in the cycle to reduce the attractiveness of these large cultivars to thieves. Alternatively, farmers who require small amounts of cash to cover daily expenses sometimes harvest a few kilograms of seaweed before the end of the farming cycle. One small-scale trader in the village was regularly visited discreetly at night by farmers who wished to sell just a few kilograms of seaweed in exchange for an immediate cash payment. These early harvests are not ideal because they often occur when the cultivar is in a period of high growth. Seaweed propagules follow a sigmoidal growth curve such that growth begins slowly, accelerates in the middle of the cycle and then plateaus (van Oort et al. 2022). In biophysical terms, harvest should ideally occur after the phase of rapid growth, but before marginal growth rates decline significantly, however this often does not align with socio-economic imperatives.

Conversely, farmers sometimes harvest later than optimum in order to maintain a supply of propagules to begin the next cycle. Such farmers expend considerable effort tending unproductive plots during unfavourable seasons in order to ensure that they have seeds available when more favourable conditions arrive. This practice is particularly widespread in Nusa Tenggara Timor (NTT), where there are fewer farmers to obtain seeds from at the start of a growing season (Langford, Waldron et al. 2022). Financial relations and decision-making in Pitu Sunggu will be discussed further in Chapter 9.

### Farm maintenance

Epiphytic algae can significantly reduce seaweed growth rates. Farmers take steps to reduce epiphytes, mud and barnacles by shaking the seaweed in the water. Regular cleaning was viewed to be important to seaweed growth. As one farmer told us, '[F]or the success of the seaweed harvest, the most important thing is that the farmer is diligent and always takes care of the seaweed. Especially by cleaning the seaweed of *tai wai* [epiphytic microalgae] regularly'.

Epiphytic molluscs are also a potentially severe problem if left untreated. In Pitu Sunggu, molluscs tend to affect outer locations more severely, so farmers with suitable sea space can transfer their seaweed to near-shore, dry season locations to avoid barnacle growth. However, as discussed in Chapter 5, not all farmers have access to these coveted locations. One farmer showed us his seaweed covered in barnacles and bemoaned that because his family didn't get a good enough location to grow seaweed, he had no choice but to continue cultivating in unsuitable locations. Farmers report that barnacles found growing on sacol can be dealt with by bringing the seaweed to shore and leaving it on land overnight. The barnacles begin to die and fall off the seaweed, which can then be returned to the sea the following day. However, this method can only be used for sacol as cottonii cannot be left out of the sea overnight or it will begin to perish. This is consistent with findings that sacol is more tolerant of dehydration than cottonii (Pang et al. 2015). Molluscs also grow on the ropes and bottles used for seaweed farming, and these are typically cleaned thoroughly at the end of a farming cycle.

Lastly, farms are affected by the growth of macroalgae – referred to by farmers as moss. Farmers report that moss seriously damages seaweed but that no effective strategies for managing this have yet been developed apart from manual removal.

Some farmers in other areas such as NTT wash their seaweed in laundry detergent to remove moss and prevent regrowth (see Langford, Turupadang et al. 2022). We observed Pitu Sunggu farmers watching this process in an online video. The farmers expressed enthusiasm for the approach and discussed trialling it in Pitu Sunggu. However, this approach pollutes marine waters so other methods should be explored.

## Propagules

The decisions farmers make with regard to propagules are crucial elements affecting farm performance. These decisions centre around two key factors: propagule size and propagule 'quality'.

### *Propagule size*

Several studies have noted the relationship between propagule size and farm yields (e.g. Neish 2008; van Oort et al. 2022). Farmers in Pitu Sunggu recognise that larger propagules generally generate higher yields. As one farmer explained, '[T]he size of the harvested seaweed will depend on the size of the propagule cultivated. If the propagule was quite big, then the harvest will also be quite big.' However, as noted above, the production strategy of farmers depends not only on yield optimisation calculations, but also on loss and risk minimisation which, in turn, depends on household circumstances and appetite for risk (Langford, Turupadang et al. 2023; Langford, Waldron et al. 2023). Farmers in Pitu Sunggu regularly use small propagules to minimise the risk of losses from breakage of seaweed off the longlines. At times when high waves are common, farmers often minimise potential losses by using smaller seeds and harvesting before the seaweed gets so big that breakages become common. This is a well-known strategy, as Neish (2008, p. 39) notes,

> Generally choppy waters tend to favor "Small-Long" strategies [small propagules grown over long time-frames] but smooth waters with strong currents favor "Large-Medium" approaches [large propagules grown over medium time-frames]. In all cases a major determining factor in strategy choice is the point at which significant propagule breakage-losses take place.

These 'small-long' strategies are also used for socio-economic reasons, including reducing the risk of theft.

The other reason farmers use small propagules is because they lack sufficient planting material. In such instances, instead of filling a subset of their ropes with large propagules they often prefer to divide their seeds amongst all of their available ropes. That is, rather than filling half their longlines with propagules of 30g in size, they might choose to fill all their longlines with propagules 15g in size. Spreading propagule use over larger areas is a risk management strategy as this practice is perceived to reduce the risk of all the seaweed in one place being destroyed.

It is also relevant to note that farmers do not directly control propagule size but give their propagules to seaweed binders who cut them to size and attach them to longlines. Binders will typically cut propagules into a size that will enable them to fill all the ropes they have been given, which maximises their payment for binding (which is paid on a per-rope basis). In this way, farmers only indirectly determine propagule size by choosing the weight of propagules and number of ropes provided to binders. We observed that binders sometimes used as little as 2kg of propagules per 25m longline or a propagule size of around 12.5g (assuming a 15cm spacing). This is far lower than the size farmers regard as ideal (6kg per longline or 37.5g per propagule) and the size recommended by government (propagules of 50–100g) (MKPRI 2019). Using small propagules on many ropes rather than large propagules on fewer ropes also increases (per unit) binding and maintenance labour per gram of propagule used. The use of this strategy suggests that farmers are trading off the lower returns from smaller propagules for the lower risk of spreading available propagules over larger areas with more diverse environmental conditions.

An alternative to this strategy would be to purchase propagules from other farmers. In some low-intensity seaweed producing areas such as in NTT, this is often not an option due to thin propagule markets in the area (Langford, Turupadang et al. 2022). In Pitu Sunggu, however, farmers have access to thicker propagule markets thanks to the high density of seaweed farmers in surrounding areas. Even so, most farmers prefer to multiply propagules themselves and only buy small quantities of seeds due to the high capital costs and the risk of losing them in production. During the course of our study, the cost of buying seaweed propagules to fill 150 longlines varied from Rp. 5–8 million (AU$500–AU$800), a considerable capital expense. If these cultivars were subsequently lost due to ice-ice disease or other factors (as is reasonably likely in the early stages of a new season), it would impose a high debt burden on most farmers. To minimise the risk of debt, most farmers only buy seeds when they need to (at the start of a new growing season or if their seaweed is completely destroyed) and then only buy small quantities, usually below the quantities required to operate at full capacity. As one farmer explained, '[I]f people have seeds, they prefer to multiply them themselves rather than buy them. Except for people who have a lot of capital.' This also represents a strategy to trade off yield for risk.

### Propagule 'quality'

The low 'quality' of propagules in sufficient supply is regarded as a major problem by analysts (Hurtado et al. 2014) and farmers (Zamroni 2018; Langford et al. 2023; 2024). Perceptions of the propagule 'quality' can however be ill-defined and require closer examination. Farmers are interested in propagule 'quality' insofar as it leads to differences in performance as measured by the growth rates and/or resilience to losses due to breakage and ice-ice disease. These differences in propagule performance result from a range of factors, including genetic material, adaptability to the environment (evidenced by colour, shape, size), degree of damage (from disease, grazing, epiphytes, mud, dehydration) or age. This section explores how these different factors combine to create propagule 'quality' as perceived by farmers.

### Genetic material

Two species of *Kappaphycus* are widely farmed globally – *Kappaphycus alvarezii* (cottonii) and *Kappaphycus striatus* (sacol). There are clear differences in the behaviour and appearance of these two species, as discussed above. However, there is very little genetic variation within the species of *Kappaphycus alvarezii*.[1] Most commercial cultivars globally have been shown to have a high level of genetic conformity despite variation in colour, size and shape (Zuccarello et al. 2006; Ratnawati et al. 2020). Some work has been undertaken to establish whether wild strains of *Kappaphycus alvarezzi* with improved production characteristics could be identified (Narvarte et al. 2022) and such work may offer new strains with improved growth rates and/or resilience to environmental stress if they are proven to perform under various (cultivated) conditions. To date, however, most cultivated cottonii have been shown to have an identical COI genotype despite wide variations in growth rate, carrageenan yields and characteristics, colour and shape (Ratnawati et al. 2020; Simatupang et al. 2021; Zuccarello et al. 2006). The connection between the limited genetic diversity and the wide range of observable features of cottonii requires further investigation to establish the extent to which seaweed growth is related to genotypic and phenotypic variation, as available evidence is yet to determine a significant role for genetics. Further research using both traditional molecular tools and new genomic tools could reveal such interactions in the future (Simatupang et al. 2021).

### Adaptation to environment

Notwithstanding the limited genetic variation in cultivated seaweed revealed to date, seaweed cultivated in different locations and conditions demonstrates vastly different growth behaviour, which changes plant colour, shape, growth rate and carrageenan content (Simatupang et al. 2021). This is due to seaweed's 'morphological plasticity' or ability to adapt to different ocean conditions. Seaweeds undergo very visible adaptations to new environmental conditions. They change colour and shape (Neish 2008) and may adapt to grow best in higher or lower salinity and temperature conditions (Araujo et al. 2014). Simatupang et al. (2021) found that propagules sourced from different locations and then cultivated together at a 'common garden' site converged in their morphology and colour, suggesting that much of the observed variation in seaweed growth is related to environmental conditions. Similarly, Neish (2008) described key changes in colour and shape that farmers can use to identify positive and negative responses of the seaweed to the environment. In line with these findings, farmers in Pitu Sunggu also described adaptation occurring over short time frames. They observed two 'types' of sacol – a 'round' shaped one and a 'branched' type, with the former appearing in plots closer to the shore and the latter in plots further off-shore. They also observed that if the locations of these two 'types' of sacol were switched, their morphologies also reverse after several farming cycles.

In addition, farmers in Pitu Sunggu observe variations in propagule 'quality' throughout the year. Good quality was indicated by a large size even with a short growing time, and clean, undamaged propagules with a vibrant colour (not yellowish or dull). Complaints about poor seed 'quality' were based on poor growth and suitability for their plots. This became a particular problem in June, when farmers widely complained about the slow growth of seaweed and high prevalence of mud, epiphytes and barnacles. Problems with production were attributed to a lack of 'quality' propagules. As one farmer told us, 'No seaweed looks healthy. They are all evenly damaged.' On one occasion when we observed a farmer preparing small, poorly coloured propagules for replanting and expressed surprise that such propagules were to be replanted, the farmer responded, 'What can I say? I have no proper seeds available. Some are big, but they are dirty. But what do you want? We are all out.'

Farmers did sometimes complain about declining yields over time – for example, one farmer complained that where his farm had previously yielded 50–60kg/longline, it now yielded only 40kg. This has been referred to as 'strain fatigue' in the literature, but this label is possibly misleading since there are a range of factors which may contribute to the decline. Propagules are clonally propagated so are genetically identical to each other, so while it is possible that the issue is internal to propagules themselves – for example, related to age, damage or infection with bacteria – it is also likely to be linked to changes in ocean conditions. It has been suggested that declining yields over time are linked to over-planting which leads to changes in conditions in coastal areas (Neish 2008). Declining yields in Pitu Sunggu are probably linked to the intensification of seaweed cultivation which increases competition for nutrients, as well as external factors such as reduced fertiliser runoff from reduced terrestrial pond fertilisation. Indeed, farmers told us that they had noticed that plots left fallow for several years often regained their productivity when planted again. As one farmer told us, 'I see that the growth of seaweed in new locations tends to be better. For example, there was a location that had been abandoned for two years, and when it was occupied again, the growth was good again.' The reasons for this are not clear but are worth exploring given the wider importance of this phenomenon (e.g. Neish 2008; Steenbergen et al. 2017).

### Damage

Another key aspect of propagule 'quality' is the absence of any damage to the seaweed. This may include damage by grazing fish (which remove new branches emerging from the thallus), damage by molluscs, the presence of mush or epiphytes or damage from being out of the ocean for long periods. With regard to the latter, government guidelines recommend that seeds should not be out of the water for more than six hours (MKPRI 2019), and as a result, farmers assert that propagules purchased from other farmers in the village are 'fresher' and grow better as a result than those sourced by collectors. It seems obvious that propagules with visible damage are not ideal for use as seeds, but, as this chapter seeks to demonstrate, farmers contend with a wide range of potential threats to their seaweed which means that they are often forced to operate under suboptimal conditions. The common complaint by farmers about a lack of 'quality' seeds is therefore closely linked to cultivation conditions overall.

## *Age*

Finally, the age of seaweed propagule is a key aspect of 'quality'. According to government guidelines, propagules should be cultivated for 25–35 days and seaweed destined for sale should be cultivated for 45–50 days to ensure sufficient carrageenan development (MKPRI 2019). When asked, most farmers in our study said that they preferred propagules that were around 25 days old for cottonii and 30 days old for sacol. As one farmer explained, '[I]t is better to use the young seedlings, they grow faster. If we use the old seedlings, they grow slower.' However, it wasn't always possible to use young seeds if growth rates had not been sufficient to support division.

### *Tissue culture propagules*

The Indonesian government currently has a programme which produces *Kappaphycus alvarezzi* propagules using tissue culturing methods. Some academics have suggested that propagules grown from tissue culture grow faster than other propagules though the evidence is mixed (Hurtado et al. 2015) and other studies have found that after accounting for ocean conditions these differences are not statistically significant (Budiyanto et al. 2019). In addition, studies to date are mostly short term (<1 year) and so are not able to track performance over the longer time frames that farmers report declining yields (3–5 years). Industry advises that tissue cultured propagules also experience declining yields after several years of cultivation (personal communication, May 2021). The mechanism by which these propagules might achieve higher growth rates are not clear, but as all propagules are genetically identical and may also experience declining growth rates, they are likely to be linked to the young age of these propagules. Further research undertaking multiyear, in situ studies of the growth of tissue cultured propagules are needed to systematically assess their performance relative to other cultivars.

## Floats

Floats are a key component of the seaweed farming apparatus because cultivars are highly sensitive to small changes in depth. Buoyancy control is therefore a critical factor in farm performance.

### *Main float*

The seaweed farm apparatus is fixed in place by anchors on the sea floor. Ropes are attached to these anchors and held at the surface by large buoys, normally jerry cans (such as empty cooking oil containers). Farmers feel that jerry cans are highly suitable for this purpose – and preferrable to large permanent buoys – because they sway easily with the current, which minimises rope breakage. Seaweed longlines are attached to the main rope and held at the surface by a large number of individual plastic floats attached to the longline. Used plastic bottles are normally used for this purpose.

### Float spacing

Seaweed is highly sensitive to light and salinity. Seaweed needs sunlight to grow, but if the sun is too intense and raises the water temperature the seaweed can be damaged. Similarly, changes in salinity at the ocean surface after rainfall create a hyposaline environment that is damaging to seaweed. The spacing of plastic bottles on the longlines creates small differences in depth across a single longline that have a strong impact on seaweed growth. In the dry season, farmers observe that seaweed propagules located closer to bottles – and therefore closer to the surface – have faster growth and are darker in colour. Propagules further from floats – in slightly deeper water – tend to be lighter in colour and smaller in size and more affected by molluscs. A different pattern emerges during the wet season. Seaweed propagules located closer to the plastic bottles and the ocean surface are often damaged by rainfall, whereas those further from the floats in slightly deeper water have better growth and less damage. As a result, in the dry season, farmers attach more plastic floats to the seaweed than in the wet season. The number of floats used also depends on the size of the propagules. Farmers typically add floats during the farming cycle as the seaweed becomes heavier. Farmers note the general pattern of using more bottles for sacol in the dry season and less for cottonii in the wet season, but in practice they closely monitor their seaweed growth and adjust float spacing accordingly. As one farmer explained,

> The color will be different, as well as the growth, if we use many bottles … We need to monitor the [seaweed] condition. If the growth of the seaweed with more bottles is good and those with less bottles seem to be stunted, that's when we will add more bottles.

Farmers also need to respond to growth of molluscs on the ropes, seaweed and bottles, which can sink the seaweed further below the surface, by adding plastic bottles to increase buoyancy.

### Float buoyancy

Farmers respond to high rainfall events in the wet season by travelling to their farms and sinking seaweed below the surface by filling bottles with water, as discussed in Chapter 6. This is one of the key benefits of using plastic bottles as floats rather than permanent floats. Other benefits include their light weight which makes them portable and allows them to sway with the currents. Farmers who fail to lower float levels after rainfall often experience substantial losses. As one farmer who had failed to attend to all of his seaweed on the day of a high rainfall commented, '[W]hen the flood hit, everything was immediately ruined. It didn't wait another day.' Efforts to develop alternatives to single-use plastic bottles in seaweed farming need to consider the multifunctionality of these devices – being light, portable and fillable.

## Rope maintenance

The maintenance of clean ropes at an appropriate tension, spacing and orientation is an often overlooked but important aspect of seaweed farming. Decisions about ropes have an impact on seaweed growth by affecting their spacing and cultivation density.

### *Rope tension*

Extension agents recommend that farmers keep rope tension high to avoid ropes becoming tangled. However, many farmers report that too much tension in the rope will contribute to seaweed breakages due to water motion, so they aim to keep them slightly looser than recommended. Farmers report that tightening loose ropes is one of the most time-consuming jobs they undertake but is important to ensure seaweed receives sufficient light and water motion.

### *Rope orientation*

Some farmers have observed that the direction of the longlines also impacts on the exposure of seaweed to water motion and adjust the orientation of the lines based on prevailing ocean conditions. Although extension agents advise them on the correct rope orientation, in practice farmers adjust orientation based on water motion in different areas. It is common practice for farmers with plots in stagnant areas close to the shore to position longlines parallel to the shore. As one farmer explained,

> If there are no waves, the longlines are positioned parallel to the shoreline, because the seaweed needs waves. Since the area there is a lack of waves, the longlines are made parallel so that the strength of the waves has more impact on the seaweed longlines.

### *Rope cleaning*

Farmers clean ropes after harvest in order to remove molluscs and epiphytic algae. Cleaning the ropes thoroughly is important to avoid injury to seaweed binders, as well as to stop the ropes becoming heavy and sinking below the surface. If farmers are very busy, they still ensure that ropes are cleaned but will save time by not cleaning the bottles – removing the molluscs from the rope is viewed as a non-negotiable activity.

*Double or single rope method*

Farmers in Pitu Sunggu use what is known as a 'double rope method' in which two ropes with seaweed attached are twisted together. Farmers report that this method means that the ropes are not as easily shaken by waves or broken, which has benefits for both rope longevity and seaweed survival. Pitu Sunggu farmers report that similar results can be achieved by using a single thicker rope. The double rope method also increases the number of propagules attached to each line, although similar results could be achieved using shorter spacings on a single, thicker rope. The use of the double rope method was viewed very favourably by farmers, and in our survey of technology uptake, it was the most frequently listed positive development in seaweed farming in recent years.

*Rope and propagule spacing*

It is recommended that propagules are attached to ropes at a spacing of 25–30cm, and that ropes are installed on-farm at a distance of at least 1m from each other (MKPRI 2019). Large farmers typically observe this rope spacing. They note that if ropes are spaced too close together they may get tangled. As one farmer put it, 'The seaweed competes for nutrition. It is better for the longlines to be free [separate]. If the position of the longlines is too close to one another, the seaweed growth will be small.' However, with increasing seaweed prices and competition over sea space, we noted that a number of farmers were planting seaweed longlines much closer together – some only 50–70cm apart. This is typical of seaweed farmers who do not have sufficient access to sea space. They economise on space by increasing the intensity of (space-saving) labour and capital. Spacing decisions are therefore made within the individual constraints of farmers.

# Farm performance

Environmental and socio-economic factors and associated decisions by farmers create differences in farm performance between locations and between farmers. Some areas are known to perform better than others at different times of the year while some farmers perform better than others in similar ocean conditions. As Neish (2008) noted, close farm monitoring and management is key to success. Farm performance is measured by farmers in the overall yields, which are closely linked to growth rates. However, processors are more interested in the characteristics of the carrageenan extracted from the seaweed. A recent study of seaweed cultivated at ten different sites in Indonesia (Simatupang et al. 2021) showed considerable variability in growth rates and carrageenan yields and characteristics. Growth rates ranged from ~0.5–2.8 kg/m of rope, carrageenan content from 9–26 per cent, gel strength from 292–735 $g/cm^2$ and viscosity from 30–139 cP. Cozzolino et al. (2023) also report large variations in carrageenan content and characteristics across different samples. At present, these different seaweed characteristics are not reflected in price differentials (Langford, Zhang et al. 2022). Consequently, farmers base decisions on growth rate alone.

Farm management in turn alters the environmental and socio-economic constraints in which seaweed farming takes place. Farm performance generates income for farmers which may alter their access to capital and sea space, thereby leading to different production decisions. Similarly, farming activities influence some of the characteristics of the surrounding ocean environment, such as nutrient levels, turbidity, pollution, epiphyte incidence and the presence of grazing wildlife. The environmental impact of seaweed farming has been partially studied but is not yet well understood. Seaweed farming changes ocean conditions by changing the amount of sunlight that reaches the sea floor, increasing detritus and altering sedimentation, secreting poisonous substances, competing for nutrients and providing a source of food for some fauna (Kelly et al. 2020). Intensive use of coastal space for seaweed farming alters fish assemblages in the area (Kelly et al. 2020) which may impact on proximate seagrass meadows and coral reefs (Kelly et al. 2020; Unsworth et al. 2018) and may contribute to plastic pollution including microplastics from on-farm bottle degradation and waste plastic bottles not properly disposed of (Hovey et al. 2023). Despite this growing body of literature, the environmental impact of seaweed farming is not well studied under diverse environmental and socio-economic conditions, and possible methods to reduce the impacts are not well tested. Some proposed approaches include observing maximum planting densities (e.g. Blankenhorn 2007) and maintaining fishing- and farming-free marine protected areas proximate to the farms (Le Gouvello et al. 2017).

Seaweed farming may also alter marine environments by changing the way that coastal people engage with them. Seaweed farming is argued to have environmental benefits as the emergence of a new livelihood activity in coastal villages could reduce pressure on ecosystems from over-fishing (Sievanen et al. 2005; Zamroni et al. 2011). In Pitu Sunggu, seaweed farming has certainly changed the use of coastal resources. The coastal areas are now intensively used for seaweed farming and fishing activities take place in the gaps between farms or further out to sea. However, the impact of these changes in the distribution and intensity of fishing activity are not well understood and vary by location. Some studies have indicated that seaweed farming has reduced destructive fishing practices (Zamroni et al. 2011). Others observed increases in fishing activity as farmers used money generated from seaweed to invest in fishing gear (Sievanen et al. 2005). Steenbergen et al. (2017) described a case of collapse of seaweed farming in a village due to excessive pressure on the environment. With government plans to expand seaweed production into new areas and to intensify existing areas, there is a critical need for further research in this area.

## Conclusion

Farmers make decisions about seaweed production in response to an elaborate combination of environmental conditions and socio-economic constraints. Many farmers are diligent and monitor their plots regularly. Farmers know that certain practices can increase seaweed growth rate. This includes using good-quality propagules of adequate size, spacing the ropes appropriately so the seaweed gets enough nutrients, tying the seaweed close enough to the surface to get enough light, regularly cleaning it of epiphytes and barnacles, carefully maintaining ropes and monitoring the seaweed daily and moving it to a new location or adjusting the planting parameters if necessary. However, there are several reasons why perfectly rational farmers may not take up 'best practices' designed to maximise growth rate, including boredom, being busy (maintaining a large number of seaweed plots or other activities at the same time) and risk avoidance (through strategies such as low investment in propagules at the start of a new season using small propagules or harvesting early to avoid the risk of losses). The preceding two chapters have explored the work of growing seaweed in Pitu Sunggu. The next chapter moves from the farm to the household to explore the way that this work and the income it brings have transformed the social lives of Pitu Sunggu residents.

## Note

1  Most studies focus on *Kappaphycus alvarezii* – *Kappaphycus striatus* has yet to be studied in detail.

## References

Araujo, Patrícia G., Ana Lívia, N. L. Ribeiro, Nair S. Yokoya, and Mutue T. Fujii. 2014. "Temperature and Salinity Responses of Drifting Specimens of *Kappaphycus alvarezii* (*Gigartinales, Rhodophyta*) Farmed on the Brazilian Tropical Coast." *Journal of Applied Phycology* 26: 1979–1988.

Blankenhorn, Sven Uli. 2007. "Seaweed Farming and Artisanal Fisheries in an Indonesian Seagrass Bed – Complementary or Competitive Usages." PhD Thesis. University of Bremen, Germany.

Budiyanto, Budiyanto, Ma'Ruf Kasim, and Sarini Yusuf Abadi. 2019. "Growth and Carrageenan Content of Local and Tissue Culture Seed of *Kappaphycus alvarezii* Cultivated in Floating Cage." *Aquaculture, Aquarium, Conservation & Legislation* 12 (1): 167–178.

Cozzolino, Daniel, Mohamad Rafi, Wahyu Ramadhan, Rudi Heryanto, Scott Waldron, and Zannie Langford. (2023) "Developing a Rapid Assessment for Seaweed Qualities." Melbourne, Australia: Australia-Indonesia Centre. https://pair.australiaindonesiacentre.org/wp-content/uploads/2023/09/Final-Report_TWP5_ENG_Developing-a-rapid-assessment-for-seaweed-qualities.pdf

Hovey, Renae, Shinta Werorilangi, Widyastuti Umar, Hendra Hasyim, Nurul Masyiah Rani Harusi, Karen F. Dexter, Radhiyah Ruhon, Zulung Zach Walyandra, and Zannie Langford. 2023. *End of Life of Plastics Used in Seaweed Aquaculture in South Sulawesi.* Melbourne, Australia: Australia-Indonesia Centre.

Hurtado, Anicia Q., Grevo S. Gerung, Suhaimi Yasir, and Alan T. Critchley. 2014. "Cultivation of Tropical Red Seaweeds in the BIMP-EAGA Region." *Journal of Applied Phycology* 26 (2): 707–18. https://doi.org/10.1007/s10811-013-0116-2

Hurtado, Anicia Q., Iain C. Neish, and Alan T. Critchley. 2015. "Developments in Production Technology of *Kappaphycus* in the Philippines: More than Four Decades of Farming." *Journal of Applied Phycology* 27 (5): 1945–1961. https://doi.org/10.1007/s10811-014-0510-4

Kelly, Emily L. A., Abigail L. Cannon, and Jennifer E. Smith. 2020. "Environmental Impacts and Implications of Tropical Carrageenophyte Seaweed Farming." *Conservation Biology* 34 (2): 326–337. https://doi.org/10.1111/cobi.13462

Langford, Alexandra, Welem Turupadang, Marclien D. R. Oedjoe, Christian Liufeto, Berty Bire, Marcel Yohanis, Yulius Suni and Scott Waldron. 2022. "Seaweed Farmer Resilience in Eastern Indonesia after COVID-19." *Australian National University Indonesia Project.*

Langford, Alexandra, Scott Waldron, Nunung Nuryartono, Syamsul Pasaribu, Boedi Julianto, Irsyadi Siradjuddin, Radhiyah Ruhon, Zulung Zach Walyandra, Imran Lapong, and Risya Arsyi Armis. 2023. *Sustainable Upgrading of the South Sulawesi Seaweed Industry.* Melbourne, Australia: Australia-Indonesia Centre.

Langford, Alexandra, Scott Waldron, Jing Zhang, Radhiyah Ruhon, Zulung Zach Walyandra, Risya Arsyi Armis, Imran Lapong, Boedi Julianto, Irsyadi Siradjuddin, Syamsul Pasaribu, and Nunung Nuryartono. (2024). "Diverse Seaweed Farming Livelihoods in Two Indonesian Villages." In *Tropical Seaweed Cultivation – Phyconomy: Proceedings of the Tropical Phyconomy Coalition Development, TPCD 1, Held at UNHAS University, Makassar, Indonesia. July 7th and 8th* edited by A. Hurtado, A. Critchley, and I. Neish. 00-00. New York: Cham, Switzerland: Springer Nature.

Langford, Alexandra, Jing Zhang, Scott Waldron, Boedi Julianto, Irsyadi Siradjuddin, Iain C. Neish, and Nunung Nuryartono. 2022. "Price Analysis of the Indonesian Carrageenan Seaweed Industry." *Aquaculture* 550 (737828). doi.org/10.1016/j.aquaculture.2021.737828

Le Gouvello, Raphaëla, Laure-Elise Hochart, Dan Laffoley, François Simard, Carlos Andrade, Dror Angel, Myriam Callier et al. 2017. "Aquaculture and Marine Protected Areas: Potential Opportunities and Synergies." *Aquatic Conservation* 27 (S1): 138–150. https://doi.org/10.1002/aqc.2821

MKPRI (Menteri Kelautan dan Perikanan Republik Indonesia) 2019. *Keputusan Menteri Kelautan dan Perikanan Republik Indonesia Nomor 1 / Kepmen-KP/2019 Tentange Pedoman umum pembudidayaan rumput laut.*

Narvarte, Bienson Ceasar V., Lourie Ann R. Hinaloc, Tom Gerald T. Genovia, Shienna Mae C. Gonzaga, April Mae Tabonda-Nabor, and Michael Y. Roleda. 2022. "Physiological and Biochemical Characterization of New Wild Strains of *Kappaphycus alvarezii* (*Gigartinales, Rhodophyta*) Cultivated under Land-Based Hatchery Conditions." *Aquatic Botany* 183 (103567): 1–10. https://doi.org/10.1016/j.aquabot.2022.103567

Neish, Iain C. 2008. "Good Agronomy Practices for *Kappaphycus* and *Eucheuma*: Including an Overview of Basic Biology." *Seaplant.net* Monograph no. HB2F 1008 V3 GAP.

Pang, Tong, Litao Zhang, Jianguo Liu, Hu Li, and Junpeng Li. 2015. "Differences in Photosynthetic Behaviour of *Kappaphycus alvarezii* and *Kappaphycus striatus* During Dehydration and Rehydration." *Marine Biology Research* 11 (7): 765–772. https://doi.org/10.1080/17451000.2015.1007876

Ratnawati, Pustika, Nova F. Simatupang, Petrus R. Pong-Masak, Nicholas A. Paul, and Giuseppe C. Zuccarello. 2020. "Genetic Diversity Analysis of Cultivated *Kappaphycus* in Indonesian Seaweed Farms Using Coi Gene." *Squalen* 15 (2): 65–72. https://doi.org/10.15578/squalen.v15i2.466

Sievanen, Leila, Brian Crawford, Richard Pollnac, and Celia Lowe. 2005. "Weeding through Assumptions of Livelihood Approaches in ICM: Seaweed Farming in the Philippines and Indonesia." *Ocean & Coastal Management* 48 (3): 297–313. https://doi.org/10.1016/j.ocecoaman.2005.04.015

Simatupang, Nova Francisca, Petrus Rani Pong-Masak, Pustika Ratnawati, Agusman, Nicholas A. Paul, and Michael A. Rimmer. 2021. "Growth and Product Quality of the Seaweed *Kappaphycus alvarezii* from Different Farming Locations in Indonesia." *Aquaculture Reports* 20: 100685. https://doi.org/10.1016/j.aqrep.2021.100685

Steenbergen, Dirk J., Cliff Marlessy, and Elisabeth Holle. 2017. "Effects of Rapid Livelihood Transitions: Examining Local Co-Developed Change Following a Seaweed Farming Boom." *Marine Policy* 82: 216–23. https://doi.org/10.1016/j.marpol.2017.03.026

Unsworth, Richard K. F., Rohani Ambo-Rappe, Benjamin L. Jones, Yayu A. La Nafie, A. Irawan, Udhi E. Hernawan, Abigail M. Moore, and Leanne C. Cullen-Unsworth. 2018. "Indonesia's Globally Significant Seagrass Meadows Are Under Widespread Threat." *The Science of the Total Environment* 634: 279–286. https://doi.org/10.1016/j.scitotenv.2018.03.315

van Oort, P. A. J., N. Rukminasari, G. Latama, A. Verhagen, and A. K. van der Werf. 2022. "The Bio Economic Seaweed Model (BESeM) for Modelling Tropical Seaweed Cultivation – Experimentation and Modelling." *Journal of Applied Phycology* 34 (5): 2627–2644. https://doi.org/10.1007/s10811-022-02799-8

Zamroni, Achmad, Khaled Laoubi, and Masahiro Yamao. 2011. "The Development of Seaweed Farming as a Sustainable Coastal Management Method in Indonesia: An Opportunities and Constraints Assessment." *Transactions on Ecology and the Environment* 150: 505–516. https://doi.org/10.2495/SDP110421

Zamroni, Achmad. 2018. "Small Scale Entrepreneurship of Seaweed in Serewe Bay, East Lombok, Indonesia: Challenges and Opportunities." *Journal of Development and Agricultural Economics* 10 (5): 165–175. https://doi.org/10.5897/JDAE2017.0874

Zuccarello, Giuseppe C., Alan T. Critchley, Jennifer Smith, Volker Sieber, Genevieve Bleicher L'honneur, and John A. West. 2006. "Systematics and Genetic Variation in Commercial Shape *Kappaphycus* and Shape *Eucheuma* (*Solieriaceae, Rhodophyta*)." *Journal of Applied Phycology* 18 (3–5): 643–651. https://doi.org/10.1007/s10811-006-9066-2

# 8 Gendered work and casual labour in the Indonesian seaweed industry

*Zannie Langford, Radhiyah Ruhon,*
*Zulung Zach Walyandra, Risya Arsyi Armis,*
*and Imran Lapong*

## Introduction

The last four chapters have explored how seaweed farming has entered and transformed village livelihoods, reorganised use of sea space, and is practised by farmers. But who are these farmers, and how are other people in coastal communities involved in seaweed value chains? There are around 62,000 marine seaweed farming households in Indonesia (BPS 2022), and much of the work discussed so far is undertaken by men working at sea, a common gendered organisation of labour. But women in seaweed farming households also work in the seaweed industry, often in and around the home drying, bagging, and selling seaweed, and tying propagules to longlines. Large numbers of casual wage labourers that typically don't have their own farms also participate in the industry by working as binders who attach seaweed to longlines. Children also participate in many tasks.

Much has been written about seaweed farming livelihoods in Indonesia, generally emphasising the benefits it provides through increased incomes (e.g. Mariño et al. 2019, Zamroni 2021; Langford et al. 2024). Participation in seaweed farming is linked to the availability of alternative livelihood activities (Andréfouët et al. 2021). In some areas farmers are highly specialised and earn cash income almost exclusively from seaweed farming (Aslan et al. 2018; Mariño et al. 2019; Langford, Turupadang et al. 2022; Langford, Waldron et al. 2023). As described in Chapter 4, today, Pitu Sunggu seaweed farmers are highly specialised in seaweed farming. Many of them also undertake marine fishing (60 per cent) or pond farming (38 per cent), while some also undertake off-farm work (11 per cent), crab fishing (10 per cent), or collecting of ocean shells (7 per cent). Very few farmers (1 per cent) produce food crops such as corn or rice. Pitu Sunggu farmers therefore are highly export orientated in their production and rely on the sales of seaweed and/or shrimp and fish for the cash they need to purchase food. This is unlike the situation in the nearby village of Laikang, where 27 per cent of seaweed farmers also grow rice and 18 per cent also grow corn. Pitu Sunggu seaweed farmers are therefore quite specialised – 77 per cent of them earn more than half of their household income from seaweed farming and 83 per cent spend more than half of their household labour on seaweed farming. The income earnt through seaweed farming contributes to seaweed farmer well-being through both material benefits as well as

DOI: 10.4324/9781003183860-11

the sense of pride, confidence, and agency that it engenders (Larson et al. 2020). Seaweed farming income is often spent on items such as motorbikes (Larson et al. 2022), housing (Mirera et al. 2020), school fees (Mirera et al. 2020), and luxury items (Aslan et al. 2022).

However, the benefits of the seaweed industry are not equally distributed amongst farmers, who have differing abilities to capitalise on new livelihood opportunities (Wright 2017; Teniwut and Teniwut 2018; Teniwut et al. 2017). Seaweed farming varies by area, seasonal patterns, profitability, and levels of risk (Langford, Waldron, Sulfahri et al. 2021). Women in the industry are involved to varying extents in different locations (Msuya and Hurtado 2017; Valderrama 2015) and engage in both paid and unpaid (i.e. own or family farm) work (Eranza et al. 2015). This chapter explores the ways that seaweed farming has transformed livelihoods in coastal communities, not only for seaweed farmers themselves, but for those they share a house with, those they employ, and those who live in coastal villages alongside them.

### Seaweed farming jobs

The focus of the last few chapters has been on-farm. However, what happens off-farm after the seaweed is harvested and brought to shore is equally important to the functioning of the industry. Seaweed is normally not sold in its 'wet', living state,[1] but dried, packed into bags, and then sold to local traders. The role of local traders in seaweed marketing will be discussed further in Chapter 9. Drying seaweed involves spreading it out on either a bamboo drying platform or on the ground in the sun (Figure 8.1). Depending on sunlight and rainfall patterns, seaweed requires anywhere from a few days to over a week to dry to a level acceptable to buyers (Stone et al. 2023). During this time it must be regularly turned to ensure even drying. If it rains, it must be covered with a tarpaulin. Chickens must be prevented from damaging it through manure, dirt, scratching, and trampling. This means that the seaweed must be monitored and occasionally tended to, especially on days with frequent rain showers. It is often women who do this work amongst other household duties.

After it is dry, seaweed must be packed into bags and stored until it is sold. Sometimes farmers store seaweed for long periods before selling it. One farmer showed us his store of seaweed stacked under his house and explained that it had been there for nearly a year because he was waiting for a good time to sell when prices are high. Others sell their seaweed straight away, especially when they have urgent expenses. When it is time to sell, the seaweed is transported to the house of a trader where it is weighed, examined, and purchased. Trader relations will be discussed further in Chapter 9.

During this drying, bagging, and selling process, farmers are also busy re-planting their seaweed for a new cycle. Part of a seaweed harvest is normally used as propagules for the next cycle, and these must be cut into appropriate sizes and attached to ropes before they can be taken to sea. As discussed in previous chapters, this is the most time-consuming work of seaweed farming. Since propagules

*Figure 8.1* Seaweed drying in the sun in front of a house, with seaweed already dried and packed into white sacks visible behind it

Source: Image by Radhiyah Ruhon, June 2022, Pitu Sunggu.

should ideally be out of the water for no more than six hours (MKPRI 2019), and our observations suggest that the average, middle-aged binder takes around one hour to complete one longline,[2] farmers rely on teams of binders to complete the work within the time frame. They hire teams of binders to work for them for a day to complete a set of longlines, often 50–100 in one session, completed by binder groups ranging in size from 5–15 people. These people are not seaweed farmers, but their labour is essential to the operation of the seaweed industry. This chapter overviews the range of people involved in these different seaweed farming jobs and the way it affects their livelihoods.

## Farmers

According to a recent commodity-specific survey (BPS 2022), there are 62,754 households farming seaweed in marine areas and 88,171 individual owner-managers. It is reported that across the country, each household typically has a man who works as the owner-manager of the farm, while there is variation in the number of women reporting as owner-managers, from 0.18 per household in Kalimantan, up to 0.58 per household in Nusa Tenggara Timor (NTT) (BPS 2022). In South Sulawesi, around one in four seaweed producing households report a woman as an owner-manager of the farm (BPS 2022). The extent to which women participate in on-farm work may influence their likelihood of self-reporting as farm owner-managers, and level of participation is often linked to production method. If the more physically demanding longline method is used, female involvement in on-farm work seems to be less common than if the off-bottom method is used

(Msuya and Hurtado 2017). Female involvement in on-farm work may also be influenced by cultural norms in different parts of Indonesia. The highest involvement of women is in Maluku, Nusa Tenggara Barat (NTB), and NTT (0.42, 0.57, and 0.58 women per household respectively), all of which have a high Christian population (Table 8.1).

Although women across Indonesia are much less likely to report themselves as owner-managers of a seaweed farm, they are heavily involved in the seaweed industry, especially in off-farm tasks. In our survey of Pitu Sunggu seaweed farmers, we asked respondents who was involved in different on- and off-farm tasks (Figure 8.2). We found that men were almost always involved in seaweed farm installation, maintenance, and harvesting (>95 per cent of households). Most men also participated in drying, bagging, and selling (~75–80 per cent of households), and often reported some involvement in seaweed binding (39 per cent of households). Women had limited involvement in seaweed farm installation, maintenance, and harvesting (<10 per cent of households) but were normally involved in seaweed drying (84 per cent of households) and seaweed binding (71 per cent of households). They were often involved in seaweed bagging and selling (43 per cent and 35 per cent of households respectively). This suggests that although women may not report working as owner-managers of the farm, they are involved in many aspects of off-farm processing and marketing necessary to the operation of the business. Similar findings have been reported elsewhere. For example, an NTT study found that both men and women contributed similar amounts of labour to seaweed farming activities (Fitriana 2017).

Seaweed farming work is therefore significantly gendered in Pitu Sunggu, although these seem to be patterns rather than strict rules. During our qualitative research we frequently asked men why they did not participate in seaweed binding work, and women why they did not participate in on-farm work. Respondents reported that while in the past there were some rules restricting women going to sea – during menstruation, for example – today, there were no definitive rules about which of these jobs men and women could do. Women could also work at sea, and

*Table 8.1* Involvement of women and men as seaweed farm owner-managers by province

|  | *Total households* | *Men per household* | *Women per household* |
|---|---|---|---|
| Sulawesi Selatan | 24,922 | 1.08 | 0.25 |
| Nusa Tenggara Timur | 10,166 | 0.99 | 0.58 |
| Maluku | 7,275 | 1.09 | 0.42 |
| Sulawesi Tenggara | 7,083 | 1.04 | 0.38 |
| Sulawesi Tengah | 2,895 | 0.95 | 0.22 |
| Jawa Timur | 2,152 | 0.96 | 0.26 |
| Kalimantan Utara | 2,075 | 1.25 | 0.18 |
| Nusa Tenggara Barat | 2,056 | 1.03 | 0.57 |
| Other provinces | 4,130 | 1.08 | 0.27 |
| Indonesia | 62,754 | 1.06 | 0.35 |

Source: Data from BPS (2022).

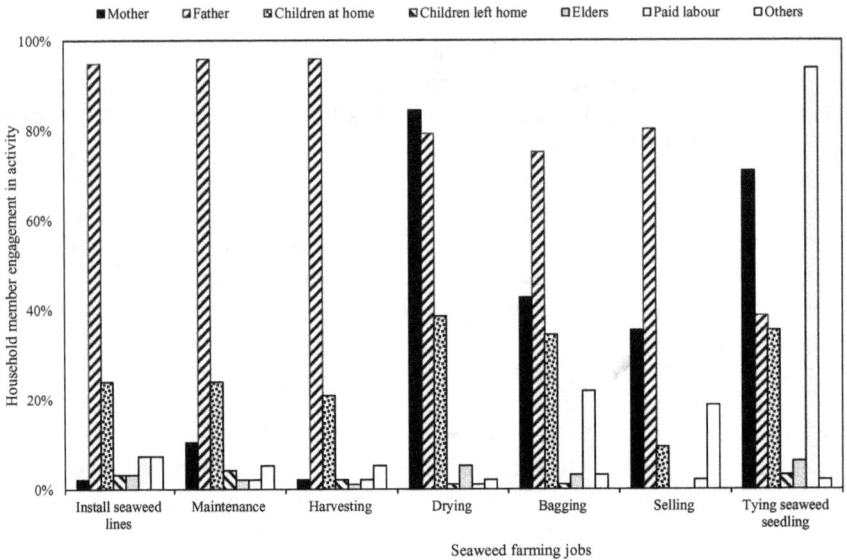

*Figure 8.2* Participation of different household members in common seaweed farming jobs
Source: Data from authors' survey.

men could also work in the household as seaweed binders. However, they said that it would normally be the case that the men would do the 'heavy' work at sea. This heavy work was outside the reach of some village residents. For example, one elderly crab fisherman reported that he could no longer farm seaweed as he was not strong enough any more. Similarly, a widowed woman lamented that her financial situation had declined since she had lost her husband as she was not physically strong enough to manage the seaweed farms herself. The work is also not suitable for heavily pregnant women or women raising small children due to the long hours spent in the sun sitting hunched in boats, and as a result, most women viewed the work as men's work.

Thus, while there are no fixed rules preventing women from working at sea, in this village it is uncommon. Harvesting work can only be undertaken by one household member at a time, since seaweed must be harvested and loaded into small boats, and when the boat is full there is no space for an additional worker. Cleaning work can be undertaken by two people, but is normally undertaken by either one or two men. The installation, maintenance, and harvesting work is undertaken under the sun in an often windless area which makes the work very hot and slow, requiring that the worker spend hours crouched in a boat bent over the side, tending to their seaweed. As a result, men were less likely to participate in seaweed binding as they were often busy with on-farm work, although many would occasionally help out with this work when they were idle. Similarly, men were partially involved in drying work including bringing harvested seaweed to a drying location and spreading it out, but it was common for women to monitor the drying process and regularly turn the seaweed over if the drying location was close to the house. As a result of this division of labour, male farmers may be more likely to see themselves as the farm manager. In addition, men typically installed the anchors claiming the farm area which often means that they see themselves as plot owners (as discussed in Chapter 5).

This is not, however, universal. As shown above, Women in NTT are more than twice as likely as those in South Sulawesi to report themselves as owner-managers of a seaweed farm (BPS 2022). In our research with NTT female farmers (Langford, Turupadang et al. 2022), one woman described how she managed a seaweed farm alone by directly replanting her seeds while at sea. In order to avoid the heavy work of harvesting and planting, rather than lifting the whole seaweed longline from the water and taking it to shore, she individually harvested seaweed bunches, took cuttings from them, and directly reattached the cuttings to the longline, thereby combining the seaweed harvesting, tying, and installation jobs (Langford, Turupadang et al. 2022). This style of work is not common practice however, as it means that she spent very long hours at sea in the sun, something that farmers try to avoid.

In Pitu Sunggu, most women preferred indoor work. When asked why they did not work outdoors tending to the shrimp ponds, one woman joked that women preferred to only go to the pond when it was time to harvest the shrimp for consumption, leaving the work of maintenance to their husbands. This division of labour is also linked to the additional household duties of women, who are often responsible for preparation of food and care of young children. This work occurs in the home, so seaweed binding can be done while also caring for young children. As binding is paid as piecework, it provides flexibility for women to undertake other household jobs. The ability for this work to be easily integrated into women's lives is one reason that it has been taken up widely and relatively unproblematically. The gendered division of labour was not seen as problematic by community members in our study, with men and women alike reporting that if they wished to undertake the others' jobs, they could.

### Children

There is some involvement of children in the seaweed industry. Respondents reported that 20–24 per cent of 'children living at home' were involved in on-farm work and 34–39 per cent of this group were involved in drying, bagging, and binding work (Figure 8.2). These figures do not necessarily represent young children however, and often include teenagers and young, unmarried adults. Many young children were not heavily involved in managing the family seaweed farm, but occasionally helped by binding or collecting small pieces of seaweed left scattered by other farmers. Participation of children in farm work in rural areas is common and may be beneficial to themselves and their families, depending on the type and level of work they undertake. According to the International Labour Organisation (ILO) (2023, n.p.):

Not all work done by children should be classified as child labour that is to be targeted for elimination. The participation of children or adolescents above the minimum age for admission to employment in work that does not

affect their health and personal development or interfere with their schooling, is generally regarded as being something positive. This includes activities such as assisting in a family business or earning pocket money outside
school hours and during school holidays. These kinds of activities contribute
to children's development and to the welfare of their families; they provide
them with skills and experience, and help to prepare them to be productive
members of society during their adult life.

The ILO defines child labour as work that 'deprives children of their childhood,
their potential and their dignity, and that is harmful to physical and mental development', including work that is 'mentally, physically, socially or morally dangerous
and harmful to children; and/or interferes with their schooling' (ILO 2023, n.p.).
The work undertaken by children in Pitu Sunggu does not appear to fill this defini-
tion. Children seem to be only lightly involved in seaweed work, and when this occurs it is generally only after school hours, and only for small amounts of time and
money. It is typically driven by the initiative and interest of the child, rather than
out of necessity for the family, and children normally keep the sums they receive
as 'pocket money'.

For example, one woman in nearby Laikang village laughed about how her
child had taken up binding for a couple of hours after school and within a few
months had earned enough for a mobile phone, which the child purchased before
immediately ceasing the binding work. Another described how their child had collected fallen pieces of seaweed from the seashore and dried them by the side of
the road and sold them for a small sum. In another example, we observed a mini
seaweed farm set up close to the shoreline using a short piece of rope and some
sticks dug into the mud. When we enquired about the purpose of this set up, we
were told that it was just 'child's play'. These forms of participation are initiated by
the children in their free time and are perceived by their parents as beneficial to the
child, since they do not interfere with their children's schooling and provide them
with positive experiences of farm work.

There may be differences in the involvement of children in binding work in
other areas. In the period of our study seaweed farmers benefited from high prices
and many farmers emphasised the importance they placed on their children's education. Some hoped that their children would become civil servants. One woman
told us that she forbade her children to participate in on-farm work because she
wanted them to study hard and gain lucrative employment in the city. Others saw
education as an important skill for navigating the modern world, even if the children were to remain in the village and work on seaweed.

**Farmer demographics**

Seaweed farmers tend to be older, with 70 per cent of farmers across Indonesia
aged 35 or over (BPS 2022). They also tend to have limited education, with 64 per
cent of farmers having a primary school education or less (BPS 2022). These figures are similar for men and women. The educational profile of farmers is closely

linked to the age profile, as older farmers belong to a generation where education was often difficult to access in rural areas or not prioritised (Figure 8.3). For residents of Pitu Sunggu, the first local primary school was established in 1976. Farmers report that those who enrolled in it without a birth certificate were assigned a birth year of 1969, based on the assumption that a student should be seven years old when they commenced their education, although many students were in fact several years older than this. This generation of seaweed farmers were approximately 54 years old in 2023. Village residents older than this typically did not attend primary school, while younger residents had increasing access to education.

Today, education is widely prioritised in the village, although the extent to which higher education is encouraged varies widely depending on the financial resources of the parent(s), the educational proficiency of the child, and the land and sea resources of the parent(s) (which affect, for example, the ability of the household to pay school fees or forgo paid labour). Some young 25–30-year-old seaweed farmers who reached adulthood before seaweed farming intensified in 2017 have managed to accumulate large areas of ocean space and have become very wealthy,

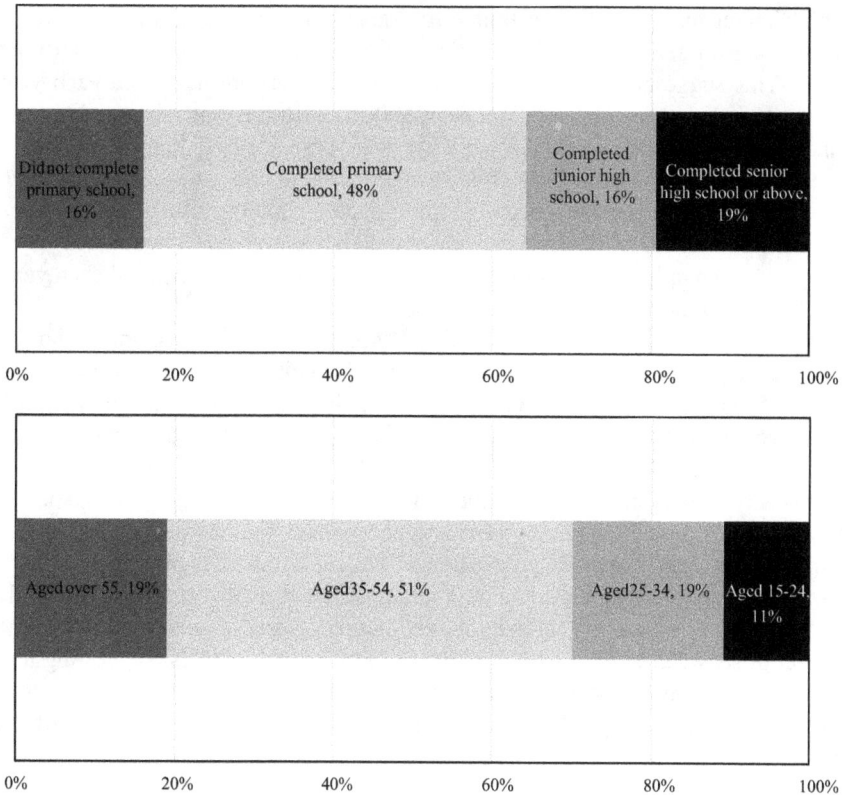

*Figure 8.3* Educational and age profile of seaweed farmers in Indonesia
Source: Data from BPS (2022).

while others have migrated from the village. Today, however, the only way youth reaching adulthood can enter seaweed farming is to inherit sea space from their family, as the sea space area is now full. Unlike the situation just ten years ago, opportunities in marine activities are defined by existing family resources.

### Wage labourers

Wage labour is one of the most significant and overlooked types of work in the seaweed industry. 94 per cent of seaweed farmers in Pitu Sunggu report using paid labour to complete binding work. This work is also undertaken by women in 71 per cent of households, men in 39 per cent of households, children living at home in 35 per cent of households, and elderly household members in 6 per cent of households (as shown in Figure 8.2). Although no formal estimates of the number of people engaged in seaweed binding work in Indonesia exists, an approximate estimate can be made assuming that Indonesia produces approximately 2 million tonnes of wet seaweed annually (see Appendix A1).

Assuming that one longline yields 40kg seaweed (this estimate is consistent with Simatupang et al. 2021 estimate for Pangkep) and 10kg of this is used for seeds, 30kg of seaweed would be harvested from each longline after a farming cycle. This suggests that around 66.7 million longlines are harvested each year (to produce 2 million tonnes of wet seaweed).[3] If each longline takes one hour to bind, this is 66.7 million hours of work per year. A 'specialised' binder who works five hours per day for on average five days per week in peak times for around two months of the year, two days per week for around six months of the year, and one day per week for four months of the year, would work around 562 hours (112 days) per year. To complete the required 66.7 million hours of work required each year, 118,000 such binders would be needed.

With around 62,000 seaweed farming households (BPS 2022), and assuming that each household also contributes one person's binding labour, this would leave around 56,000 casual binders working in the industry. This is likely to be a conservative estimate as it assumes successful yields from every binding cycle (while as Chapters 6 and 7 showed the need to harvest and replant in response to weather events is common) and because it does not account for the changing of the species cultivated which occurs twice a year and necessitates farmers dedicating several farming cycles for propagule production alone. The estimate is therefore very approximate due to the lack of data on binding activities available. However, while approximate, this back-of-the-envelope estimation gives a sense of the scale of the impact of seaweed farming *outside* seaweed farming households, suggesting that binding work may support the livelihoods of nearly as many non-farming households as farming households. It is therefore a highly significant component of the livelihoods generated from seaweed farming that should be incorporated into research, policy, and development programmes focused on the industry.

Seaweed binding work is often undertaken by groups of 5–10 seaweed binders who specialise in the work. In Pitu Sunggu these groups work in spaces under or outside people's homes, though in other places such as Laikang they work in shelters constructed along the coast, where they are close to seaweed farmers coming in from the sea and can enjoy the cool sea breezes as they work. The creation of seaweed binding jobs is widely viewed as one of the positive benefits of the seaweed industry. As discussed in Chapter 5, it is partially a result of the unequal distribution of sea space, which has created a cohort of very large farmers who are not able to undertake their binding work themselves. It is also a result of the time-sensitive nature of the work, since seaweed propagules should be returned to the sea within six hours or they will begin to perish. This means that a seaweed farmer must arrange in advance for a binding group to work on their seaweed for a particular day, and on that day must rise early in the morning, travel to their seaweed farm, bring the seaweed longlines to shore, separate the propagules from the ropes, select a proportion of them for use as seeds, and transport the seeds and ropes to a binding group for reattachment. They can then rest while this binding is undertaken and return in the afternoon to collect the lines and install them back in the sea.

Over the course of our study, as the prices paid for seaweed increased and many farmers intensified their production, demand for binding labour increased and in many cases was not able to be met. As one farmer told us, 'The binders here are overwhelmed with work. Lots of farmers want them to tie their seeds. Usually, we look for available labour in the village, but none are available.' Some farmers reported that they drew on their kinship connections to access the labour required. Others reported travelling outside the village to binding groups further from the sea, thus spreading the reach of seaweed farming benefits further inland.

*Figure 8.4* A group of seaweed binders (known locally as *passio*) working on a load of seaweed

Source: Image by Radhiyah Ruhon, November 2022, Pitu Sunggu.

Binding labour has traditionally been sourced from the inland hamlet of Bonto Sunggu, but as seaweed farming has intensified and demand has grown, farmers have sought labour further afield. One farmer told us that since at least 2019, binding labour has been sourced mostly from outside the village. In 2022, as farming intensified further with rising prices, several people established businesses acting as intermediaries between farmers and binding groups in inland villages.

One such intermediary lives in Pitu Sunggu, where he noticed high demand for workers, but also has family in an inland village where there was low employment. He started a service where he transported the seeds from the seaweed farmers to binders in inland villages, and returned the completed longlines at the end of the day. He charges Rp. 6,000/rope for the service, pays binders Rp. 4,500/rope, and retains the rest as a transportation and finder's fee. He said that the service was viewed very positively by the residents of the inland village: 'women there don't have jobs, so they are grateful to be brought jobs. So they can also experience the benefits of seaweed.' Or as another farmer told us, 'many people are affected by the economy because of seaweed. Not only in the coastal areas, but even near the mountains.' It appears that some binders had been previously unemployed, while others had worked peeling cashews or cleaning rice, but wages from seaweed binding was much higher than for these other jobs.

Some farmers prefer to use specialised binder groups (rather than family members or neighbours) because they are able to tie many ropes in a single day. Other farmers continued to use local binders wherever possible, as they felt it was important to give jobs to their neighbours. As one described, 'We just give the jobs to the people here, so they can also benefit'. Indeed, many binders we spoke to were grateful for the work. One woman with two children and no husband spoke at length about her gratitude for the work, reporting that the income helped her regain her self-esteem since she no longer had to rely on relatives for her survival. Women, widows, elderly people, and people with disabilities were frequently involved in this work as they had few other options. As one woman told us, 'This is my only job, nothing else. If I don't do this, I eat nothing.' The work benefits women because they are able to earn an income while still meeting their obligations around the house, so it has been taken up relatively uncontentiously, providing women with greater autonomy. In Pitu Sunggu there are also several cases of people with disabilities such as physical mobility limitations and blindness working as seaweed binders.

Increasing labour shortages and costs for binding, however, has affected seaweed farmers. With growing demand for labour during 2022, many seaweed binding groups took the opportunity to renegotiate their remuneration. Prices rose from Rp. 4,000/longline to Rp. 6,000/longline and the terms of labour were also questioned. For example, whether it included twisting the longlines together to create the 'double-ropes', whether the longline was 25m or 28m in length, and whether the binders were permitted to keep any leftover seeds at the end of the day. It is common practice in many binding groups to keep leftover seaweed, which they dry and sell (Figure 8.5). These leftovers can represent a sizeable additional income stream and may incentivise the use of smaller propagule sizes.

*Figure 8.5* Binder dries leftover seaweed on roadside

Source: Image by Radhiyah Ruhon, March 2022, Pitu Sunggu.

Some binders would charge Rp. 4,000/rope if they kept the seeds, and Rp. 5,000/rope if not. Others would charge an additional Rp. 1,000/rope to include twisting of the double ropes As some binders negotiated higher rates, others also negotiated matching rates from farmers for services on better terms (e.g. on leftover seaweed). Farmers who urgently needed binding labour had no choice but to agree to these terms. As one farmer said, 'it is because the seeds really need to be tied, meanwhile there is no binding service available around, so like it or not, I must accept the price offered. I need it.' Farmers complained that rising binding wages put pressure on their margins, and that once the price of binding labour went up, it would not go down again if prices fell. Indeed, this concern seemed to have actualised after market prices halved at the end of 2022 (Langford, Zhang et al. 2022).

Nonetheless, farmers rely on seaweed binders to manage their business, and are anxious to remain on good terms with seaweed binders. Similarly, binders seek to remain on good terms with farmers, to an extent that depends on their personal circumstances. Some binders in vulnerable financial circumstances go to great lengths to maintain their access to work. One elderly woman explained that because she needed the work and knew that she was slow (only completing around 5–7 longlines in an eight-hour day), she sought to maintain good relations with farmers and demonstrate her gratitude by returning leftover propagules. Other binders demanded wage increases as seaweed prices rose. Overall, people viewed this wage work as a key contributor to well-being in the village. As one person told us, 'It empowers the community. Starting from the binding of seaweed, many unemployed people get jobs. That is one of the main developments in this area to empower the community … Especially women and widows.'

Despite the large number of wage labourers working in the seaweed industry and the potential importance of this work to inland livelihoods, relatively few studies of the social impact of seaweed farming have explicitly examined these workers. Further studies in this area would probably reveal the diverse impacts of this work in these non-seaweed farming locations.

### Seaweed gatherers

Finally, we observed a number of people working as what might be described as 'seaweed gatherers'. These were people who did not farm seaweed themselves, but would collect seaweed which had fallen off the longlines of others. These included fishermen, who would often intentionally hang gill nets to catch fallen seaweed or use push nets to collect it (Figure 8.6), as well as people – often women – who would walk along the shore to collect seaweed as it washed ashore. This was common in both Laikang and Pitu Sunggu villages. In Laikang, which sits on a large sandy bay with clearer waters, we also found individuals who would go 'diving' for seaweed, searching for it along the ocean floor and collecting it into their boat. The yields from this activity vary according to the method used: fishers using nets or diving gear in combination with boats collected significant quantities of seaweed and found this to be a profitable activity. In contrast, those who gathered seaweed washing up on the shore tended to collect smaller quantities and earned less income

*Figure 8.6* Left: Seaweed gathering from along the shoreline; and Right: Using push nets and a boat

Source: Image by Zulung Zach Walyandra, in September and October 2022, Pitu Sunggu.

from this practice. The latter were often less-well off and did not have capital (in the form of boats and nets) to deploy, but managed to earn small amounts of income from the activity. These activities were particularly commonly undertaken after a period of high waves, when seaweed breakages were common.

## Conclusion

Much has been written about the social impact of the seaweed industry. While early research emphasised seaweed as a source of income for low-income households, more recent research highlights the range of social benefits brought to seaweed farming households. This chapter expanded on this by exploring labour dimensions disaggregated by the demographic groups of men, women, and children. It also highlighted the large number of casual wage labourers working in the industry and documented ways that casual binding work has spread the impact of seaweed farming beyond coastal villages into nearby inland areas. Further research into these casual labourers is an important aspect in understanding the broader social impact of seaweed farming in Indonesia. The next chapter explores the role of traders and financiers in Pitu Sunggu.

## Notes

1 However, some companies are changing this with the establishment of centralised, quality controlled drying facilities (Langford, Turupadang, and Waldron 2023).
2 Fast binders, including children and younger women, can sometimes bind seaweed at twice this rate, while older binders take up to 1 hour 30 minutes per longline. The time taken also depends on the length of the line (which is normally 25m, but sometimes up to 28m), whether the binder is simultaneously 'twinning' the lines (i.e. twisting two ropes together to form a 'double-rope'), and the size, species, and morphology of the propagules (with cottonii reportedly easier to bind than sacol).
3 Some seaweed areas use off-bottom, raft and cage methods: the purpose of this calculation is to provide an approximation of the scale of seaweed binding work, not to accurately represent the diversity of farming systems and yields across Indonesia.

## References

Andréfouët, Serge, I. Made Iwan Dewantama, Eghbert Elvan Ampou. 2021. "Seaweed Farming Collapse and Fast Changing Socio-Economic Ecosystems Exacerbated by Tourism and Natural Hazards in Indonesia: A View from Space and from the Households of Nusa Lembongan Island." *Ocean and Coastal Management* 207 (105586): 1–8.

Aslan, La Ode Muhammad, Rustam Supendy, Siti Aida Adha Taridala, Harafin Hafid, Wa Ode Sifatu, Zalili Sailan, and La Niampe. 2018. "Income of Seaweed Farming Households: A Case Study from Lemo of Indonesia." *IOP Conference Series. Earth and Environmental Science* 175 (1): 12221. https://doi.org/10.1088/1755-1315/175/1/012221

Aslan, La Ode Muhammad, Nur Isiyana Wianti, Siti Aida Adha Taridala, Manat Rahim, Ruslaini, and Wa Ode Sifatu. 2022. "The Debt Trap of Seaweed Farmers: A Case Study from Bajo Community in Bungin Permai, Indonesia." In IOP Conference Series. *Earth and Environmental Science* 1033 (1): 12056. https://doi.org/10.1088/1755-1315/1033/1/012056

BPS (*Badan Pusat Statistik*). 2022. Hasil Survei Komoditas Perikanan Potensi Rumput Laut 2021 Seri 2. *Badan Pusat Statistik*. https://www.bps.go.id/publication/2022/08/ 29/269de33babc6e3d52bbae5b6/hasil-survei-komoditas-perikanan-potensi-rumput-laut-2021-seri-2.html

Eranza, Datu Razali Datu, James M. Alin, Arsiah Bahron, and Roslinah Mahmud. 2015. "Determinants of Women's Participation in Seaweed Farming in the Regency of Jeneponto, South Sulawesi, Indonesia." *Mediterranean Journal of Social Sciences* 6 (5): 43–45. https://doi.org/10.5901/mjss.2015.v6n5s5p43

Fitriana, Ria. 2017. "Gendered Participation in Seaweed Production – Examples from Indonesia." *Asian Fisheries Science* 30: 245–264. https://doi.org/10.33997/j.afs.2017.30. S1.013

ILO (International Labour Organisation) 2023. "What Is Child Labour?" *International Labour Organisation*. https://www.ilo.org/ipec/facts/lang--en/index.htm

Langford, Alexandra, Welem Turupadang, Marclien D. R. Oedjoe, Christian Liufeto, Berty Bire, Marcel Yohanis, Yulius Suni and Scott Waldron. 2022. "Seaweed Farmer Resilience in Eastern Indonesia after COVID-19." *Australian National University Indonesia Project.*

Langford, A., W. Turupadang, and S. Waldron. (2023). "Interventionist Industry Policy to Support Local Value Adding: Evidence from the Eastern Indonesia Seaweed Industry." *Marine Policy* 151 (105561). doi.org/10.1016/j.marpol.2023.105561

Langford, Alexandra, Scott Waldron, Nunung Nuryartono, Syamsul Pasaribu, Boedi Julianto, Irsyadi Siradjuddin, Radhiyah Ruhon, Zulung Zach Walyandra, Imran Lapong, Risya Arsyi Armis. 2023. *Sustainable Upgrading of the South Sulawesi Seaweed Industry*. Melbourne, Australia: Australia-Indonesia Centre.

Langford, Alexandra, S. Waldron, Sulfahri, and H. Saleh. 2021. "Monitoring the COVID-19-Affected Indonesian Seaweed Industry Using Remote Sensing Data." *Marine Policy* 127 (104431): 1–10. doi.org/10.1016/j.marpol.2021.104431

Langford, Alexandra, Scott Waldron, Jing Zhang, Radhiyah Ruhon, Zulung Zach Walyandra, Risya Arsyi Armis, Imran Lapong, Boedi Julianto, Irsyadi Siradjuddin, Syamsul Pasaribu, and Nunung Nuryartono. (2024). "Diverse Seaweed Farming Livelihoods in Two Indonesian Villages." In *Tropical Seaweed Cultivation – Phyconomy: Proceedings of the Tropical Phyconomy Coalition Development, TPCD 1, Held at UNHAS University, Makassar, Indonesia. July 7th and 8th* edited by A. Hurtado, A. Critchley, and I. Neish. 00-00. New York: Cham, Switzerland: Springer Nature.

Langford, Alexandra, Jing Zhang, Scott Waldron, Boedi Julianto, Irsyadi Siradjuddin, Iain C. Neish and Nunung Nuryartono. 2022. Price Analysis of the Indonesian Carrageenan Seaweed Industry. *Aquaculture* 550(737828). doi.org/10.1016/j.aquaculture.2021.737828

Larson, Silva, Natalie Stoeckl, Mardiana E. Fachry, Muhammad Dalvi Mustafa, Imran Lapong, Agus Heri Purnomo, Michael A. Rimmer, and Nicholas A. Paul. 2020. "Women's Well-being and Household Benefits from Seaweed Farming in Indonesia." *Aquaculture* 530 (735711). https://doi.org/10.1016/j.aquaculture.2020.735711

Larson, Silva, Natalie Stoeckl, Michael A. Rimmer, and Nicholas A. Paul. 2022. "Understanding Feedback Relationships Between Resources, Functionings and Well-Being: A Case Study of Seaweed Farming and Artisanal Processing in Indonesia." *Ambio* 51 (4): 914–925. https://doi.org/10.1007/s13280-021-01581-3

Mariño, Mónica, Annette Breckwoldt, Mirta Teichberg, Alfred Kase, and Hauke Reuter. 2019. "Livelihood Aspects of Seaweed Farming in Rote Island, Indonesia." *Marine Policy* 107 (103600). doi.org/10.1016/j.marpol.2019.103600

Mirera, D. O., A. Kimathi, M. M. Ngarari, E. W. Magondu, M. Wainaina, and A. Ototo. 2020. "Societal and Environmental Impacts of Seaweed Farming in Relation to Rural

Development: The Case of Kibuyuni Village, South Coast, Kenya." *Ocean & Coastal Management* 194: 105253. https://doi.org/10.1016/j.ocecoaman.2020.105253

MKPRI (Menteri Kelautan dan Perikanan Republik Indonesia). 2019. *Keputusan Menteri Kelautan dan Perikanan Republik Indonesia Nomor 1 / Kepmen-KP/2019 Tentang Pedoman umum pembudidayaan rumput laut.*

Msuya, Flower E. and Anicia Q. Hurtado. 2017. "The Role of Women in Seaweed Aquaculture in the Western Indian Ocean and South-East Asia." *European Journal of Phycology* 52 (4): 482–494. https://doi.org/10.1080/09670262.2017.1357084

Nuryartono, N., S. Waldron, A. Langford, Sulfahri, K. Tarman, S. H. Pasaribu, U. J. Siregar, and M. F. D. Lusno. 2020. *A Diagnostic Analysis of the South Sulawesi Seaweed Industry.* Melbourne, Australia: Australia-Indonesia Centre.

Simatupang, Nova Francisca, Petrus Rani Pong-Masak, Pustika Ratnawati, Agusman, Nicholas A. Paul, and Michael A. Rimmer. 2021. "Growth and Product Quality of the Seaweed *Kappaphycus alvarezii* from Different Farming Locations in Indonesia." *Aquaculture Reports* 20: 100685. https://doi.org/10.1016/j.aqrep.2021.100685.

Stone, Serafina, Zannie Langford, Imran Lapong, Risya Arsyi Armis, Zulung Zach Walyandra, Radhiyah Ruhon, Annie Wong, Boedi Julianto, Irsyadi Siradjudding, and Scott Waldron. 2023. "Technology Uptake by Smallholder Farmers: The Case of the Indonesian Seaweed Industry." *Journal of Agribusiness in Developing and Emerging Economies.* https://doi.org/10.1108/JADEE-01-2023-0011

Teniwut, Wellem Anselmus, and Roberto M. K. Teniwut. 2018. "Minimizing the Instability of Seaweed Cultivation Productivity on Rural Coastal Area: A Case Study from Indonesia." *Aquaculture, Aquarium, Conservation & Legislation* 11 (1): 259–71.

Teniwut, Wellem Anselmus, Yuliana K Teniwut, Roberto M. K Teniwut, and Cawalinya L Hasyim. 2017. "Family vs Village-Based: Intangible View on the Sustainable of Seaweed Farming." IOP Conference Series: *Earth and Environmental Science* 89 (1): 12021. https://doi.org/10.1088/1755-1315/89/1/012021.

Valderrama, D., J. Cai, N. Hishamunda, N. Ridler, I. Neish, A. Hurtado, . . . J. Fraga. 2015. "The Economics of Kappaphycus Seaweed Cultivation in Developing Countries: A Comparative Analysis of Farming Systems." *Aquaculture Economics & Management*, 19, 251–277. doi:10.1080/13657305.2015.1024348

Wright, Emily C. 2017. "The Upshot of Upgrading: Seaweed Farming and Value Chain Development in Indonesia". Master's Thesis. University of Hawai'i.

Zamroni, A. 2021. "Sustainable Seaweed Farming and Its Contribution to Livelihoods in Eastern Indonesia." IOP Conference Series: *Earth and Environmental Science*, 718: 12099. Bristol: IOP Publishing. https://doi.org/10.1088/1755-1315/718/1/012099

# 9 Seaweed marketing

## Village-based traders as financial and market intermediaries

*Zannie Langford, Radhiyah Ruhon,*
*Zulung Zach Walyandra, Risya Arsyi Armis,*
*and Imran Lapong*

### Introduction

The previous chapters have explored how participation in the seaweed industry has transformed the livelihoods of farmers in Pitu Sunggu. Through their work, farmers have transformed village landscape and use of sea space (Chapters 4 and 5), maintained production through concentrated efforts to address the biophysical challenges of farming (Chapters 6 and 7) and incorporated various forms of labour in ways that have altered the livelihoods of men, women, children and vulnerable people (Chapter 8). In this chapter, the focus shifts from the people involved in the production of seaweed to those involved in marketing it for sale: local and regional traders. The chapter takes an actor-based and local-level approach to examining the structures, conduct and performance of the Indonesian seaweed marketing system.

Local traders act as intermediaries between village-based farmers and city-based traders and processors. They are residents of the village who buy seaweed from farmers, pack and redry it if necessary (Figure 9.1) and transport it to Makassar for sale. Their work is important because they must mediate between farmers, who offer seaweed for sale with varying degrees of quality (as measured mainly by dirt and moisture content, but also by characteristics of the carrageenan it contains), and buyers in Makassar who often – but not always – apply strict criteria for quality (Komarek et al. 2023; Mulyati et al. 2020; Mulyati and Geldermann 2017). They also mediate between farmers and global networks of production and capital in other ways, by providing farmers with access to the financial services necessary to buy important inputs for seaweed farming, as well as, in some parts of Indonesia, other household goods such as fridges (e.g. Aslan et al. 2022; Fausayana et al. 2014; Prasetia et al. 2022; Zamroni 2018, 2021). This chapter completes our examination of livelihood transformations in Part II of this book with this study of the work that takes the seaweed out of the village and into global value chains.

### Senior traders

At the commencement of our study there were three well-established village-based traders in Pitu Sunggu. These 'senior traders' included two men and one woman.[1] All were well-established when we first entered Pitu Sunggu, purchasing almost all

DOI: 10.4324/9781003183860-12

*Figure 9.1* Left: Repacking of seaweed at trader locations in Pitu Sunggu; and Right: Repacking warehouse in Laikang

Source: Image by Zannie Langford, Pitu Sunggu and Laikang, May 2022.

of the seaweed produced in the village and selling it to regional buyers in Makassar. Each had started buying seaweed several years previously and had built up
their seaweed trading businesses over time. One senior trader told us,

> Initially I started buying seaweed on a small scale, starting with one or two
> people ... Starting from that, I became a small-scale trader and I sold the
> seaweed back to a large trader [in the village] ... at first it took me one month
> to collect a ton of seaweed ... As time went by, over about a year, some peo-
> ple from the island began to sell their seaweed to me [and] ... from there,
> many farmers started selling their seaweed to me. I sold my seaweed to a
> bigger trader until he offered to teach me to sell it directly to the company [in
> Makassar] ... After I started selling seaweed directly to companies, I started
> to assist farmers by providing capital, such as ropes and other things.

This trader reported that when they first started trading, they faced considerable
risk due to low profit margins and high levels of risk associated with improperly
dried seaweed.

> In the past, yields were small, and profits were slim. The profit from the sale
> of seaweed was only 200 Rupiah per kg, and at the beginning the farmers did

not know the level of dryness required, so the quality of the seaweed was not good and I received little profit from sales … to be honest, I almost stopped working as a seaweed trader twice because our efforts were not matched by the results we got.

This trader reported that after several years, they too introduced a new trader to the practice of trading seaweed, teaching them how to sell seaweed to the same company in Makassar. This trader has since also grown their business to become one of the senior traders in the village.

These senior traders have traditionally had close working relationships with each other, each offering similar prices to farmers. They say this is a professional courtesy to each other, by providing 'transparency' about seaweed prices and not competing for customers by undercutting each other on price. Over time, these traders have each built close connections with their own cohorts of farmers, and many farmers have sold exclusively to the same trader for many years. Senior traders also provide financial services to farmers in the form of loans, cash or seaweed farming inputs, such as propagules and ropes. These goods and services are one way that these traders can compete with each other for clients. The loans they offer range in size from around Rp. 1–50 million (~ AU$100–5,000), have no fixed repayment date and attract no interest payments. The lenders do require, however, that indebted farmers sell their seaweed exclusively to them.

In addition to borrowing from traders, many farmers also deposited their savings with village-based traders rather than with banks. Many farmers had had negative experiences saving money in banks. As the amounts saved were small, they found that their savings were eroded by fees and had become discouraged from using banks. As one farmer explained,

I just keep my money at home … because now the bank does not pay interest … In the past I had saved money in the bank, but when I went to take it out, it turned out [my account] was minus Rp. 50,000. I thought they would pay a profit, but no – they just took the money.

Instead of saving money in banks, many village residents save their money at home or buy gold as a store of value. Many seaweed farmers save their money with traders by depositing their seaweed and not collecting their payment. As one farmer told us,

I left it [the proceeds from a recent sale] there on purpose … I intentionally save it there … I am saving it for capital. … I need that money, but I do not yet want to take it … There are some people who always save their money [with the trader] for up to a year … I have never kept money in the bank.

Saving with traders is not without risks. One farmer told us that on one occasion he had been saving money with a crab trader under a similar system, but the trader went bankrupt and never returned his money. Another told us of an incident with a seaweed trader when he intentionally did not take payment on delivery of a large load of seaweed, but when he returned to take the money, the trader refused to pay him. As he said,

> There are also dangers. There was also an incident with me, 1.5 tons [of seaweed], until now it has not been paid for ... It's gone ... they had a receipt for me, but [when they looked for it] there was no receipt there. Maybe his men took it. Then the trader said that everything had been paid for.

The loans provided by traders are well suited to farmers because they are informal, do not require any paperwork, provide instant access to cash, have flexible loan terms and repayment dates. However, the loans come with a requirement to sell seaweed exclusively to the trader until the loan is repaid.

Farmers in Pitu Sunggu vary in their opinion of this system. Some express a heartfelt gratitude to 'their' trader for all the loans that they had received over the years and feel that to negotiate on price alone would be ungrateful. They see these loans as a kindness which should be remembered, even if their trader pays them a lower price for their seaweed than other traders are offering. As one farmer who always loyally sold to the same trader told us,

> I am happy if I ask [for a loan] and receive it ... if they say yes, I am very grateful. Even though the money is borrowed, I think of it as a gift ... So, I never haggle [on price when selling] ... 1,000 or 2,000 [Rupiah/kg (~ AU\$0.10–0.20/kg)] [price difference] is meaningless to me ... I never ask the price when I bring goods ... What is 2,000 [Rupiah]? I am also a poor person, but trust is worth more than 1,000 or 2,000 [Rupiah]. Because if I behave like that, if I ask questions first, or if I don't wholeheartedly give my goods to that person, it means that I don't trust them. So, if for example the price offered to me is low, but I already want to give the seaweed to them, it is impossible for me not to give it. So it's better not to ask [the price] ... There is a sense of shame. That is, you could say that the value of oneself is more important than money. I also need money, but even if it's a lot of money, I care more about my name.

As the farmer above expresses, the idea of haggling on price, or even enquiring about price, is contrary to his personal values. Many farmers expressed similar views. They had received loans from traders at times when they desperately needed them, not only for seaweed inputs but also for personal costs such as weddings and for their children's education, and they saw these loans as initiating a personal relationship that would be violated by haggling on price. These farmers often boasted to us that, for example, 'I never ask the price. I just bring it directly and straight away I say, "just take it".'

Indeed, this easy access to capital was one of the reasons that many people thought that seaweed was a more favourable livelihood than other options. As one farmer told us,

> Whatever we ask for at the traders, they will definitely give it, even if it's not for seaweed capital, they still give it … What I feel is, of all the jobs, it's easiest if you work on seaweed because it is so easy to get money from traders.

Traders lend money for seaweed and non-seaweed related purchases, but they only lend money to seaweed farmers (non-seaweed farmers cannot request loans for non-seaweed related purchases). Farmers noted that loans from traders are preferable because unlike banks or pawnshops, they charge no interest and have no fixed repayment date, so that they can easily work repayment around seaweed farming cycles.

Many farmers and other community members use other forms of debt, either instead of or in addition to loans from traders. It is common for farmers to have multiple loans, including: large, long-term loans from banks (Rp. 5–25 million (~ AU\$500–2,500) over 2–3 years at 4–5 per cent interest); small, short-term loans from pawnshops (Rp. 1–2 million (~ AU\$100–200) over 4 months at 25–30 per cent interest); loans from co-operatives; and short-term loans from friends and family to help them meet repayments on their other loans (our respondents indicated that amounts borrowed from kin were typically less than Rp. 1 million (~ AU\$100) and repaid within a month or two). Loans from traders were viewed very favourably as they could be large (up to Rp. 50 million (~ AU\$5,000)), incurred no interest, had no fixed repayment schedule or loan term and did not prevent the borrower from borrowing again while the loan was outstanding, either from the trader or from other sources. As a result, seaweed trader loans were viewed favourably by many farmers, and they understood that their access to these loans was based on continued loyalty to the same trader. As one farmer told us, this access to loans from traders was invaluable and worth more than small differences in price:

> For me, there are no drawbacks (to the system) … for others, when they sell a lot of seaweed they usually think about whether they want to move traders, or what they want to do … but for me, I don't, because I think that time is too long and someday we will need help again.

However, there were those who expressed resentment of the system. This resentment focused on the buying practices of the traders. Firstly, some farmers complained about what they saw as anti-competitive pricing. As one farmer explained, the seaweed traders 'developed very quickly compared to farmers, so they have many advantages … they communicate with each other to lower the price, so the price of seaweed decreases'. They knew that prices were higher in Makassar and many calculated that the price difference exceeded the cost incurred by the traders in transporting the seaweed. These farmers resented the collusion between senior traders which kept their profit margins high. Secondly, some farmers resented being forced to sell to a trader they had borrowed from when the price offered by this trader was lower than that of other traders. Many farmers told us that a price difference of around Rp. 500/kg (~ AU$0.05) was common between traders (which would mean a price discount of 1–2 per cent, depending on market prices). Some farmers resented being forced to sell to traders at lower prices because they had outstanding loans. Some also complained that traders discouraged them from repaying their debts, with one farmer telling us that they had frequently visited a trader to try to make repayments but had been repeatedly told that the trader was not able to receive the funds at that time. As one farmer told us, 'often we want to pay it off but the traders don't want us to ... [they are] afraid that people will move to sell their seaweed [to someone else]'.

Traders acknowledged that this is a fair reflection of how the system works. However, they emphasised that their prices were negotiable. As one trader told us, 'I have communicated this to all the seaweed farmers, if there is a difference in price and you want the price to be increased like the purchase price of other traders, you can convey it'. They noted that they would not necessarily raise the price, but that it could be discussed. The outcome of such negotiations would depend on the farmer, including the quantity and quality of seaweed they sell, as well as the capacity of the trader to receive it. However, as described above, many farmers do not feel comfortable discussing prices in this way. Traders also emphasised their role in providing financial services to farmers as a social good. They explained that unlike banks, they did not charge interest, as one trader put it, 'if I give cash worth 50 million, then 50 million will also be returned ... This does not constitute usury. If you take a loan at a bank with high interest, that is what causes usury.' Several farmers corroborated this, noting that this trader could provide instant loans of up to Rp. 50 million (~ AU$5,000) at no interest and with no fixed repayment date.

Traders further highlighted the help that they gave farmers in times of need. One trader told us, 'I help with capital for a wedding, to build a house, to buy a vehicle ... there are also those that I help to buy land, and later the money will be paid using seaweed'. Traders stressed that they are flexible with the repayments and did not 'burden' farmers with repayments. One trader said,

When it's harvest time, I usually ask the farmer who I gave the capital to, should I take only part of the proceeds from selling the seaweed, or should I take all of it? ... I don't want to burden the seaweed farmers; it depends on the farmers.

Traders also told us that some farmers took advantage of this flexibility and failed to repay their loans. As one trader explained there is a risk of loan default: 'there are some farmers whose children are grown but the debt has not yet been paid off, many are like that ... they never pay annually and the debts pile up, and now it's been six years'.

Traders felt that they provided a valuable service to farmers and absorbed the risk of default. As a result, some felt resentful when farmers that they had lent money to sold their seaweed to other traders without first repaying the loan. One senior trader said,

> Sometimes there are farmers who sell their seaweed to other traders because there is a difference of Rp. 500. Actually, I don't have a problem with that and I don't force them, but since I was the one who provided the capital, it would be nice if it could be sold to me and they just ask me to match the price ... If I get a farmer like that, I can't be bothered to help meet his needs [by providing loans]. It gives me the impression that these farmers only come to me when needed and when the seaweed is harvested they sell it to other traders. We should both benefit.

The trader noted that '[I] never [force people to pay their debts], as long as the seaweed is always sold to me. However, if the farmers stop selling me their seaweed, I often urge them to pay off their debts immediately.' For the most part this system of using interest-free loans to capture supply has been enforced successfully in this way up to now. However, recently, the entrance of new traders to the village has begun to change the system.

**New traders: breaking the oligopsony**

In the two months from December 2021 to January 2022, three new traders started operations in Pitu Sunggu, all of which were also village residents and seaweed farmers who had entered the seaweed trading business to take advantage of rising seaweed prices. However, these traders proceeded cautiously, as they did not want to create conflict with the senior traders by poaching their suppliers. As one new trader explained,

> I rarely ask other people [to sell to me], except my family. But if a farmer offers me [seaweed] I will take it, because most of the farmers [can't sell to me because they] already have debt ... I haven't dared to look for more farmers because they are already attached [to another trader]. I will only buy seaweed if a farmer calls me.

These traders hoped to increase the number of farmers that supplied them and employed a range of strategies to do so.

The first strategy new traders employed was to appeal to their family to sell them seaweed. Most residents of Pitu Sunggu put a high value on family and help each other in times of need. They prefer to sell to family wherever possible, so traders with relatives with seaweed farms benefit from this relationship. New traders sought to oblige family members to sell to them wherever possible, and began their operations using small sales from family members.

The second strategy they employed was to offer higher prices than the senior traders. Senior traders told us that the margins they made on selling seaweed to Makassar ranged from Rp. 1,000–1,800/kg (~ AU\$0.10–\$0.18). One of them told us that if they earned a margin of Rp. 1,500/kg, they considered that to be 'pretty good'. They noted that these margins were much higher than they were some 10–15 years ago when they were establishing themselves, when they had margins of just Rp. 200/kg (~ AU$0.02). Junior traders were prepared to make smaller margins, but initially they were hesitant to publicly offer higher prices than senior traders because they wanted to avoid conflict with them. One new trader told us,

> I feel guilty with old traders if I increase the purchase price ... I don't want to be [competitive] ... so far I have only followed their price and have not dared to increase the price.... I don't want any fuss ... Hopefully there will be no problems in the future.

However, by July 2021 the new Pitu Sunggu traders, together with several other new traders who had commenced operation in the neighboring village, had begun offering higher prices than the senior traders. Two traders in the neighbouring village in particular had begun aggressively raising prices to attract farmers. This was perceived positively by farmers who had previously been critical of the 'cooperation' between the senior traders, as it was viewed as a fairer distribution of profits between traders and farmers. However, farmers who remained loyal to their senior trader were critical of those who moved sales to other traders, viewing them as disloyal and short-sighted. They valued the financial services that the senior trader provided them with in the form of loans or savings and did not consider small price differences a good reason to move. This was especially the case as prices were rising significantly each week at the time, so differences of Rp. 500–1,000/kg (~ AU $0.05–0.10) did not seem significant to them.

The third way that new traders sought to capture supply was by purchasing small volumes of seaweed and paying on delivery. Whereas senior traders tended to deal in large quantities at the end of the harvest cycle, and would often delay payment, new traders sought to capture supply by buying small quantities of seaweed and making payment on delivery. One new trader told us that he was often visited at night by farmers who urgently needed a little cash to meet short-term needs. These farmers often had debt problems with other traders, so sought to conceal these sales by visiting the new traders discretely. In these instances, new traders had to hold sufficient cash in hand to make immediate payment. They would aggregate these small sales until they had sufficient volumes to sell a load of seaweed to Makassar.

The fourth strategy used by traders to capture supply was to accept low quality seaweed. They were hesitant to reject poorly dried seaweed because they were eager to keep farmers as suppliers. As one trader explained,

> [in a sack of seaweed] sometimes the top is dry but the bottom is moist. I've found this kind of thing on seaweed from a farmer twice ... [I have] not yet [told him]. At first I unloaded the first sack and then the second sack, I didn't rebuke him at that time because the second purchase might not be like that but the result was still the same. I don't know what to do next ... but I do not want to let him move to another trader. We can fix the quality, we can dry it again.

These traders had to be very careful employing this strategy, as if they overpaid for dried seaweed with a high moisture content, they could easily incur losses. However, as prices rose dramatically through 2021, traders did not need to be particularly risk averse, as losses incurred through shrinkage could be made up through rising prices within a few weeks. This situation has not continued however, and prices have since fallen (as described in Chapter 2).

Finally, these traders, like senior traders before them, sought to capture supply by providing loans. They would buy ropes for farmers and provide them on credit, and then seek to discourage repayment of the loans. As one new trader told us,

> For me, the farmers don't need to pay me for the ropes in the form of money, they only need to deposit their seaweed to me. Most farmers, if they have paid off the rope, they move to other traders … So I prefer that farmers don't repay their loans quickly. The important thing is that they bring their dried seaweed to me … it's better not to be paid first, in my opinion. Because the farmer is attached to me. If paid off, farmers are not bound to me, so they are free to sell to other traders.

In this way, new traders sought to emulate the strategies of large traders by developing a supplier base that bound suppliers to them using debt relations.

Senior traders viewed the activity of new traders as a threat. They criticised the practice of increasing prices without consultation with the senior traders. One senior trader said,

> Now the competition is getting tougher with other traders. It used to be good, when the company set a price, all the purchase prices [in the village] had to be the same … but now there is no 'transparency' in determining the purchase price at the local trader level.

As a result, one of the senior traders had started matching the prices of the new traders. This suggests that seaweed buying in Pitu Sunggu may be becoming more competitive as more traders have begun to operate and have disrupted the cooperation that existed between traders. Although many farmers continue to sell to their traders, other options have opened up that resemble those used in areas of more intensive production. In Laikang, Takalar, for example, farmers report that in a two-year period they may sell to up to fifteen traders. Traders in Laikang offer two prices – a higher price for farmers without debt, and a lower price (reduced by around Rp. 1,000/kg (~ AU$0.10)) for farmers with debt. In this way, the system of loan provision in Laikang acts as a de facto interest-bearing system, set within a more competitive marketing system. As a result, farmers in Laikang widely view the financial services provided by traders less positively than those in Pitu Sunggu. One Laikang farmer told us that he preferred to borrow money from a local pawnshop rather than from traders, reinterpreting the motto of the pawnshop as he quoted it back to us: 'solve problems without [creating] problems'. Increased competition between buyers on price benefits farmers, but it may also erode the ability of traders to flexibly provide financial services, which may have unintended negative impacts.

## Managing quality

Junior and senior traders alike must carefully manage seaweed quality to protect their profit margins. Seaweed is normally not sold in its 'wet', living state, but dried to produce 'raw dried seaweed' (RDS). The moisture content of seaweed has a critical impact on the value of the dried seaweed product, since it effectively changes the quantity of 'weed solids' that the buyer actually receives. As a result, drying practices are a key issue for the industry. They are a major source of risk for traders because seaweed may shrink (lose moisture weight) during shipping and storage, meaning that the effective quantity drops over time (Mulyati et al. 2020; Mulyati and Geldermann 2017; Langford, Turupandang et al. 2023; Langford, Waldron et al. 2023). Our study undertook a nationwide analysis of seaweed prices over a ten-year period and found no consistent correlation between price and moisture content, even after controlling for long-term trends and seasonal patterns (Langford et al. 2022).

Indonesian national standards for dried seaweed specify that RDS should have a maximum moisture content of 38 per cent and an impurity content of 3 per cent (including dirt, sand, barnacles, etc.) (Standar Nasional Indonesia 2018). Figure 9.1 illustrates how changes in the moisture and dirt content affect the total volume of seaweed which must be bought to acquire the same amount of actual weed content. In order to obtain one kilogram of weed solids, at the national standard moisture and dirt content of 38 per cent moisture and 3 per cent dirt, a trader needs to buy 1.67kg of seaweed. However, if the farmer is diligent and sells seaweed with only 2 per cent dirt content and 35 per cent water content, one kilogram of weed solids will be found in only 1.59kg of seaweed, such that the farmer will lose 5 per cent of potential revenue. On the other hand, if the farmer does not dry the seaweed well enough and sells the seaweed at 5 per cent dirt content and 45 per cent moisture content, he will be paid for 2kg of seaweed – an increase in sales value of 20 per cent – unless the trader reduces the price paid for the seaweed. Traders normally reduce prices by around Rp. 2,000/kg (~ AU\$0.20) (4–10 per cent depending on the current prices) if the moisture content does not meet the standards and may reject the seaweed if the moisture content is excessively high. This means that farmers have an incentive to dry their seaweed to the minimum acceptable moisture reduction – and no drier – as they will lose revenue for doing so.

Our study of farmers' use of drying technologies (Stone et al. 2023) showed that many farmers were aware of improved drying technologies, and had been encouraged to use these in the past. These include oven-technologies (similar technologies had previously been used in the area for rice) and hang-drying techniques (which many farmers had previously trialled). However, farmers had little interest in these technologies for two reasons. First, they felt they were ineffective – ovens were expensive to operate, and the hang-drying method made the seaweed very difficult to remove from the lines once dry. Second, they felt that hang-drying in particular dried the seaweed *too well* – that is, it created drastic reductions in moisture

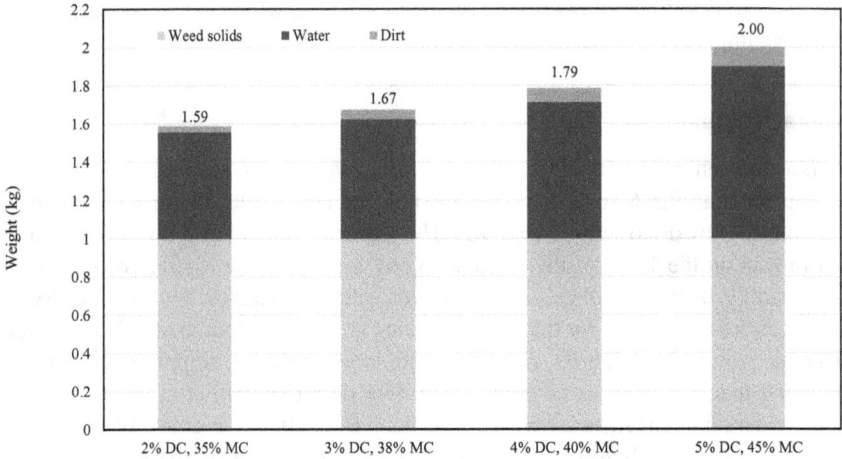

*Figure 9.2* Weight of seaweed required to retrieve one kilogram of seaweed solids under varying levels of moisture content (MC) and dirt content (DC)

*Source:* Authors' schematic.

content – and therefore seaweed weight – which was not adequately compensated for by an increase in price.

In our household survey (Langford et al. 2024), most farmers did not think that 'quality' had a strong impact on prices. In Pitu Sunggu, 27 per cent of farmers told us that they did not think that moisture affected the price of seaweed 'at all', while 68 per cent thought that it would impact the price 'a little bit'. Only 4 per cent of farmers said they thought that moisture impacted the price 'quite a bit', and no farmers thought that it would impact the price a lot. These figures are similar for dirt content, where 34 per cent of farmers believed it had no impact, 59 per cent a little impact, 6 per cent quite a bit of impact and 1 per cent a lot of impact. Interestingly, these results are different from those in the same survey in Laikang, where farmers are much more likely to report that quality indicators have a significant impact on price. This probably reflects the differences in marketing systems in these areas. As described above, Laikang is a much larger seaweed-producing area with many (>15) traders, has a more market-based system of borrowing and price formation and there is less reliance on close connections between traders and farmers.

Pitu Sunggu farmers varied in the care that they took preparing their dried seaweed. Some farmers expended considerable effort drying and cleaning their seaweed. One of the largest farmers in the village showed us a purpose-built sieve that he had constructed to clean the seaweed. He told us that he always cleaned the seaweed carefully to remove all of the dirt, because he didn't want to be seen as being dishonest in claiming to sell a 100kg sack of seaweed, but which in fact contained several kilograms of dirt. He believed that his farming success was a reward for his honest practices, as he said, '[P]erhaps we always have a good harvest because, you know, we always sell only the clean product'. There are a number

*Figure 9.3* Farmer diligently cleaning seaweed for sale
Source: Image by Radhiyah Ruhon, July 2022, Pitu Sunggu.

of like-minded farmers in Pitu Sunggu. On one occasion we observed one farmer spending several days carefully cleaning a sack of seaweed (Figure 9.3).

Other farmers took little care over quality. One farmer was frequently observed during the dry season harvesting his seaweed from muddy waters close to the shore, and then immediately laying it out on the pier for drying without first washing the mud off. Other farmers noted that the muddy seaweed would not only have high dirt content but also high moisture content due to poor drying (since the mud would impair evaporation of the water). The farmers pointed out the muddy seaweed had been drying on the pier for ten days, but was still not dry. Farmers criticised this practice as they worried that the bad practices of this farmer would cause a drop in price for everyone if it gave seaweed from the region a bad reputation. In economic terms, information asymmetries between sellers and buyers on quality attributes leads to cases of adverse selection that devalues the market to the detriment of the good actors. Overall, farmers in Pitu Sunggu exhibit a mix of individualistic and collective behaviours which result in dried seaweed of different moisture and dirt content being produced.

## Regional traders: Makassar warehouses

Makassar warehouses are larger, inter-regional traders that aggregate product from small village traders and sell to processors or exporters, or export themselves. Local traders in Pitu Sunggu sell to warehouses or carrageenan processors in Makassar, so are responsive to the prices and quality requirements of these buyers. As described above, traders must manage the moisture content of the seaweed they buy to avoid losses. For example, if they purchase 100kg of seaweed at Rp. 30,000/kg (~AU$3.00), they have paid 3 million Rupiah for the sack. However, if this sack contained even the slightly higher moisture content of 40 per cent, after shrinking to the required moisture content of 38 per cent (either during storage or after trader redrying), the sack will weigh only 94.3kg. If sold to a buyer in Makassar at a higher price of Rp. 31,500/kg (~ AU$3.15) the revenue will be 2.97 million Rupiah (~ AU$297) – a loss, even before accounting for expenses incurred. Given the small margins, it is critical that traders estimate and account for moisture content accurately.

To avoid losses, senior traders in the village strictly cut prices when standards were not met. These traders reported that they could easily tell the moisture content of seaweed brought to them just by looking at it. As one experienced trader told us, 'I never learned [formally], only from experience, I already know the water content by just holding the seaweed without using a measuring instrument'. Other techniques for assessing moisture content include squeezing it, looking for the presence of salt crystals (which traders report often indicate a moisture content below 38 per cent) and breaking open the thallus of the seaweed to check for moisture inside. Traders report that they do not have trouble accurately assessing the moisture content of the seaweed in this way. It seems likely that traders would have extensive experience making these judgements as they repeatedly see their seaweed being measured and sometimes discounted at the next stage in the chain. However, the preparation and bagging of the product remains highly variable and any subjective measurement is highly prone to inaccuracy. There may be opportunities to develop low-cost measurement devices that could support accurate assessments (e.g. Cozzolino et al. 2023).

Although price margins appear slim, it appears that traders regularly buy seaweed from farmers that is better than the minimum quality standard in terms of moisture content required by Makassar buyers. In this case, traders benefit by lowering shrinkage rates, but do not receive higher prices from Makassar buyers as a reward for exceeding quality standards. As a result, like farmers, traders also aim to supply seaweed to warehouses that meets, but does not exceed, the quality standards set by them. Traders in both Laikang and Pitu Sunggu told us that it was common practice for local traders to mix too-wet and too-dry seaweed to achieve the commercially optimal moisture content. As one person told us, '[T]he traders look for good quality seaweed and then mix it with the bad ones … Like my nephew, he takes the seaweed in Tarakan [North Kalimantan], where the seaweed is a little wet, then he mixes it with the dry seaweed here.' Often this means mixing sacol and cottonii species together, because sacol has a thicker stem that is harder to dry.

Traders reported that the extent to which excessive moisture or dirt content would lead warehouses to reject seaweed or reduce prices varied based on the demand for seaweed at the time. Major carrageenan producers accept orders from buyers in advance, and so must buy sufficient seaweed to fill their existing orders, which sometimes means accepting seaweed at increased prices and reduced quality. One local Pitu Sunggu trader observed that when seaweed was in short supply, quality parameters were relaxed, whereas when supply was sufficient (as indicated by lower prices), more stringent requirements were enforced:

> when the price of seaweed goes down, the factory has a lot of wishes or imposes a lot of standards … I often bring seaweed with a good level of dryness, but the company still considers it wet … It's very different when the price of seaweed goes up – the company immediately accepts whatever seaweed I bring to sell to them.

The price-grade conditions in warehouses are passed back down the chain. If village-based traders observe that quality criteria have been relaxed, they too can relax quality criteria for their purchases from farmers. As a result of this transmission of standards, farmers observed that as prices rose through 2021–2022, quality criteria relaxed and traders were not as strict with them. This suggests that 'quality' issues widely reported by industry are closely linked to warehouse signals, rather than to a lack of farmer capacity to meet quality standards.

Pitu Sunggu traders told us that they normally sell to the same warehouse, but occasionally sell to two or three others. Traders select warehouses for several reasons. One reason is price, however, village-based traders reported that price differences between warehouses tend to be small (Rp. 200–500/kg (~ AU\$0.02–\$0.05)) and as such this is not the main factor driving their choice of buyer. The main reason given by all village-based traders is that they have personal contacts at the warehouse which mean that have a more positive selling experience in those locations. Trust between Makassar warehouses and village-based traders facilitates smoother transactions. This is valued especially by senior traders where they are known, their staff are treated well, their seaweed is processed quickly, payment is smooth and timely and quality requirements are clear and enforced consistently.

Newer traders also prefer to sell at warehouses where they are known, although for slightly different reasons. They also note that they are 'treated better' at warehouses in which they have existing contacts, meaning that the transactions are processed quickly. One new trader told us he was made to wait for many hours with his seaweed before it was weighed when he attempted to sell at a new location. He felt angry and humiliated by the experience and did not return to that buyer. Sometimes if the new traders had seaweed which was not of a high quality, they would select a buyer that they knew often accepted lower quality seaweed. One new trader explained that they often returned to one of their preferred buyers because they 'still accept moist seaweed – the inspection is not very strict'. In general, however, village-based traders were anxious to avoid being rejected at the warehouse so sought to ensure quality standards were met. Other reasons for warehouse selection include the convenience of the warehouse location, which allowed for shorter travel times, and the speed of payment, which ranged from a few days to several weeks. Some warehouses also offer loans to village-based traders through the warehouse receipt system (see Permani et al. 2023), and in such cases, the choice of buyer is dictated in advance as traders are obliged to sell to their lender, mirroring the system of lending between village-based traders and farmers.

## Conclusion

The marketing system described in this chapter transmits price and other signals from Makassar buyers to village-based traders to seaweed farmers, about the quantity and quality of seaweed demanded. When demand is strong, buyers offer higher prices and also relax quality criteria. The system appears to function effectively in triggering supply responses in terms of production volumes, but less well for quality characteristics. Indeed, the system incentivises local traders to mix high and low quality seaweed together (to meet minimum standards at highest weight), and incentivises farmers to end the drying process, not when the seaweed is fully dry, but when it meets minimum standards. In the absence of significant price-grade differentials, neither local traders nor farmers seek to maximise the quality of the seaweed they deliver. Furthermore, they have no incentive to use technologies that increase quality, such as improved drying techniques and technologies. It may be possible for warehouses to improve the quality of the product they receive by using buying schedules where price differentials both reward and discount for quality, but this would impose increased transaction costs, especially for quality measurement which, if done off-farm, would delay payment to farmers.

Local traders capture supply from farmers by offering financial and procurement services. Farmers appear to be relatively satisfied with the arrangement and many are grateful for these services. Points of dissatisfaction for farmers include lower prices perceived to arise from collusion between traders.

## Note

1 To avoid identifying the traders, they will be referred to in gender neutral terms in this chapter, including in quote translations.

## References

Adharini, Ratih Ida, Eko Agus Suyono, Suadi, Anes Dwi Jayanti, and Arief Rahmat Setyawan. 2019. "A Comparison of Nutritional Values of *Kappaphycus alvarezii*, *Kappaphycus striatum*, and *Kappaphycus spinosum* from the Farming Sites in Gorontalo Province, Sulawesi, Indonesia." *Journal of Applied Phycology* 31 (1): 725–730. https://doi.org/10.1007/s10811-018-1540-0

Aslan, La Ode M., Nur Isiyana Wianti, Siti Aida Adha Taridala, Manat Rahim, Ruslaini, and Wa Ode Sifatu. 2022. "The Debt Trap of Seaweed Farmers: A Case Study from Bajo Community in Bungin Permai, Indonesia." IOP Conference Series. *Earth and Environmental Science* 1033 (1): 12056. https://doi.org/10.1088/1755-1315/1033/1/012056

Cozzolino, Daniel, Mohamad Rafi, Wahyu Ramadhan, Rudi Heryanto, Scott Waldron, and Zannie Langford. (2023) "Developing a Rapid Assessment for Seaweed Qualities". Melbourne, Australia: Australia-Indonesia Centre. https://pair.australiaindonesiacentre.org/wp-content/uploads/2023/09/Final-Report_TWP5_ENG_Developing-a-rapid-assessment-for-seaweed-qualities.pdf

Fausayana, Ine, Darmawan Salman, M. Saleh. S. Ali, Rahim Darma, Sitti Nurani Sirajuddin, and Akyyar. 2014. "Lending Models of Seaweed Farming of Bajo Community."

*Australian Journal of Basic and Applied Sciences*, 8 (7): 434–440. https://www.cabdirect.org/cabdirect/abstract/20143305036

Komarek, A., E. R. Cahyadi, J. Zhang, A. Fariyanti, B. Julianto, R. Arsyi, I. Lapong, A. Langford, S. Waldron, and M. Grist. 2023. *Increasing Incomes in Carrageenan Seaweed Value Chains in Takalar, South Sulawesi*. Melbourne, Australia: Australia-Indonesia Centre.

Langford, Alexandra, Scott Waldron, Nunung Nuryartono, Syamsul Pasaribu, Boedi Julianto, Irsyadi Siradjuddin, Radhiyah Ruhon, Zulung Zach Walyandra, Imran Lapong, and Risya Arsyi Armis. 2023. *Sustainable Upgrading of the South Sulawesi Seaweed Industry*. Melbourne, Australia: Australia-Indonesia Centre. Available at https://pair.australiaindonesia-centre.org/research/sustainable-upgrading-of-the-south-sulawesi-seaweed-industry-2

Langford, Alexandra, Scott Waldron, Jing Zhang, Radhiyah Ruhon, Zulung Zach Waly-andra, Risya Arsyi Armis, Imran Lapong, Boedi Julianto, Irsyadi Siradjuddin, Syamsul Pasaribu, and Nunung Nuryartono. (2024). "Diverse Seaweed Farming Livelihoods in Two Indonesian Villages." In *Tropical Seaweed Cultivation – Phyconomy: Proceedings of the Tropical Phyconomy Coalition Development, TPCD 1, Held at UNHAS University, Makassar, Indonesia. July 7th and 8th* edited by A. Hurtado, A. Critchley, and I. Neish. 00-00. New York: Cham, Switzerland: Springer Nature.

Langford, Alexandra, Jing Zhang, Scott Waldron, Boedi Julianto, Irsyadi Siradjuddin, Iain C. Neish, and Nunung Nuryartono. 2022. "Price Analysis of the Indonesian Carrageenan Sea-weed Industry." *Aquaculture* 550 (737828). doi.org/10.1016/j.aquaculture.2021.737828

Mulyati, Heti and Jutta Geldermann. 2017. "Managing Risks in the Indonesian Seaweed Supply Chain." *Clean Technologies and Environmental Policy* 19 (1): 175–189. https://doi.org/10.1007/s10098-016-1219-7

Mulyati, H., J. Geldermann, and T. Kusumastanto. 2020. "Carrageenan Supply Chains in Indonesia." IOP Conference Series. *Earth and Environmental Science* 414 (1): 12013. https://doi.org/10.1088/1755-1315/414/1/012013

Nor, Adibi M., Tim S. Gray, Gary S. Caldwell, and Selina M. Stead. 2020. "A Value Chain Analysis of Malaysia's Seaweed Industry." *Journal of Applied Phycology* 32 (4): 2161–2171. https://doi.org/10.1007/s10811-019-02004-3

Nuryartono, N., S. Waldron, A. Langford, Sulfahri, K. Tarman, S. H. Pasaribu, U. J. Siregar, and M. F. D. Lusno. 2020. *A Diagnostic Analysis of the South Sulawesi Seaweed Industry*. Melbourne, Australia: Australia-Indonesia Centre.

Permani, Risti, Yanti N. Muflikh, Fikri F. Sjahruddin, Nunung Nuryartono, Scott Waldron, Alexandra Langford, and Syamsul Pasaribu. 2023. "The Policy Landscape and Supply Chain Governance of the Indonesian Seaweed Industry: A Focus on South Sulawesi." Melbourne, Australia: Australia-Indonesia Centre.

Picaulima, Simon, Syahibul Hamid, Anna Ngamel, and Roberto Teniwut. 2016. "A Model for the Development of the Seaweed Agro Industry in the Southeast Maluku District of Indonesia." *Eurasian Journal of Business and Management* 4 (4): 46–55. https://doi.org/10.15604/ejbm.2016.04.04.005

Prasetia, Hafiizh, Firman Zulpikar, Jeverson Renyaan, Muhammad Safaat, and Ary Mauliva Hada Putri. 2022. "The Impact of Covid 19 on Seaweed Smallholder Farmers in Nusa Tenggara Barat, Indonesia." *Omni-Akuatika* 18 (S1): 96–100. https://doi.org/10.20884/1.oa.2022.18.S1.986

Rahim, M., L. O. M. Aslan, S. A. A. Ruslaini, N. I. Taridala, A. Wianti, Budiyanto Nikoyan, and H. Hafid. 2019. "Livelihood Features of Seaweed Farming Households: A Case Study from Bungin Permai Village, South Konawe, South East (SE) Sulawesi, Indonesia."

IOP Conference Series: *Earth and Environmental Science* 370 (1): 12025. https://doi.org/10.1088/1755-1315/370/1/012025

Standar Nasional Indonesia (2018). "Rumput laut kering." ICS 67.120.30. *Badan Standardisasi Nasional.*

Stone, Serafina, Zannie Langford, Imran Lapong, Risya Arsyi Armis, Zulung Zach Walyandra, Radhiyah Ruhon, Annie Wong, Boedi Julianto, Irsyadi Siradjudding, and Scott Waldron. 2023. "Technology Uptake by Smallholder Farmers: The Case of the Indonesian Seaweed Industry." *Journal of Agribusiness in Developing and Emerging Economies.* https://doi.org/10.1108/JADEE-01-2023-0011

Zamroni, Achmad. 2018. "Small Scale Entrepreneurship of Seaweed in Serewe Bay, East Lombok, Indonesia: Challenges and Opportunities." *Journal of Development and Agricultural Economics* 10 (5): 165–175. https://doi.org/10.5897/JDAE2017.0874

Zamroni, Achmad. 2021. "Sustainable Seaweed Farming and Its Contribution to Livelihoods in Eastern Indonesia." IOP Conference Series: *Earth and Environmental Science* 718: 12099. https://doi.org/10.1088/1755-1315/718/1/012099

Zamroni, Achmad and Masahiro Yamao. 2012. "An Assessment of Farm-to-Market Link of Indonesian Dried Seaweeds: Contribution of Middlemen Toward Sustainable Livelihood of Small-Scale Fishermen in Laikang Bay." *African Journal of Agricultural Research* 7 (30). https://doi.org/10.5897/AJAR11.1059

# Conclusion

*Zannie Langford*

The Indonesian seaweed industry has grown rapidly over the last twenty years, bringing socio-economic and environmental transformations to coastal villages across the country, particularly in Eastern Indonesia. This book has explored how local people in Indonesia work to produce this commodity: how they have reorganised claims to the sea, developed strategies for growing seaweed in highly variable ocean conditions, reorganised labour use both within the household and across the region, and worked to buy and sell seaweed and transport it to Makassar for processing or export. The work of these local people has been supported by industry and government actors within Indonesia and abroad. This book has sought to link these two domains together. Part I explored the global carrageenan industry, telescoping down from the global to national and provincial levels, while Part II explored in detail the livelihood transformations which have taken place to supply this global industry. This chapter reflects on the content of the book and offers policy recommendations. It describes the major findings of each chapter, and how findings can inform policy.

## The development of the Indonesian carrageenan seaweed industry

The *Preface* of this book described the early development of the carrageenan seaweed industry, and the long process undertaken by industry actors to successfully establish production in Indonesia and the Philippines. It explored the key role played by carrageenan processors in establishing seaweed farming in Indonesia, and how the introduction of low-cost, semi-refined carrageenan production technologies reshaped the industry by enabling the proliferation of low-cost processing facilities. It also highlighted the measures taken by industry and government to expand the industry to new locations, which provides insights for other tropical regions and countries globally seeking to build a seaweed industry. Significant industry or government support is required to provide the knowledge, tools and marketing systems required by farmers to develop this activity in new locations.

The *Introduction* explored global seaweed production. Cultivating plants in the ocean is appealing to people interested in sustainable development, because it seems to offer the potential to shift farming across a new frontier (the ocean), and to

DOI: 10.4324/9781003183860-13

reduce pressure on land resources. It also has the potential to provide ecosystem services and new livelihood opportunities for coastal households. Despite this interest, there is relatively little understanding of what the global seaweed industry actually looks like, which leads to a gap between ideals and reality. *Chapter 1* introduced fundamental information on the global industry. It highlighted that seaweed farming is a relatively new activity globally, and that current global marine seaweed production is dominated by just six species of seaweed. Around 55 per cent of seaweeds are grown to produce the food seaweeds used for nori, wakame and kombu, with 40 per cent grown to produce hydrocolloid containing seaweeds, including *Eucheuma* and *Kappaphycus* for carrageenan production, and *Gracilaria* for agar production. Currently, there is very limited cultivation of other species, owing in part to a lack of understanding of how to grow and process them efficiently, and of their environmental requirements and impact. While there is considerable worldwide excitement about the potential for seaweed to store carbon, reduce methane emissions and increase food security, this needs to be checked against the feasibility of producing those seaweeds in different environmental, geographic, socio-economic and policy contexts.

*Chapter 1: The global carrageenan market* set the macro level scene for other chapters of the book. It compared carrageenan with other major hydrocolloids used for gelling, thickening and stabilising purposes in foods, cosmetics and pharmaceuticals such as gelatin, cellulose gum, pectin, xanthan gum and Arabic gum. Each of these hydrocolloids performs different functions and is used in specialised combinations to provide goods with specific textures and consistencies. These hydrocolloids can complement and substitute for each other to different extents in different applications. The main use of carrageenan currently is in the food and beverage industry, with nearly half of all carrageenan being used in meat and dairy applications. More than half of the market is a lower purity form of carrageenan known as semi-refined carrageenan (SRC) used in a range of low value, non-food applications including pet food, but it is increasingly used in food products. Most carrageenan used is *kappa-carrageenan* and *iota-carrageenan*, derived from the species of seaweed grown in Indonesia. Indonesia dominates global carrageenan seaweed production and the majority is processed in China, or by Chinese companies operating in Indonesia. There has been a trend towards increased processing in countries of origin, a process that has been expedited by Chinese direct investment in the Indonesian seaweed processing sector. The benefits for domestic value adding and employment have to be weighed up against the negative effects of competition on local processors, and pose questions for government on policies towards foreign direct investment (FDI), including on tax and environmental standards. The data and analysis presented in the chapter suggest that carrageenan markets are likely to continue to grow, given demand settings for processed food globally and in developing countries, but also in a wide range of other products to form a diversified market. While the hydrocolloid sector is highly competitive, carrageenan has sought-after performance characteristics. This bodes well for the governments, companies and households of Indonesia that have invested in seaweed production or plan to.

*Chapter 2: The Indonesian seaweed industry* analysed the industry at a national meso level, between the global and provincial levels. The industry is relatively new, and has only had significant commercialisation since the 1990s. The early development of the industry was driven largely by multinational companies seeking to develop a new production base. The production base has become increasingly atomised. Seaweed farming is completely dominated by small, individual households that make their own management decisions and innovations, with a dearth of corporatised seaweed farms or estates. Households deal with local traders with whom they have close personal or reciprocal relationships but there is a dearth of formal, contract governance systems. Attempts have been made to corporatise the production and marketing systems but smallholder-based production systems and 'spot' marketing systems are more efficient. These pro-poor characteristics are also a source of international competitive advantage. Indonesia has for many years been the world's largest producer of carrageenan seaweeds. Indonesian policy aims to consolidate or expand this position as a seaweed producer and, furthermore, to increase the share of it used for domestic processing to meet value adding and employment objectives. Government has sought to expedite the process through the tools of industry policy (export bans, subsidies for domestic processing) with adverse effects. However, Indonesia is meeting the objective of building a domestic processing base through FDI, especially from China. Chinese-invested companies in Indonesia account for perhaps half of the actual output of Indonesia's carrageenan processing sector and companies in China process virtually all of Indonesia's exports of raw dried seaweed. While government has sought to play a mediating role, industry development and conduct is driven by powerful forces that directly and profoundly affect local governments and households.

*Chapter 3: The South Sulawesi seaweed industry* explored the provincial industry and value chain. Even given statistical qualifications (Appendix 1) the industry appears to have grown strongly over the last two decades. The chapter described how seaweed produced in this region is combined with production from other areas in Eastern Indonesia as it makes its way to Makassar for processing, packing and export. The chapter demonstrated how value is added to seaweed at different points of its transformation and highlighted the central role of South Sulawesi in the Indonesian seaweed industry. Lessons from the growth and development of the leading province may be applicable to other areas seeking to grow seaweed industries.

*Chapter 4: Export commodity frontiers and the transformation of village life* explored how the village of Pitu Sunggu has been transformed by successive waves of export commodity 'booms', leading to rapid and long-lasting environmental and socio-economic changes. The residents of Pitu Sunggu are almost entirely dependent on export commodity production for their livelihoods: very little food is grown there. The village landscape was transformed from mixed cropping to field rice production in the early twentieth century, from field rice to wet rice during the Green Revolution of the 1970s, from wet rice to shrimp and fish farming from the 1980s and, most recently, the sea from communal fishing grounds to seaweed farms from the early 2000s. These export-oriented transformations have led to increased incomes for many residents. Nowadays, most of them earn the majority of their income through either seaweed farming, pond farming of shrimp and fish or fishing for blue swimmer crabs and market fish. Although people consume wild-caught marine fish and farmed fish and shrimp, most of these products are destined for export markets. Pitu Sunggu residents are therefore impacted heavily by global markets, demand and prices. This has created a situation of high but often volatile incomes for farmers, including periods of 'boom and bust' as production of different commodities has expanded and contracted in response to market signals and changing government incentives (such as those affecting the availability of fertiliser). These booms and busts have also affected migration patterns. People return to the village when prices and incomes are good (as has been the case with seaweed farming in recent years) and travelled abroad or within Indonesia to work when livelihood options were limited. This chapter highlighted the historical importance of export crops to livelihoods in this coastal village, and the close relationship between export commodity prices and social and economic organisation.

*Chapter 5: From communal access to private ownership: negotiating rights to the sea* explored how seaweed farmers in Pitu Sunggu transformed their offshore sea area tenure from communal fishing grounds into parcels of individually 'owned' plots of sea space. It traced how the first seaweed farmers established plots and sought to exclude fishers from their farming areas, and how this was contested for many years. Over time, as more and more people established plots, resistance decreased, until eventually most people who had resisted seaweed farming had become seaweed farmers themselves, and the rights of individuals to claim exclusive ownership of sea space by installing farm markers became widely recognised. It described how today seaweed farming plots in Pitu Sunggu are rented, bought and sold for high prices, as well as the types of conflicts that persist between different seaweed farmers, between farmers and fishers and over encroachment of seaweed farms on boat lanes. It highlighted how these use informal arrangements which, while not formally recognised by government, are widely recognised and enforced within Pitu Sunggu. It also cited examples of other seaweed farming areas which have established similar rules for sea use. In some cases, local rules limit the number of plots an individual farmer may own, the locations which may be claimed for farming and the hours during which farms may be operated. In Pitu Sunggu, many farmers have claimed large areas of sea space, and rely on casual wage labourers to keep their areas productive. The system of property rights has therefore led to the creation of a labour market and, in turn, this labour market is essential to valorising the claims that seaweed farmers have on the sea. Without their work, it would be impossible to farm such large areas. The chapter highlights how seaweed farming has required a reconfiguration of local understandings of sea space access and rights in ways that have benefited some community members and disadvantaged others, such as crab fishers who were excluded from what was previously crab fishing grounds.

*Chapter 6: Environmental and socio-economic constraints to marine farming* and *Chapter 7: Farmer decision-making in the Indonesian seaweed industry* explored the range of environmental and socio-economic factors affecting farmer decision-making and farm performance. Chapter 6 described how ocean conditions such as surface water temperature, salinity, light penetration, water motion, nutrient levels, turbidity, ephiphytes, grazing wildlife and pollution impact upon seaweed growth. These conditions create distinct seasonal patterns of cultivation that vary between areas. Chapter 6 also examined how socio-economic factors such as access to sea space, access to labour, access to capital, seaweed market prices and risk of theft also constrain the production opportunities available to individual farmers. Chapter 7 explored how farmers respond to these environmental and socio-economic constraints and opportunities by altering their production strategies in order to optimise production throughout seasonal changes in ocean conditions and to manage the risk of ice-ice disease, including by varying the species and location of seaweed planted and the timing of it throughout the year, their farm maintenance patterns, propagule use, float use and rope use. It showed that many farmers use a strategy of geographic diversification, in which they farm multiple plots in different locations in the sea, which both minimises the risk of total seaweed loss due to ice-ice disease in any one plot, and provides flexibility to move seaweed between plots in response to different growth rates in different locations.

These chapters also described propagule use dynamics, particularly the issue of propagule 'quality' which is widely identified as a major problem facing farmers. It showed that in Pitu Sunggu, propagule 'quality' is primarily a seasonal problem. The issue is not a lack of genetic material for propagation as is sometimes assumed, but poor environmental growing conditions at certain times of year which mean that farmers have trouble producing any seaweed at all, including to use as propagation material. Propagule 'quality' generally improves as oceanic conditions become more suitable to seaweed cultivation in different seasons of the year. There is also a long-term decline in farm productivity noted by farmers, however, this is likely to be at least in part a result of changing ocean conditions rather than an issue of declining propagule 'quality' as is sometimes assumed. As seaweed farming in Pitu Sunggu has intensified over time, more seaweed competes for nutrients in the waters off-shore of the village and as such it is likely that declining yields are linked to the increasing area under cultivation – and possibly also to declining fertiliser use in coastal shrimp ponds.

These chapters also explored the role of farmer decision-making in farm productivity, showing how successful farmers monitored their farms almost daily and made small changes in farm apparatus in response to their observations. For example, farmers increased the number of plastic bottle floats if they noted that seaweed growing closer to the floats grew better than that further away (which was slightly deeper). They reduced rope tension in response to increased water motion to reduce breakages and increased or reduced propagule size in response to risks of breakages from waves in different seasons. They harvested seaweed early to prevent theft or breakage, and moved seaweed between locations in response to epiphytes, mud or grazing by fish. They brought seaweed to shore for cleaning if epiphytic growth was excessive, or harvested and replanted undamaged portions of it if damage was excessive. Farmers make a range of decisions designed to maximise yield, but also – and often more commonly – to minimise the risk of losses, which are frequent in certain seasons, due to ice-ice disease and from breakages from wave motion. The chapter explored how diligent farm management practices are a key component of the performance of seaweed farms.

The chapters also explored how socio-economic factors affect farm management, including how a lack of access to sea space leads some farmers to plant at much higher densities than is recommended, how access to binding labour often affects harvest and planting timing, how binder decision-making affects the size of the propagules they bind, how the risk of theft at times of high prices affects the size of propagules farmers use and the timing of harvest. It also showed how seaweed prices drive both increasing uptake of seaweed farming by new farmers, and intensification of farming by existing farmers. The chapters showed how farm performance and management decisions are closely linked to socio-economic factors and, how these must be considered in monitoring performance of existing farms.

*Chapter 8: Gendered work and casual labour in the Indonesian seaweed industry* explored the range of different types of labour that are employed in the seaweed industry in Pitu Sunggu. It highlighted the gendered nature of on- and off-farm jobs in the industry, with men primarily involved with on-farm work, and women involved in the many preparatory and post-harvest activities. It also highlighted the role of casual wage labourers in the industry, often large numbers of people from inland villages who work as seaweed binders. These binders are often people with few other livelihood opportunities, such as women with caring responsibilities, widows and people with disabilities. In addition, many people find work 'gathering' seaweed that has broken off farming ropes and washed ashore. This chapter highlighted the way that seaweed is incorporated into the livelihoods of a diverse range of households, in both seaweed-farming and non-seaweed-farming villages, and suggested that people who do not farm seaweed but work in the industry indirectly should also be considered stakeholders.

*Chapter 9: Seaweed marketing: village-based traders as financial and market intermediaries* explored the local seaweed marketing system in Pitu Sunggu, in which local traders receive seaweed from farmers and sell it to Makassar-based warehouses. It described how senior traders in the village capture seaweed supply by providing financial services to farmers, but how with rising seaweed prices, new entrants to the industry are trying to disrupt this structure. Farmers and traders alike report that they are able to accurately assess seaweed moisture levels, but do not necessarily strive to deliver higher quality characteristics, including moisture. It highlighted how price signals from Makassar are transmitted to traders and farmers which appears to be an effective method of generating supply responses, but not of increasing seaweed quality, which is of primary interest to processors. Both traders and farmers have incentives to meet minimum quality standards, but not to exceed them. If their seaweed is drier than minimum standards, it will often be mixed with wetter seaweed to increase weight and sales revenue. This chapter suggested that farmers and local traders have the capacity to supply better qualities of seaweed, but that this would require a buying schedule with higher price-grade differentials, and the incentives for all the actors in the chain to use it.

## Policy recommendations

The chapters of this book telescoped down from macro- to meso- and micro-level analysis, with an emphasis on socio-economic and environmental factors in the decision-making of farmers. Several policy implications follow from this:[1].

*Policy recommendation 1: Development of alternative seaweed products should be subject to economic feasibility analysis*

Chapter 1 reviewed the global market for carrageenan. The Indonesian government aims to diversify the market for carrageenan seaweed products including producing foods, fuels, fertilisers, animal feeds and carbon capture. In addition, it has tried to develop new products from other seaweed species found in Indonesia. These products do not yet have a market, and because of this, it is unclear whether such a market can be established or it is commercially viable for households and companies to service such a market. It is recommended that biophysical research and development into alternative seaweed markets, products and species is integrated with market research and economic feasibility analysis. A technoeconomic analysis such as this could guide research and development programmes and policies by providing insights into the most economically feasible product development options.

*Policy recommendation 2: Consider the environmental impacts of FDI in processing*

The Indonesian government has sought to improve processing technology in-country by incentivising FDI in seaweed processing plants. However, these plants were approved with liberal tax and investment terms, and carrageenan processing can have a considerable environmental impact. Now that Indonesia has established a significant carrageenan processing base, it is recommended that future investments are carefully evaluated across social, economic and environmental criteria to maximise benefits to Indonesia.

*Policy recommendation 3: Carefully consider the costs and benefits of industry policy*

After the boom that occurred during the course of this study, seaweed prices began a significant decline in October 2022 (Langford et al. 2022). During periods of price decline in the Indonesian seaweed industry, it has frequently been suggested that there should be pricing intervention to support farmers. During periods of price increases, it has also been suggested that domestic processors should be supported by capping prices and, indeed, such a policy was implemented in Nusa Tenggara Timor (NTT) in 2022 (Langford et al. 2023). These two types of price interventions serve different purposes. Care should be taken when intervening in markets, however, as interventions often have unintended impacts. Whether such impacts justify welfare benefits to farmers or investors (processors), under what conditions and over what timeframes, should be carefully considered before any price or trade interventions are announced or implemented.

*Policy recommendation 4: Marine spatial planning*

The findings in Chapter 5 have important policy implications for marine zoning. As the Indonesian government seeks to expand seaweed farming into new areas of eastern Indonesia, these insights suggest that advanced sea-space planning and marine zoning could support more equitable access to seaweed farming areas. Emerging production areas could, for example, pre-emptively set boundaries for sea space for not just seaweed farming, but also boat lanes, fishing activities, environmental reserves (particularly near sensitive marine ecosystems such as coral reefs and seagrass meadows and fish habitats) and tourism. Other factors that could be considered include whether limits on farm areas are desirable, and whether formal recognition of use claims by farmers would be desirable. The workshopping of marine planning and zoning areas in advance of wide establishment of seaweed farming in new regions is appropriate. In existing regions, such work is complicated by existing claims to sea space by community members, and any attempt at intervention should be undertaken through close consultation with community members to avoid generating hardship and unrest.

*Policy recommendation 5: Focus on propagule distribution rather than solely on genetics*

The problem of propagule 'quality' is not solely one of genetics. Seaweed is clonally propagated so that much of the variation in propagule quality is closely linked to the oceanographic conditions in which they are grown. When farmers complain about propagule quality, they complain about the availability of propagules or that they are small, old, exhibit colour that indicates poor health or are damaged. In some locations, the establishment of propagule nurseries in areas with good year-round growing conditions can support improved propagule availability, however the issue of quality will remain largely linked to seasonal variation in the suitability of different coastal areas to seaweed farming. Efforts to support supply of quality propagules should focus on production and distribution of quality seaweed propagules, rather than solely on technological developments.

*Policy recommendation 6: Focus on plastics end-of-life management rather than plastic float alternatives*

Plastic floats are used widely in seaweed farming because they are highly suitable for the purpose. They are light so they are easy to transport, they move easily with currents so reduce breakages and can be filled with water to sink seaweed below the surface of the ocean to minimise losses after rain. Any attempt to reduce the use of these floats would need to offer a suitable alternative to them with similar features at low cost. At present, no such options exist. However, significant improvements in plastic management could be made by supporting farmers' end-of-life management of plastics. Many used bottles from seaweed farming are at present thrown into the sea or river, or left to decompose on vacant land. The rate of plastic degradation increases over time, such that used plastic bottles may represent a greater threat to marine ecosystems than those used on farms (because bottles used on farm are newer, and therefore degrade more slowly than older bottles which have been discarded). Improved end-of-life management of bottles therefore represents an important first step in improving plastic management from seaweed farming. It is recommended that end-of-life plastics management be prioritised as a key step towards reducing the contribution of seaweed farming to ocean plastics.

*Policy recommendation 7: Build recognition of the importance of local knowledge into farm extension programmes*

The government extension system would benefit from the incorporation of local knowledge into research, extension and training services. While several important technological gains have been taken up (such as the use of bamboo drying platforms for part of the drying process, and the use of the twin-rope method to reduce breakages), advice or technologies that are general in nature and not adapted to local realities are unlikely to be taken up. Farmers are intimately tuned in to their local environment and can assess whether technologies are suitable and beneficial. Closer engagement with farmers and their complex decision-making processes – as outlined in this book – can support more effective farm extension services.

*Policy recommendation 8: Review aid distribution system to reduce conflicts between farmers*

Government aid has caused significant conflict between seaweed farmer group members. The aid provided is often only sufficient for a few of the group members, and distribution between members is often uneven. This has led to frustration by many farmers and a lack of interest in joining farmer groups or working cooperatively. In addition, the government assistance provided is often not fit for purpose. For example, programmes may disseminate the wrong type of boat or boat motor or the wrong size or quality of rope. A review of government and group programmes directed at the seaweed industry (and indeed other industries) is recommended.

*Policy recommendation 9: Develop seaweed growth models and remote sensing technologies to build an understanding of seaweed growing locations and yields and the impact of climate change*

There is a lack of data on seaweed growing patterns and how these relate to local ocean conditions. This could be addressed through new technologies, such as by combining analysis of satellite imagery of farm locations with in-situ measurement devices to develop a model of seaweed growth under different growing conditions. Such developments could provide important data on seaweed growth and current production areas which are necessary to guide the development of a more sustainable and resilient seaweed industry, which may include improvements in seaweed quality and production under environmentally sustainable conditions. These developments will also be important for building resilience strategies in the face of climate change. Farmers experience frequent seaweed losses due to prolonged rainfall and high temperature events. A closer understanding of seaweed production quantities, locations and growing conditions could support more informed industry planning.

*Policy recommendation 10: Recognise casual labourers as a key group of seaweed industry stakeholders*

A range of demographic groups work in the seaweed production sector. Some are men, women and children in seaweed farming households, while others are from non-seaweed farming households, including casual labourers. These casual workers should be recognised as important stakeholders in the industry, as they include vulnerable populations such as widows, people with disabilities and women with caring responsibilities who may have no other source of income. Casual labourers working in the industry should be included in statistics, analysis and policy on the industry in order to consider the impact of development on the industry, including in non-seaweed farming areas.

*Policy recommendation 11: Quality improvements require processor-led price signals*

Farmers and village-based traders are highly attuned to the quality (primarily moisture and dirt content) requirements of processors and warehouses, but seek to avoid supplying seaweed which exceeds requirements as they would suffer decreased revenues because of the lower weights, which is the basis of payment. As such, local traders regularly mix high and low quality seaweed, and farmers avoid drying their seaweed for too long. Improved price quality signals could incentivise the supply of higher quality seaweed.

*Policy recommendation 12: Financial services for farmers could be diversified*

Farmers often use financial services from a number of different sources, including from village-based traders, bank loans, cooperatives, pawnshops and family members. Each of these sources has different loan sizes, interest rates and repayment terms, and farmers select each of these to meet their needs. Interest rates from some sources are high, while in Pitu Sunggu, village-based traders do not charge interest but require farmers to sell seaweed to them directly, sometimes at lower than market prices. Farmers prefer to use debt rather than to save their money in banks as bank charges are high relative to the amounts they seek to save. Consequently many farmers also 'save' money with traders. Digitally mediated financial products have been suggested as a way of giving farmers' access to credit. Although this is an innovative method, our research suggests that farmers have access to credit from a range of sources which suit their needs, but could benefit from a better range of savings products.

## Conclusion

This book has explored how an export commodity – carrageenan seaweed – has transformed livelihoods in one village of Indonesia. Many people have benefited from the higher incomes available through the industry, but the distribution of benefits has been uneven. The industry has had both positive and negative social and environmental impacts as it has drastically transformed the use of sea space by coastal villages. It is hoped that the research and recommendations in this book will contribute to the development of the seaweed industry – globally, in Indonesia and in other countries – and improve the livelihoods of coastal communities that participate in it.

## Note

1  Additional policy recommendations from AIC projects are outlined in our policy brief (Waldron et l. (2023) and main project report (Langford, Waldron et al. (2023).

# References

Langford, Alexandra, Welem Turupadang, and Scott Waldron. 2023. "Intterventionist Industry Policy to Support Local Value Adding: Evidence from the Eastern Indonesia Seaweed Industry." *Marine Policy* 151 (105561). doi.org/10.1016/j.marpol.2023.105561

Langford, Alexandra, Scott Waldron, Nunung Nuryartono, Syamsul Pasaribu, Boedi Julianto, Irsyadi Siradjuddin, Radhiyah Ruhon, Zulung Zach Walyandra, Imran Lapong, and Risya Arsyi Armis. 2023. *Sustainable Upgrading of the South Sulawesi Seaweed Industry*. Melbourne, Australia: Australia-Indonesia Centre. https://pair.australiaindonesiacentre.org/research/sustainable-upgrading-of-the-south-sulawesi-seaweed-industry-2

Langford, Alexandra, Jing Zhang, Scott Waldron, Boedi Julianto, Irsyadi Siradjuddin, Iain C. Neish and Nunung Nuryartono. 2022. "Price Analysis of the Indonesian Carrageenan Seaweed Industry." *Aquaculture* 550 (737828). doi.org/10.1016/j.aquaculture.2021.737828

Waldron, Scott, Nunung Nuryartono, Alexandra Langford, Syamsul Pasaribu, Kustiariyah, Tarman, Ulfah J. Siregar, Muhammad Farid Dimjati Lusno, Sulfahri, Boedi Sarjana Julianto, Siradjuddin, Ruhon, Radhiyah, Walyandra, Zulung Zach Walyandra, Lapong, Irsyadi Muhammad Imran, Risya Arsyi Armis, Eugene Sebastian, Helen Brown, Fadhilah Trya Wulandari, Hasnawati Saleh, and Steve Wright. 2022. *Policy Brief: Sustainable Upgrading of the South Sulawesi Seaweed Industry*. Melbourne, Australia: Australia-Indonesia Centre. https://pair.australiaindonesiacentre.org/wp-content/uploads/2022/11/SIP-1-EN-ONLINE.pdf

# Appendix 1

## Reconciling Indonesian seaweed industry statistics from different sources

*Zannie Langford, Radhiyah Ruhon, and Scott Waldron*

Various agencies produce the following statistical series relevant to the seaweed industry in Indonesia:

- Seaweed production volume and value
- The number of households that produce seaweed
- Seaweed and carrageenan exports and imports
- Carrageenan processing capacity and production

This appendix outlines reporting of these statistics, the differences in reporting between agencies and the possible degree of inaccuracy. The first section provides available recent data from the Ministry of Marine Affairs and Fisheries (KKP), the Ministry of Industry (Kemenperin 2022) and Statistics Indonesia (*Badan Pusat Statistik* (BPS)) and the Presidential Decree on the seaweed industry of 2019 (PER-PRES 33–2019). The second section compares and evaluates this data.

### Available statistics

#### Ministry of Marine Affairs and Fisheries (KKP and DKP)

*Production volumes*

The KKP collects seaweed production data across Indonesia and reports on an annual basis.

The KKP uses a 'bottom up' statistical collection method to estimate seaweed production, where local level DKP enumerators estimate production, area and number of participating households at local levels which are then aggregated up to national levels. Official data reports volumes of wet seaweed, disaggregated by main species. As is the case for many other commodities in other countries, the seaweed statistics are subject to inaccuracy at every step of this process. It is simply not possible for already overstretched government departments or indeed village officials to measure the weights, or indeed areas, cultivated in highly atomised industries like seaweed. The yields used derive production from estimates of quantities of dry seaweed provided by farmers and cannot account for regional, seasonal or other variations.

NATIONAL PRODUCTION (2021)

In 2021, the KKP reported to the Food and Agriculture Organisation of the United Nations (FAO) that Indonesia produces 7,058,254 tonnes of carrageenan-bearing seaweeds (*Kappaphycus alvarezii* ('cottonii'), *Kappaphycus striatus* ('sacol') and *Eucheuma denticulatum* ('spinosum') and 1,915,313 tonnes of the agar-bearing seaweed species *Gracilaria* (FAO 2023). In a separate source, the KKP reported that in 2021 Indonesia produced 7,063,738 wet tonnes of carrageenan seaweed (consisting of 4,734,868 wet tonnes of cottonii and 2,328,870 wet tonnes of spinosum) and 1,903,035 wet tonnes of *Gracilaria* (KKP 2022). The figures reported by the KKP to the FAO (2023) and separately by the KKP (2022) are similar but not identical.

PRODUCTION BY PROVINCE (2020)

The KKP (2022) reports statistics that disaggregate production to the provincial level, but the most recent year reported is 2020 (Table A1.1). For reference, this source (KKP 2022) also reports national level production statistics of 6,060,552 tonnes of cottonii, 2,030,244 tonnes spinosum and 1,456,730 tonnes of *Gracilaria*, as well as small quantities of *Sargassum* (80,662 tonnes) and *Caulerpa* (232 tonnes) species. The provincial data does not disaggregate production by species, but pertains to total seaweed production (Table A1.1).

AREA UNDER PRODUCTION (2020)

The KKP also reports area under production for the top ten seaweed producing provinces (Table A1.1).

## South Sulawesi Ministry of Marine Affairs and Fisheries (KKP SulSel)

Data collected by the KKP at the provincial level disaggregates between pond and marine seaweed production and includes estimates of the number of seaweed farming households (Table A1.2).[1] A comparison between cultivated area and production suggests that yields vary enormously (from 2 tonnes per ha to 164 tonnes per ha), implying a large statistical anomaly. The average cultivation area ranges from 0.6ha per household to 4.6ha per household.

### *Ministry of Industry (Kemenperin 2022)*

*Domestic seaweed processing*

The Ministry of Industry reported that in 2021 Indonesian processors produced 8,531 tonnes of alkali treated cottonii (ATC), 14,102 tonnes of semi-refined carrageenan (SRC), 2,423 tonnes of refined carrageenan (RC) and 3,911 tonnes of agar.

*Table A1.1* KKP statistics on seaweed production and area by province

| Province | 2020 seaweed production volume (tonnes) | 2020 production area (ha) | Species cultivated |
|---|---|---|---|
| Sulawesi Selatan | 3,442,076 | 40,317 | cottonii, *Gracilaria* |
| Nusa Tenggara Timur | 2,158,903 | 7,907 | cottonii, spinosum |
| Sulawesi Tengah | 927,787 | 15,927 | cottonii, spinosum |
| Jawa Timur | 699,236 | 217 | cottonii, *Gracilaria* |
| Nusa Tenggara Barat | 677,111 | 14,097 | spinosum, *Gracilaria* |
| Kalimantan Utara | 523,258 | 7,428 | cottonii |
| Sulawesi Tenggara | 272,325 | 4,380 | cottonii, spinosum |
| Sulawesi Utara | 247,024 | 1,843 | cottonii, spinosum |
| Maluku | 191,489 | 4,197 | cottonii |
| Sulawesi Barat | 94,187 | 818 | spinosum, cottonii |
| Jawa Barat | 87,275 | | |
| Maluku Utara | 81,555 | | |
| Jawa Tengah | 80,458 | | |
| Banten | 70,555 | | |
| Kalimantan Timur | 28,067 | | |
| Bali | 17,145 | | |
| Gorontalo | 8,671 | | |
| Lampung | 5,460 | | |
| Kepulauan Riau | 4,095 | | |
| Papua | 905 | | |
| Papua Barat | 607 | | |
| DKI Jakarta | 117 | | |
| Kalimantan Barat | 93 | | |
| Kepulauan Bangka Belitung | 22 | | |
| Indonesia | 9,618,421 | 102,254 | |

Source: KKP (2022).

Using conversion ratios reported in Langford et al. (2023), this is the equivalent of 744,980 wet tonnes of seaweed consumed by the domestic processing sector (Table A1.3). These products are then either exported or consumed by the domestic processing sector.

*Raw seaweed exports*

Based on Customs statistics that are likely to be relatively accurate, the Ministry of Industry reported that Indonesia exported 187,504 tonnes of RDS in 2021. Assuming a 7.5:1 wet to dry ratio, this is the equivalent of 1,406,280 tonnes of (all species of) wet seaweed in 2021. In 2020, 168,364 tonnes of RDS were exported (equivalent of 1,263,780 tonnes wet seaweed). Adding statistics derived from the Ministry of Industry, sectors downstream from production (domestic processors and exporters) used an equivalent of 2,151,260 tonnes of wet seaweed in 2021 and 2,008,760 tonnes in 2020.

*Table A1.2* Seaweed production, cultivation area and number of households involved in major producing areas in South Sulawesi in 2020

| 2020 | Pond seaweed production | Marine seaweed production (tonnes) | Marine cultivation area (ha) | Marine cultivation households | Marine seaweed farming yield (tonnes per ha) | Marine seaweed farming average household cultivation (tonnes per household) | Marine seaweed farming average cultivation area (ha per household) |
|---|---|---|---|---|---|---|---|
| Wajo | 42,262 | 433,817 | 5,577 | 2,525 | 78 | 172 | 2.2 |
| Pangkep | 10,653 | 410,299 | 3,431 | 5,165 | 120 | 79 | 0.7 |
| Takalar | 50,757 | 358,360 | 3,200 | 4,002 | 112 | 90 | 0.8 |
| luwu | 322,100 | 290,432 | 5,548 | 1,260 | 52 | 231 | 4.4 |
| Jeneponto | | 231,754 | 3,249 | 7,125 | 71 | 33 | 0.5 |
| Bulukumba | 861 | 191,389 | 7,085 | 3,173 | 27 | 60 | 2.2 |
| Bone | 116,961 | 184,910 | 2,045 | 2,494 | 90 | 74 | 0.8 |
| Luwu Timur | 150,892 | 147,820 | 904 | 559 | 164 | 264 | 1.6 |
| Bantaeng | | 86,285 | 3,521 | 3,822 | 25 | 23 | 0.9 |
| Luwu Utara | 175,430 | 38,419 | 829 | 301 | 46 | 128 | 2.8 |
| Palopo | 92,873 | 30,860 | 503 | 469 | 61 | 66 | 1.1 |
| Pinrang | 1,032 | 13,211 | 3,620 | 787 | 4 | 17 | 4.6 |
| Sinjai | 33,051 | 12,925 | 405 | 698 | 32 | 19 | 0.6 |
| Barru | | 680 | 139 | 133 | 5 | 5 | 1.0 |
| Selayar | | 642 | 267 | 361 | 2 | 2 | 0.7 |
| Gowa | 102 | | | | | | |
| Total | 996,975 | 2,431,802 | 40,322 | 32,874 | | | |
| Average | | | | | 93 | 79 | 1.6 |

Source: DKP Prov. SulSel (2021).

*Table A1.3* Quantities of seaweed processed in Indonesia

| 2021 | Wet seaweed to dry product ratio[2] | Produced | Wet tonne equivalent |
|---|---|---|---|
| ATCC | 22.5:1 | 8,531 | 191,947.5 |
| SRC | 30:1 | 14,102 | 423,060 |
| RC | 37.5:1 | 2,423 | 90,862.5 |
| Agar | 10:1 | 3,911 | 39,110 |
| **Total** | | | 744,980 |

Source: Kemenperin (2022).

## Agricultural Census 2013

Unlike the KKP statistics, which are collected by a production-oriented line bureau based on bottom-up methods and extrapolation, agricultural censuses organise large teams of enumerators to collect statistics more directly. National agricultural censuses are conducted every ten years, with the support of the FAO. Results are often used to 'smooth out' annual production statistics, including on a retrospective basis. The last agricultural census conducted in Indonesia was for 2013 and 2023 statistics had not been released at the time of writing. The Agricultural Census 2013 (*Sensus Pertanian*) was the first source of data to directly count the number of seaweed farming households in Indonesia. It recorded the total number of seaweed farming households (marine and brackishwater) at 66,115 in 2013 (Table A1.4). Other data (on production and cultivation) was not reported.

***Presidential Decree***

The Presidential Decree (PERPRES 33–2019) estimated that in 2018, 267,800ha were cultivated with seaweed, and this was cultivated by households with an average plot of 1ha, giving an estimated total number of 267,800 seaweed farming households.

*BPS 2021 Survey data*

The BPS (2022a) conducted a survey that estimated national seaweed production at 5 million tonnes, significantly lower than the KKP estimate of 9.6 million

*Table A1.4* Number of seaweed farming households in Indonesia by province

| Province | Number of seaweed farming households |
|---|---|
| Sulawesi Selatan | 22,293 |
| Nusa Tenggara Timur | 8,985 |
| Sulawesi Tenggara | 8,524 |
| Maluku | 7,415 |
| Sulawesi Tengah | 5,758 |
| Jawa Timur | 3,213 |
| Bali | 2,432 |
| Kalimantan Lltara | 1,625 |
| Nusa Tenggara Barat | 1,513 |
| Maluku Utara | 996 |
| Banten | 627 |
| Sulawesi Barat | 495 |
| Jawa Tengah | 477 |
| Kalimantan Timur | 369 |
| Sulawesi Utara | 325 |
| Kepulauan Riau | 250 |
| Jawa Barat | 175 |
| Gorontalo | 173 |
| Kalimantan Selatan | 128 |
| Lampung | 116 |
| Papua Barat | 110 |
| Papua | 69 |
| Other | 47 |
| Total | 66,115 |

Source: BPS (2013).

tonnes (Table A1.5). This is the only survey of seaweed farmers conducted since the 2013 census. It involved interviewing 7,075 farmers, the selection of which (by region) was guided by the results of the 2013 census. The survey relied on farmers' reports of the number of longlines operated and the weight of each longline at harvest.

Notably, BPS (2022a) differentiated between production which was sold as a final product (mostly in dry, but also small amounts in wet form) and production which was used as an input to the next farming cycle. In a revealing statistic, they reported that 36 per cent of all seaweed production in Indonesia was used to propagate a new cycle, with only 62 per cent sold (Table A1.6).

This is an important development as if production is estimated based on the amount of seaweed harvested, without accounting for the amount used as propagules, the amount of seaweed recorded as harvested will be much higher than the amount of seaweed recorded as sold. For example, if a farmer starts with 10kg of seaweed seeds and grows it until it reaches 30kg, then harvests it, the harvest would be recorded as 30kg when in fact only 20kg of new material has been produced. Of the 30kg harvested, 10kg will be used for seed for the next cycle, and only 20kg will be sold and available for downstream sectors. While other agricultural sectors also involve similar statistical issues (carry-over seeds for crops or replacement

*Table A1.5*  Seaweed production in Indonesia in 2020 by province

| 2020 | Total marine production (tonnes) | Proportion of total national marine production | Total pond production (tonnes) | Proportion of total national pond production |
|---|---|---|---|---|
| Sulawesi Selatan | 1,409,700 | 30% | 222,601 | 63% |
| Nusa Tenggara Timur | 1,037,875 | 22% | 30 | 0% |
| Kalimantan Utara | 441,152 | 9% | | |
| Nusa Tenggara Barat | 402,687 | 9% | | |
| Sulawesi Tengah | 393,458 | 8% | 26,436 | 8% |
| Sulawesi Tenggara | 382,218 | 8% | | |
| Maluku | 262,850 | 6% | | |
| Jawa Timur | 144,697 | 3% | 6,947 | 2% |
| Sulawesi Utara | 35,807 | 1% | | |
| Maluku Utara | 35,508 | 1% | | |
| Sulawesi Barat | 28,257 | 1% | | |
| Kalimantan Timur | 20,787 | | 39,352 | 11% |
| Bali | 19,361 | | | |
| Banten | 10,591 | | | |
| Lampung | 10,119 | | | |
| Jawa Tengah | 9,536 | | 11,420 | 3% |
| DKI Jakarta | 9,039 | | | |
| Kepulauan Riau | 5,041 | | | |
| Papua Barat | 1,571 | | | |
| Jawa Barat | | | 44,366 | 13% |
| Other | 450 | | | |
| Total | 4,660,704 | | 351,152 | |
| Grand total | | | 5,011,856 | |

Source: Data from BPS (2022a).

livestock), the large proportion of material used for multiplication in seaweed constitutes a significant statistical issue. Table A1.6 shows the amount of marine seaweed used for different purposes as reported in BPS 2022a.

*Analysis of satellite imagery*

We used high resolution satellite imagery to measure the area under production in Pangkep Regency from 2018 to 2020 (see Langford et al. 2021). We found that the total area cultivated at any time during 2020 was 782ha, but that on average 244ha was under cultivation in any given month. These areas are actual areas included in cultivation plots, so they exclude areas between plots.

## Comparison of available statistics

Table A1.8 summarises the data available from different sources.

### Number of households

Several sources report statistics on the number of households engaged in seaweed farmers (Table A1.9). The most accurate estimate is from census data. The 2023

*Table A1.6* Final use of seaweed produced by province

| 2020 | Used as propagules | Sold as propagules | Sold wet | Sold dry | Other | Total production |
|---|---|---|---|---|---|---|
| Lampung | 1,709 | 11 | 25 | 8,372 | 2 | 10,119 |
| Kepulauan Riau | 1,681 | 23 | 1 | 3,327 | 9 | 5,041 |
| DKI Jakarta | 1,313 | 113 | 34 | 7,578 | 1 | 9,039 |
| Jawa Tengah | 3,911 | 64 | 5,561 | 0 | 0 | 9,536 |
| Jawa Timur | 54,819 | 7,962 | 19,447 | 61,828 | 641 | 144,697 |
| Jawa Barat | 0 | 0 | 0 | 0 | 0 | 0 |
| Banten | 5,098 | 100 | 4,179 | 264 | 950 | 10,591 |
| Bali | 11,538 | 274 | 29 | 7,507 | 13 | 19,361 |
| Nusa Tenggara Barat | 135,463 | 6,295 | 1,597 | 258,774 | 558 | 402,687 |
| Nusa Tenggara Timur | 450,237 | 19,388 | 2,651 | 552,232 | 13,367 | 1,037,875 |
| Kalimantan Timur | 5,365 | 255 | 51 | 14,615 | 501 | 20,787 |
| Kalimantan Utara | 107,769 | 502 | 2,120 | 330,400 | 361 | 441,152 |
| Sulawesi Utara | 5,120 | 103 | 0 | 30,397 | 187 | 35,807 |
| Sulawesi Tengah | 138,252 | 5,791 | 7,421 | 239,608 | 2,386 | 393,458 |
| Sulawesi Selatan | 484,822 | 16,306 | 15,933 | 852,258 | 40,381 | 1,409,700 |
| Sulawesi Tenggara | 141,960 | 4,575 | 2,342 | 231,207 | 2,134 | 382,218 |
| Sulawesi Barat | 5,927 | 681 | 307 | 18,329 | 3,013 | 28,257 |
| Maluku | 62,108 | 601 | 623 | 198,663 | 855 | 262,850 |
| Maluku Utara | 6,846 | 133 | 37 | 28,367 | 125 | 35,508 |
| Papua Barat | 210 | 70 | 120 | 1,163 | 8 | 1,571 |
| Other | 94 | 24 | 6 | 289 | 37 | 450 |
| Total | 1,624,242 | 63,271 | 62,484 | 2,845,178 | 65,529 | 4,660,704 |

Source: BPS (2022a). BPS also reports the total number of seaweed producing households, as shown in Table A1.7 by species. For full results see BPS (2022a).

*Table A1.7*  Number of seaweed farming households in Indonesia by type of cultivation

| Number of farming households growing different types of seaweed | Total marine farming households | Total pond farming households |
|---|---|---|
| Lampung | 101 | 0 |
| Kepulauan Riau | 197 | 0 |
| DKI Jakarta | 93 | 0 |
| Jawa Barat | 0 | 454 |
| Jawa Tengah | 304 | 149 |
| Jawa Timur | 2,152 | 132 |
| Banten | 354 | 0 |
| Bali | 749 | 0 |
| Nusa Tenggara Barat | 2,056 | 0 |
| Nusa Tenggara Timur | 10,166 | 10 |
| Kalimantan Timur | 219 | 211 |
| Kalimantan Utara | 2,075 | 0 |
| Sulawesi Utara | 286 | 0 |
| Sulawesi Tengah | 2,895 | 191 |
| Sulawesi Selatan | 24,922 | 3,110 |
| Sulawesi Tenggara | 7,083 | 0 |
| Sulawesi Barat | 804 | 0 |
| Maluku | 7,275 | 0 |
| Maluku Utara | 858 | 0 |
| Papua Barat | 103 | 0 |
| Others | 62 | 0 |
| **Total** | 62,754 | 4,257 |

Source: Data from BPS (2022a).

*Table A1.8*  Summary of data collected from different sources

| | Volume | | | Area under cultivation | | | Number of households involved | | |
|---|---|---|---|---|---|---|---|---|---|
| | *National* | *Sulsel* | *Pangkep* | *National* | *Sulsel* | *Pangkep* | *National* | *Sulsel* | *Pangkep* |
| KKP (2020) | ✓ | ✓ | | ✓ | ✓ | | | | |
| DKP Sulsel (2020) | ✓ | ✓ | ✓ | ✓ | ✓ | ✓ | ✓ | ✓ | ✓ |
| Ministry of Industry (2020) | ✓* | | | | | | | | |
| Agricultural Census (2013) | | | | | | | ✓ | ✓ | |
| BPS seaweed survey (2020) | ✓ | ✓ | | | | | ✓ | ✓ | |
| Presidential Decree (2018) | | | | ✓ | | | ✓ | | |
| Satellite data (2020) | | | | | | ✓ | | | |

* Derived from export and processing statistics.

census data is not yet available. However, 2013 census data reports that there are 66,115 seaweed farming households in Indonesia (marine and pond). This aligns closely with BPS 2021 survey estimates of 67,011 seaweed farming households in Indonesia, including 62,754 marine farming households and 4,257 pond farming households. National (KKP/FAO) production statistics estimated that production volumes in 2013 and 2020 were similar (9.3 million tonnes in 2013 and 9.6 million tonnes in 2020). Unless there were large changes in the scale of household production in that period, it seems reasonable that the number of households employed in the industry would be similar. The South Sulawesi DKP reports that there are 32,874 marine seaweed farming households in South Sulawesi, around 32 per cent higher than the BPS estimate of 24,922. While broad convergence between these sources provides some confidence in the statistics, reports in the Presidential Decree of 267,800 seaweed farming households appear highly overstated, most probably because they are based on inaccurate data on cultivated area and average size of household plots.

The estimates of the number of people working in the industry are much more inconsistent than the number of households, probably due to different assumptions about who is included as a worker in the industry (Table A.10 and as discussed in Chapter 10). Consequently, the number of households involved in seaweed farming is preferred as the basis of analysis.

### Area under cultivation

The KKP estimates that 102,254ha were cultivated for seaweed farming in 2020 (of a potentially suitable farming area of 12,123,383ha (KKP 2022)). The Presidential

*Table A1.9* Estimated number of seaweed farming households in Indonesia from different sources

|  | National | | South Sulawesi | |
| --- | --- | --- | --- | --- |
|  | *Marine* | *Pond* | *Marine* | *Pond* |
| Agriculture census 2013 | 66,115 | | 22,293 | |
| BPS seaweed survey (2020) | 62,754 | 4,257 | 24,922 | 3,110 |
| DKP Sulsel (2020) | – | | 32,874 | – |
| Presidential Decree (2018) | 267,800 | | – | |

*Table A1.10* Estimated number of seaweed farmers in Indonesia from different sources

|  | National | | South Sulawesi | |
| --- | --- | --- | --- | --- |
|  | *Marine* | *Pond* | *Marine* | *Pond* |
| Agriculture census 2013 | – | | – | |
| BPS seaweed survey (2020) | 88,176 | 5,211 | 33,331 | 3,878 |
| DKP Sulsel (2020) | – | | 98,621 | – |
| Presidential Decree (2018) | – | | – | |

Decree Nomor 33–2019 estimated that a much larger area of 267,800ha was used for seaweed farming in 2018 of a potentially suitable area of 1,510,223ha. The SulSel DKP estimated that in 2020, 40,322ha were used for seaweed farming by 32,874 households. Neither the Agricultural Census 2013 nor the 2021 BPS survey reported the area under cultivation. Our recent satellite imagery analysis estimated the area under production in Pangkep in 2020. Table A1.11 reports the data from these three sources.

Estimating the area under production is difficult and is currently undertaken based on estimates provided by selected farmers. Satellite imagery provides a more precise calculation because it estimates the area under production, excluding gaps between plots. We found that the total area cultivated at any time in Pangkep during 2020 was 782ha (see Langford et al. 2021 for methodology). In the same year, KKP estimated that 3,431ha were cultivated at any time. This suggests that actual production areas may be greatly overestimated at the provincial and national level – in Pangkep in 2020 by a factor of approximately six.

### Production data

Seaweed production estimates derived from data from different sources are inconsistent, as shown in Table A1.12. KKP statistics are based on volumes sold, so exclude amounts used as propagules, which is appropriate.

The Ministry of Industry data suggests that 2.0 million tonnes of seaweed entered supply chains in 2020, while in the same year BPS put this figure at 3.3 million tonnes (excluding amounts used as propagules). The KKP estimated total production at 9.6 million tonnes wet seaweed equivalent sold in the same year. As the Ministry of Industry estimate is based on actual export data, at the national level it is likely to be the most accurate.

*Table A1.11* Estimated area under seaweed cultivation from different sources (ha)

|  | National | South Sulawesi | Pangkep |
| --- | --- | --- | --- |
| KKP (2020) | 102,254 | 40,322 | 3,431 |
| Presidential Decree (2018) | 267,800 | – | – |
| Satellite imagery (2020) | – | – | 782 |

*Table A1.12* Estimated volume of seaweed production from different sources

|  | Total volume of production (tonnes) | | | |
| --- | --- | --- | --- | --- |
|  | National | | South Sulawesi | |
|  | Marine | Pond | Marine | Pond |
| KKP RI (2020) | 8,090,796 | 1,456,730 | 3,442,076 | |
| KKP Sulsel (2020) | – | | 2,431,802 | 996,975 |
| Ministry of Industry (2020) | 2,008,760 | | – | |
| BPS seaweed survey (2020) | 2,907,662 (volume sold) | 351,152 | 868,191 (volume sold) | 222,601 |

If the estimates above on the use of wet seaweed equivalent in the Indonesian carrageenan domestic processing sector and for exports are realistic (2,151,260 tonnes), the KKP statistics (of 9,547,526 tonnes of marine and pond production) are overstated by a factor of 4.8.

## Summary

The inconsistencies revealed in this section highlight the difficulties of providing an accurate statistical depiction of the Indonesian seaweed industry. National export statistics recorded by Customs are likely to be accurate, figures are less likely to be accurate at provincial levels, where inter-provincial or inter-island trade can be more porous. The most accurate sources of data on the domestic industry appear to be agricultural census data (updated every ten years), BPS (2022a) household survey data, Ministry of Industry data and South Sulawesi DKP household participation data. The data presented in Presidential Decree Nomor 33–2019 and KKP production volume and cultivation area data appear to be greatly overstated.

After taking into account methods and discrepancies, realistic estimates include: 25,000–33,000 marine seaweed farming households in South Sulawesi and 62,000 across Indonesia; around 2 million tonnes wet seaweed production nationally; and a much smaller cultivation area than that estimated by the KKP, in the range of 42,000 ha nationally.

This analysis suggests that, similarly to other agricultural industries, data collection and reporting processes for seaweed should be improved. While this may not be administratively feasible, this would ideally involve collaboration from all government agencies collecting data on the seaweed industry (DKP/KKP, BPS, Ministry of Industry, Ministry of Agriculture). On the production side, this would involve better methods for estimating cultivation areas (based on actual farm boundaries or seaweed producing areas as a whole), more accurate (and perhaps variable) incorporation of assumptions on yield coefficients, explicitly stating whether production includes or excludes propagules and refined assumptions through which estimates of the number of households involved in seaweed are used to reach estimates of the number of people involved in seaweed farming. Ideally, it would also capture data on the number of people from non-seaweed farming households who work in the industry as casual labourers (for more on this, see Chapter 8).

## Notes

1 Note that Maros has been removed from the dataset due to a known data anomaly. In official data, an additional 16 farmers from Maros are recorded.
2 Carrageenan product conversion rates as reported in Langford et al. 2023. Agar conversion rate assumes 50 per cent weed content and 20 per cent yield.

## References

BPS (*Badan Pusat Statistik*) 2013. *Sensus Pertanian 2013*. https://st2013.bps.go.id
BPS (*Badan Pusat Statistik*) (2022a). Hasil Survei Komoditas Perikanan Potensi Rumput Laut 2021 Seri 2. *Badan Pusat Statistic*. https://www.bps.go.id/publication/2022/08/29/

269de33babc6e3d52bbae5b6/hasil-survei-komoditas-perikanan-potensi-rumput-laut-2021-seri-2.html

BPS (*Badan Pusat Statistik*) (2022b). *Statistical Yearbook of Indonesia 2022*. https://www.bps.go.id/publication/2022/02/25/0a2afea4fab72a5d052cb315/statistik-indonesia-2022.html

DKP Prov. SulSel. 2021. *Laporan Statistik Perikanan Sulawesi Selatan 2020*. Makassar: Dinas Kelautan dan Perikanan Provinsi Sulawesi Selatan.

FAO (Food and Agriculture Organisation of the United Nations). 2023. FishStatJ (software for FAO'S Fisheries and Aquaculture statistics). https://www.fao.org/fishery/en/statistics/software/fishstatj

KKP (Kementerian Kelautan dan Perikanan) 2022. Evaluasi kinerja dan rencana kebijakan budidaya rumput laut 2021–2023. *KKP 2022*.

Kementerian Perindustrian Republic Indonesia (Kemenperin). 2022. Kebijakan Industri Pengolahan Rumput Laut. *Kementerian Perindustrian 2022*.

Langford, A., S. Waldron, Sulfahri and H. Saleh. 2021. Monitoring the COVID-19-Affected Indonesian Seaweed Industry Using Remote Sensing Data. *Marine Policy 127* (104431): 1–10. doi.org/10.1016/j.marpol.2021.104431

Langford, A., W. Turupadang, and S. Waldron. 2023. Interventionist Industry Policy to Support Local Value Adding: Evidence from the Eastern Indonesia Seaweed Industry. *Marine Policy* 151 (105561). doi.org/10.1016/j.marpol.2023.105561

Presidential Decree 33–2019 (*Peraturan Presiden Republik Indonesia Nomor 33 Tahun 2019 tentang Peta Panduan Pengembangan Industri Rumput Laut Nasional Tahun 2018–2021*).

# Appendix 2

## Indonesian seaweed-related policies

*Fikri Firmansyah Sjahruddin, Yanti N. Muflikh,
Scott Waldron, and Risti Permani*

To outline the complex Indonesian policy landscape, Figure A2.1 presents a hierarchy of Indonesian laws and regulations based on the provisions stated in Law No. 12/2011. The hierarchy or pyramid consists of seven levels, ordered in level of precedence and from national level down to the regional level. In principle, regulations and policies made at lower levels should not contradict the policies at higher levels.

The policies made by the central, provincial, and district governments that impact directly or indirectly on seaweed are listed in Table 2.1. The list of policies are drawn from a much more detailed study conducted on the policy landscape and supply chain governance for seaweed in Indonesia conducted by Permani et al. (2023). Sixty-seven policy documents were collected based on desktop research

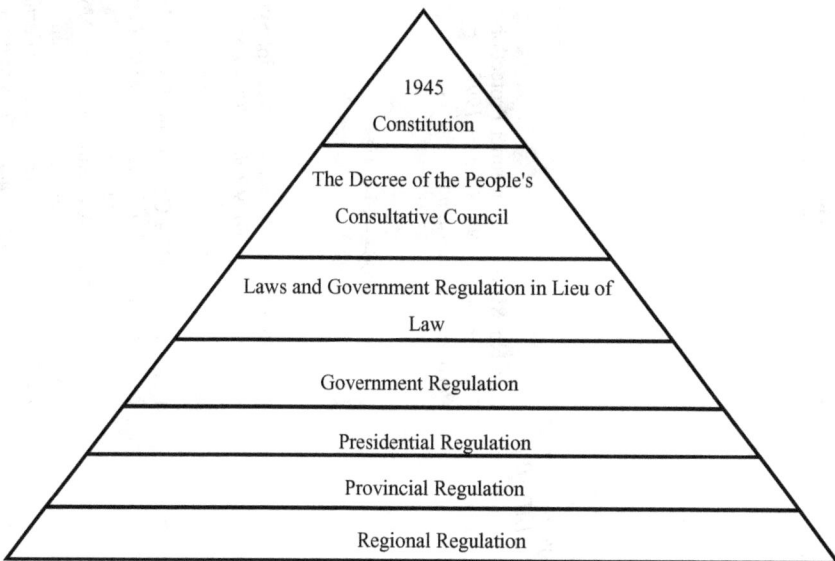

*Figure A2.1* The hierarchy of laws and regulations in Indonesia

Source: Schematic by authors based on heirarchy described in Law 12 Article 7 the Year 2011.

*Table A2.1* Indonesian seaweed-related policies

| Constitution | Policy | Description |
| --- | --- | --- |
| 1945 Constitution<br>*Undang Undang Dasar 1945*<br>The Decree of the People's Consultative Council<br>*Ketetapan Majelis Permusyarawatan Rakyat – TAP MPR* | | The Constitution of the Republic of Indonesia outlining the basic principles and structure of the Indonesian government.<br>A major constitutional document produced by the People's Consultative Council (*Majelis Permusyarawatan Rakyat* (MPR)) in Indonesia that provides guidelines and directions for the government and Indonesian institutions. |
| Laws<br>*Undang-undang* | Law No. 8/1999 – Consumer Protection<br>*Undang-undang No. 8 Tahun 1999 Perlindungan konsumen* | This law comprises all protections to ensure legal certainty for consumers. Seaweed consumers are protected by this law from illegal sellers' or traders' activities. For example, sellers who do not fulfil the agreement. |
| | Law No. 31/2004 – Fisheries<br>*Undang-undang No. 31 Tahun 2004 – Perikanan* | In Article 7, verse 5 explanation, seaweed is included as one of the fisheries resources. This law is amended by Government Regulation in Lieu of Law 2/2022. |
| | Law No. 9/2006 – Warehouse Receipt System<br>*Undang-undang No. 9 Tahun 2006 – Sistem resi gudang* | Policy base of Indonesian warehouse receipt system. In this policy, seaweed is not specifically listed as a commodity. This law was amended by Law No. 9/2011. |
| | Law No. 27/2007 – Coastal and Small Islands Management<br>*Undang-undang No. 27 Tahun 2007 – Pengelolaan wilayah pesisir dan pulau-pulau kecil*<br>Law No. 9/2011 – Amendment of Law No. 9 The Year 2006 Regarding Warehouse Receipt System<br>*Undang-undang No. 9 Tahun 2011 – Perubahan atas Undang-undang Nomor 9 Tahun 2006 tentang sistem resi gudang* | In Article 23, verse 2, aquaculture is listed as one of the small islands' utilisations. This law was amended by Government Regulation in Lieu of Law 2/2022.<br>Some articles and verses in Law No. 9/2006 regarding warehouse receipt system have been revised. |

Law No. 18/2012 – Food
*Undang-undang No. 18 Tahun 2012 – Pangan*

Food security is defined as the condition food necessities are fulfilled from an individual to a national level. Seaweed is an edible product and a material resource that contributes to national food security.

Law No. 1/2014 – Amendment of Law No. 27 The Year 2007 Related to Coastal and Small Islands Management
*Undang-undang No. 1 Tahun 2014 – Perubahan atas Undang-undang No. 27 Tahun 2007 tentang pengelolaan wilayah pesisir dan pulau-pulau kecil*

National defence and security were added in the same verse where aquaculture activity was listed. This law was amended by Government Regulation in Lieu of Law 2/2022.

Law No. 7/2014 – Trade
*Undang-undang No. 7 Tahun 2014 – Perdagangan*

Seaweed products are traded domestically and overseas. Trade is regulated by Law No. 7/2014 (amended by Law 11/2020 – Omnibus Law).

Law No. 33/2014 – Halal Product Assurance
*Undang-undang No. 33 Tahun 2014 – Jaminan produk halal*

As seaweed products are consumed in Indonesia, halal assurance of seaweed products should follow Law No. 33/2014. This law was amended by Government Regulation in Lieu of Law 2/2022.

Law No. 3/2014 – Industry
*Undang-undang No. 3 Tahun 2014 – Perindustrian*

Seaweed products are processed at a range of industrial levels. Local industries are bound by this law. Law 11/2020 – Omnibus Law amends the industry law.

Law No. 23/2014 – Local Governance
*Undang-undang No. 23 Tahun 2014 – Pemerintahan daerah*

This law authorises the division between local and central government authority. Seaweed farms and industries are often regulated at the provincial or district level. Law 11/2020 – Omnibus Law amends this industry law.

Law No. 32/2014 – Marine Affairs
*Undang-undang No. 32 Tahun 2014 – Kelautan*

In Article 17, verse 2, point c, it is stated that the government manages fisheries resources and facilitates the establishment of fisheries industries that can improve the livelihoods of aquaculture farmers and fishers. This law was amended by Government Regulation in Lieu of Law 2/2022.

*(Continued)*

*Table A2.1* (Continued)

| Constitution | Policy | Description |
| --- | --- | --- |
| | Law No. 7/2016 – Fishers, Aquaculture Farmers, and Salt Farmers Protection and Empowerment<br>*Undang-undang No. 7 Tahun 2016 – Perlindungan dan pemberdayaan nelayan pembudidaya ikan dan petambak garam* | In Article 3, it is stated that aquaculture farmers, fishers, and salt farmers are facilitated to improve, sustain, and enhance their capacity, and they are protected from risks associated with their high-risk business (amended by Law 11/2020 – Omnibus Law). |
| | Law No. 11/2020 – Cipta Kerja (Omnibus Law)<br>*Undang-undang No. 11 Tahun 2020 – Cipta Kerja* | This law is replaced by Government Regulation in Lieu of Law 2/2022 which was later amended by Law No. 4/2023. Then, the law was decreed by Law No. 6/2023. |
| Government regulations<br>*Peraturan Pemerintah* | Government Regulation No. 28/2004 – Food Safety, Quality, and Nutrition<br>*Peraturan Pemerintah No. 28 Tahun 2004 - Keamanan, mutu dan gizi pangan* | Implementation of sanitation in best agriculture practices are stated in Article 3. Best farming practices are maintained by activities such as minimising chemical use and pollutants. This maintenance is stated in Article 4. This regulation was replaced by Government Regulation No. 86/2019. |
| | Government Regulation No. 14/2015 – National industry Master Plan the Year 2015–2035<br>*Peraturan Pemerintah No. 14 Tahun 2015 – Rencana induk pengembangan industri nasional tahun 2015–2035* | Seaweed products were included as one of the priority food industries for Indonesian industry development 2015–2019. |
| | Government Regulation No. 17/2015 – Food Security and Nutrition<br>*Peraturan Pemerintah No. 17 Tahun 2015 – Ketahanan pangan dan gizi* | In Article 34, point a, it is stated that small enterprises are to be supported by incentives to enlarge their farms and improve their food products. |
| | Government Regulation No. 86/2019 – Food Safety<br>*Peraturan Pemerintah No. 86 Tahun 2019 – Keamanan pangan* | This decree amends the previous decree in order to revise the seaweed Harmonised System (HS) codes stated. The decree similarly commands seaweed and other fisheries products to acquire export/import documents within the quarantine process to ensure the quality and to avoid pests and diseases being imported or exported into or out of the country. |

| | | |
|---|---|---|
| | Government Regulation No. 27/2021 – Maritime Affairs and Fisheries Implementation<br>*Peraturan Pemerintah No. 27 Tahun 2021 – Penyelenggaraan bidang kelautan dan perikanan* | This regulation covers aquaculture activities conducted in Indonesian Fisheries Management Area, Article 2. |
| Presidential regulations<br>*Peraturan Presiden* | Presidential Regulation No. 3/2017 – National Action Plan in Accelerating National Fisheries Industries Development<br>*Peraturan Presiden No.3 Tahun 2017 – Rencana aksi percepatan pembangunan industri perikanan nasional* | The national government action plan to accelerate the development of fisheries industries between 2017 and 2019 which involved aquaculture farmers. |
| | Presidential Regulation No. 33/2019 – The Road Map of National Seaweed industry Development 2019–2021<br>*Peraturan Presiden Republik Indonesia Nomor 33 Tahun 2019 – Peta panduan (Road map) pengembangan industri rumput laut nasional tahun 2019–2021* | The national government road map to develop seaweed industries from 2019–2021. |
| | Presidential Regulation No. 18/2020 – Mid-Term National Development Plan 2020–2024<br>*Peraturan Presiden No. 18 Tahun 2020 – Rencana pembangunan nasional jangka menengah tahun 2020–2024* | The national government development plan for five years (2020–2024). |
| | Presidential Regulation No. 10/2021 – Investment Business Fields<br>*Peraturan Presiden No.10 Tahun 2021 – Bidang usaha penanaman modal* | Foreign investors can undertake business in Indonesia if the investment is worth more than 10 billion Rupiahs. |
| | Presidential Regulation No. 49/2021 – Amendment of Presidential Decree No. 10 the Year 2021 on Investment Business Fields<br>*Peraturan Presiden No.49 Tahun 2021 – Perubahan atas Peraturan Presiden No.49 Tahun 2021 tentang bidang usaha penanaman modal* | This regulation amends Article 2 and Article 6 of Presidential Regulation No. 10/2021 related to the requirements for capital investment. |
| Presidential instruction<br>*Instruksi Presiden* | Presidential Instruction No. 7/2016 – National Fisheries Industries Development Acceleration<br>*Instruksi Presiden No. 7 Tahun 2016 – Instruksi presiden tentang percepatan pembangunan industri perikanan nasional* | The President mandated to accelerate the development of national fisheries industries to improve fishers and aquaculture livelihoods for fish traders and processors. This instruction was intended to create more jobs in the industry. |

*(Continued)*

*Table A2.1* (Continued)

| Constitution | Policy | Description |
|---|---|---|
| Ministerial regulation *Peraturan Menteri* | MoF Regulation No. 176/2009 – Machinery, Goods, and Services Import Duty Exemptions for Construction and Industrial Development for Capital Investment<br>*Peraturan Menteri Keuangan No. 176 Tahun 2009 – Pembebasan bea masuk atas impor mesin serta barang dan bahan untuk pembangunan atau pengembangan industri dalam rangka penanaman modal.* | Import duty exemptions for national industry capital investment. For example, in Article a, it is stated that machinery that is not produced in Indonesia can be purchased duty free if it is imported. Seaweed processors may benefit from this regulation. |
| | MoI Regulation No. 144/2010 – The Roadmap of Luwu Industrial Core Competency Development<br>*Peraturan Menteri Perindustrian No. 144 Tahun 2010 – Peta panduan (road map) pengembangan kompetensi inti industri Kabupaten Luwu* | The roadmap of Luwu District industrial core competency 2011–2015. In Article 2, it is stated that Luwu core competency is a seaweed processing industry. |
| | MoI Regulation No. 145/2010 – The Road Map of Development of Industrial Core Competencies of Palopo<br>*Peraturan Menteri Perindustrian No. 145 Tahun 2010 – Peta panduan (road map) pengembangan kompetensi inti industri Kota Palopo* | The roadmap of Palopo District industrial core competency 2011–2015. In Article 2, it is stated that Palopo core competency is a seaweed processing industry. |
| | MoI Regulation No. 147/2010 – The Road Map of Development of Industrial Core Competencies of Maluku Tenggara<br>*Peraturan Menteri Perindustrian No. 147 Tahun 2010 – Peta panduan (road map) pengembangan kompetensi inti industri Kabupaten Maluku Tenggara* | The roadmap of Maluku Tenggara District industrial core competency 2011–2015. In Article 2, it is stated that Maluku Tenggara core competency is a seaweed processing industry. |
| | MMAF Regulation No. 39/2011 – Organisation and Working Procedure the Seaweed Research and Development Centre<br>*Peraturan Menteri Kelautan dan Perikanan Republik Indonesia No. 39 Tahun 2011 – Organisasi dan tata kerja organisasi dan tata kerja loka penelitian dan pengembangan budidaya rumput laut* | The ministry established the Seaweed Research and Development Centre and the structure for the organisation in this regulation. |
| | MoI Regulation No. 115/2011 – The Road Map of Development of Industrial Core Competencies of Sumbawa<br>*Peraturan Menteri Perindustrian No. 115 Tahun 2011 – Peta panduan (road map) pengembangan kompetensi inti industri Kabupaten Sumbawa* | The roadmap of Sumbawa District industrial core competency 2012–2015. In Article 2, it is stated that Sumbawa core competency is a seaweed processing industry. |

| Regulation | Description |
| --- | --- |
| MMAF Regulation No. 7/2013 – Certificate of Legal Origin for Seaweed<br>*Peraturan Menteri Kelautan dan Perikanan Republik Indonesia No. 7 Tahun 2013 – Sertifikat asal rumput laut* | Everyone who exports seaweed overseas is required to provide a seaweed origin certificate if required for the destination country. |
| MMAF Regulation No. 6/2014 – The Organisation and Hierarchy in The Technical Implementation Unit of Freshwater Aquaculture, Brackish Water Aquaculture, Marine Aquaculture<br>*Peraturan Menteri Kelautan dan Perikanan Republik Indonesia No. 6 Tahun 2014 – Organisasi dan tata kerja balai unit pelaksana teknis perikanan budidaya air tawar, perikanan budidaya air payau, dan perikanan budidaya air laut* | The ministry established the Technical Implementation Unit of Freshwater Aquaculture, Brackish Water Aquaculture and Marine Aquaculture, and the structure and organisational requirements are outlined in this regulation. |
| MMAF Regulation No. 49/2014 – Aquaculture Business<br>*Peraturan Menteri Kelautan dan Perikanan No. 49 Tahun 2014 - Usaha pembudidayaan ikan* | This regulation manages aquaculture businesses in Indonesia including permits and requirements. |
| MMAF Regulation No. 3/2015 – Delegation of Authority to Grant Business Licenses in The Aquaculture Sector in Accordance with The Framework of A One-Stop Integrated Service to The Head of Investment Coordinating Body<br>*Peraturan Menteri Kelautan dan Perikanan No. 3 Tahun 2015 – Pendelegasian wewenang pemberian izin usaha di bidang pembudidayaan ikan dalam rangka pelayan terpadu satu pintu kepada Kepala Badan Koordinasi Penanaman Modal* | The minister of MMAF delegates the authority to issue business permits in aquaculture to the Investment Coordinating Board with the right of substitution. For example, if the investment for aquaculture comes from overseas or if the aquaculture is located more than 12 nautical miles from the shoreline, the Investment Coordinating Board issues the business permits. |
| MMAF Regulation No. 18/2016 – Assurance of Risk Protection for Fishers, Fish Farmers, and Salt Farmers<br>*Peraturan Menteri Kelautan dan Perikanan Republik Indonesia No. 18 Tahun 2016 – Jaminan atas perlindungan atas resiko kepada nelayan, pembudidaya ikan dan petambak garam.* | In Article 2, it is stated that this regulation forms the basis of a risk protection guarantee in the form of insurance for fishers, aquaculture, and salt farmers. |
| MMAF Regulation No. 17/2017 – Organisation and Working Procedure the Fisheries Recovery Resources Agency<br>*Peraturan Menteri Kelautan dan Perikanan Republik Indonesia No. 17 Tahun 2017 – Organisasi dan tata kerja Balai Riset Pemulihan Sumber Daya Ikan* | This regulation restructures the organisation of the Fish Resources Restoration Research Centre. This regulation was replaced in 2020 by MMAF Regulation No. 80/2020. |

*(Continued)*

*Table A2.1* (Continued)

| Constitution | Policy | Description |
| --- | --- | --- |
| | MMAF Regulation No. 50/2017 – Type of Commodities Subject to Compulsory Quarantine Check of Quality and Product Safety<br>*Peraturan Menteri Kelautan dan Perikanan No. 50 Tahun 2017 – Jenis komoditas wajib periksa karantina ikan mutu dan keamanan hasil perikanan* | Seaweed is listed as one of the commodities required to be checked in quarantine facilities. HS codes are given for seaweed commodities, e.g. *Eucheuma spinosum* 1212.21.11; *Eucheuma cottonii* 1212.21.12. |
| | MMAF Regulation No. 8/2018 – The Procedures of Customary Law Management Area Establishment in Coastal and Small Island Spatial Management<br>*Peraturan Menteri Kelautan dan Perikanan Republik Indonesia No. 8 Tahun 2018 – Tata cara penetapan wilayah kelola masyarakat hukum adat dalam pemanfaatan ruang di wilayah pesisir dan pulau-pulau kecil* | Area managed by local or indigenous people can be officially acknowledged in the national spatial plan. The processes and requirements are stated in this regulation. |
| | MMAF Regulation No. 18/ 2018 – Amendment of Ministry of Marine Affairs and Fisheries Regulation No. 50 The Year 2017<br>*Peraturan Menteri Kelautan dan Perikanan No. 18 Tahun 2018 – Perubahan Peraturan Menteri Kelautan dan Perikanan No. 50 Tahun 2017 tentang jenis komoditas wajib periksa karantina ikan mutu keamanan dan hasil perikanan* | This regulation replaced MMAF Regulation No. 50/2017, however HS codes for seaweed commodities did not change. |
| | MMAF Regulation No. 17/2019 – The Requirements and Procedures to Issue Processing Eligibility Certificate<br>*Peraturan Menteri Kelautan dan Perikanan No. 17 Tahun 2019 – Persyaratan dan tata cara penerbitan sertifikat kelayakan pengolahan* | In Article 2, it is stated that processors are required to implement good practices and standard sanitising procedures. |
| | MoF Regulation No. 128/2019 – Gross income Deduction for Work, Internship, and/or Training in Order to Competency Based Human Resources Coaching and Development<br>*Peraturan Menteri Keuangan No. 128 Tahun 2019 - Pemberian pengurangan penghasilan bruto atas penyelenggaraan kegiatan praktik kerja, pemagangan, dan/atau pembelajaran dalam rangka pembinaan dan pengembangan sumber daya manusia berbasis kompetensi tertentu* | Tax deduction that can be implemented in seaweed industries development in Indonesia. |

| Policy | Description |
|---|---|
| MoCSME Regulation No. 5/2020 – Ministry of Cooperatives and Small to Medium Scales Enterprises Strategic Plan 2020–2024 / *Peraturan Menteri Koperasi dan Usaha Kecil dan Menengah No. 5 tahun 2020 – Rencana strategis Kementerian Koperasi dan Usaha Kecil Menengah 2020 - 2024* | Small enterprises are mentioned in the strategic plan. Those enterprises include seaweed businesses or seaweed farms. |
| MoMIA Regulation No. 6/2020 – Coordinating Ministry for Maritime and investments Affairs Strategic Plan 2020–2024 / *Peraturan Menteri Koordinator Kemaritiman dan Investasi No. 6 Tahun 2020 – Rencana strategis Kementerian Koordinator Kemaritiman dan Investasi 2020–2024* | Two main strategic plans that aim to accelerate investment and improve conditions that support investment. |
| MoI Regulation No. 15/2020 – Ministry of industry Strategic Plan 2020–2024 / *Peraturan Menteri Perindustrian No. 15 Tahun 2020 - Rencana strategis Kementerian Perindustrian 2020–2024* | Food and pharmaceutical industries become priorities in the development of national industries 2020–2024. |
| MMAF Regulation No. 17/2020 – Ministry of Marine Affairs and Fisheries Strategic Plan 2020–2024 / *Peraturan Menteri Kelautan dan Perikanan No. 17 Tahun 2020 - Rencana strategis Kementerian Kelautan dan Perikanan 2020–2024* | Seaweed production capacity and seedlings distribution are listed in the strategic plan. |
| MoT Regulation No. 33/2020 – Goods and Terms and Conditions for Goods Which Can Be Stored in The Warehouse Receipt System / *Peraturan Menteri Perdagangan No. 33 Tahun 2020 - Barang dan persyaratan barang yang dapat disimpan dalam sistem resi gudang* | Seaweed is one of the commodities that can access the warehouse receipt system. |
| MMAF Regulation No. 50/2020 – The Implementation of Indonesian Qualification Framework in The Seaweed Processing industry / *Peraturan Menteri Kelautan dan Perikanan Republik Indonesia No. 50 Tahun 2020 – Penerapan kerangka kualifikasi nasional indonesia bidang industri pengolahan rumput laut* | The government sets national qualifications for workers in seaweed processing industries. |

(Continued)

*Table A2.1* (Continued)

| Constitution | Policy | Description |
|---|---|---|
| | MMAF Regulation No. 55/2020 – The Procedures, Requirements, and Establishment of Aquaculture Area<br>*Peraturan Menteri Kelautan dan Perikanan Republik Indonesia No. 55 Tahun 2020 – Tata cara, persyaratan, dan penetapan kawasan budidaya perikanan* | The government sets standards for areas to be allocated as aquaculture zones. |
| | MMAF Regulation No. 57/2020 – Amendment of Ministry of Marine Affairs and Fisheries Regulation No. 17 The Year 2020 Regarding Ministry of Marine Affairs and Fisheries Strategic Plan 2020–2024<br>*Peraturan Menteri Kelautan dan Perikanan No. 57 Tahun 2020 - Perubahan atas Peraturan Menteri Kelautan dan Perikanan No. 17 Tahun 2020 Tentang Rencana Strategis Kementerian Kelautan Dan Perikanan Tahun 2020–2024* | The amendment for Ministry of Marine Affairs and Fisheries Strategic Plan 2020–2024. |
| | MMAF Regulation No. 70/2020 – The Organisation and Working Procedure of the Institute for Mariculture Research and Fisheries Extension<br>*Peraturan Menteri Kelautan dan Perikanan Republik Indonesia No. 70 Tahun 2020 - Organisasi dan tata kerja Balai Besar Riset Budidaya Laut dan Penyuluhan Perikanan* | The organisation of the Institute for Mariculture Research and Fisheries Extension was restructured. |
| | MMAF Regulation No. 84/2020 – The Organisation and Working Procedure of The Research institute for Seaweed Culture<br>*Peraturan Menteri Kelautan dan Perikanan Republik Indonesia No. 84 Tahun 2020 – Organisasi dan tata kerja Loka Riset Budidaya Rumput Laut* | The organisation of the Research Institute for Seaweed Culture was restructured. |
| | MoF Regulation No. 96/2020 – Amendment of Ministry of Finance Regulation No. 11 The Year 2020 Regarding The Implementation of Government Regulation No. 78 The Year 2019 on Tax Allowance for Particular Sector Investments and/or in Particular Regions<br>*Peraturan Menteri Keuangan No. 96 Tahun 2020 – Perubahan atas Peraturan Menteri Keuangan No.11 Tahun 2020 tentang pelaksanaan Peraturan Pemerintah No. 78 Tahun 2019 tentang fasilitas pajak penghasilan untuk penanaman modal di bidang-bidang usaha tertentu dan/atau di daerah-daerah Tertentu* | Tax allowances that can be accessed by stakeholders working in seaweed industries. |

| | | |
|---|---|---|
| | MoF Regulation No. 153/2020 – Gross Income Deduction for Particular Research and Development Sectors in Indonesia *Peraturan Menteri Keuangan No. 153 Tahun 2020 – Pemberian pengurangan penghasilan bruto atas kegiatan penelitian dan pengembangan tertentu di indonesia* | Tax deductions that can be accessed by stakeholders working in seaweed industries. |
| | MoT Regulation No. 14/2021 – Amendment to the Regulation of the Minister of Trade No. 33 The Year 2020 Concerning Goods and Requirements for Goods that Can Be Stored in The Warehouse Receipt System *Peraturan Menteri Perdagangan Republik Indonesia No. 8 Tahun 2018 – Perubahan atas peraturan Menteri Perdagangan No. 33 Tahun 2020 tentang barang dan persyaratan barang yang dapat disimpan dalam sistem resi gudang* | The amendment for commodities that can be stored and access the warehouse receipt system. Seaweed commodities can still access the receipt system. |
| | MMAF Regulation No. 22/2021 – Development of Fishery Management Plans and Fishery Management Area Institutions in The Fishery Management Areas of The Republic of Indonesia *Peraturan Menteri Kelautan dan Perikanan No. 22 Tahun 2021 – Penyusunan rencana pengelolaan perikanan dan lembaga pengelola perikanan di wilayah pengelolaan perikanan negara republik indonesia* | Plans to develop Fishery Management Areas (FMAs) and their management institutions. Seaweed farms are located in FMAs. Thus, they will also be managed by FMA. |
| | MMAF Regulation No. 27/2021 – Non-Commercial Fisheries Capture and/or Aquaculture Within The Indonesian Fisheries Management Area *Peraturan Menteri Kelautan dan Perikanan No. 27 Tahun 2021 – Penangkapan ikan dan/atau pembudidayaan ikan di wilayah pengelolaan perikanan negara republik Indonesia yang bukan tujuan komersial* | Standards for non-commercial aquaculture activities. |
| Ministerial decree *Keputusan Menteri* | MMAF Decree No. 2/2007 – Good Fish Farming Practices *Keputusan Menteri Kelautan dan Perikanan No, 2 Tahun 2007 - Cara budidaya ikan yang baik* | This ministerial decree sets standards for best aquaculture practices in Indonesia. |

(*Continued*)

*Table A2.1* (Continued)

| Constitution | Policy | Description |
| --- | --- | --- |
| | MMAF Decree No. 1/2019 – The General Guidelines for Seaweed Farming<br>*Keputusan Menteri Kelautan dan Perikanan No, 1 Tahun 2019 – Pedoman umum pembudidayaan rumput laut* | Potential development and requirements for seaweed aquaculture. Seaweed farming and harvesting methods, environmental management, human resources, supervisory, monitoring, and the evaluation of seaweed aquacultures are set outlined in this guideline. |
| Provincial regulations *Peraturan Provinsi* | South Sulawesi Regulation No. 9/2009 – Sulawesi Selatan Spatial Plan 2009–2029<br>*Peraturan Daerah No. 9 Tahun 2009 – Rencana tata ruang wilayah provinsi sulawesi selatan tahun 2009–2029* | This regulation was replaced by the South Sulawesi Regulation No. 3/2022. Originally, this regulation set the spatial planning for South Sulawesi Province 2020–2029. |
| | South Sulawesi Regulation No. 7/2018 – South Sulawesi Industrial Development Plans 2018–2038<br>*Peraturan Daerah No. 7 Tahun 2018 – Rencana pembangunan industri Provinsi Sulawesi Selatan 2018–2038* | Seaweed industries are to become priorities for South Sulawesi Province industry development. |
| | South Sulawesi Regulation No. 2/2019 – Coastal and Small Islands Zonation Plan of South Sulawesi Province 2019–2039<br>*Peraturan Daerah No. 2 Tahun 2019 – Rencana zonasi wilayah pesisir dan pulau-pulau kecil Provinsi Sulawesi Selatan Tahun 2019–2039* | This regulation set the coastal and small islands zonation plan 2019–2039. However, it was replaced by South Sulawesi Regulation No. 3/2022. |
| | South Sulawesi Regulation No. 3/2022 – Spatial Planning of South Sulawesi Province 2022–2041<br>*Peraturan Daerah Sulawesi Selatan No. 3 Tahun 2022 – Rencana tata ruang wilayah Provinsi Sulawesi Selatan tahun 2022–2041* | This regulation legalises the Spatial Planning of South Sulawesi Province 2022–2041 and replaces the previous regulation South Sulawesi's Governor Regulation No. 9/2009. |
| | South Sulawesi's Governor Regulation No. 3/2018 – The Organisation and Hierarchy of The Marine Office Branch of South Sulawesi Marine Affairs and Fisheries Office<br>*Peraturan Gubernur Sulawesi Selatan No. 3 Tahun 2018 – Organisasi dan tata kerja cabang dinas kelautan pada Dinas Kelautan dan Perikanan Provinsi Sulawesi Selatan* | The governor sets the organisational structure for the Department of Marine Affairs and Fisheries branch offices. |

| | | |
|---|---|---|
| | Nusa Tenggara Timur's Governor Regulation No. 39/2022 – Fisheries Commodity Trading System<br>*Peraturan Gubernur Nusa Tenggara Timur No. 39 Tahun 2022 – Organisasi dan tata kerja cabang dinas kelautan pada Dinas Kelautan dan Perikanan Provinsi Sulawesi Selatan* | The governor can ban dried seaweed from being traded outside the province. |
| Regional regulations<br>*Peraturan Kabupaten* | Luwu Utara's Head District Regulation No. 23/2008 – Seaweed Aquaculture Zonation in Luwu Utara Marine Territory<br>*Peraturan Bupati Luwu Utara No. 23 Tahun 2008 - Zonasi budidaya rumput Laut di wilayah perairan Kabupaten Luwu Utara* | This regulation tried to reduce conflict between seaweed farmers and other groups. Thus, it regulates and allocates areas specifically for seaweed farming. The unit area for seaweed farming is defined as a 1ha square and 3m deep at the lowest tide. The space between each square should be at least 25m. Seaweed squares cannot be allocated in water more than 4 miles from the shore. |
| | Bone's Head District Regulation No. 2/2013 – Bone Regency Spatial Management Plan 2012–2032<br>*Peraturan Bupati Bone No. 2 Tahun 2013 – Rencana tata ruang wilayah Kabupaten Bone tahun 2012–2032* | This regulation sets up the spatial management plan for Bone District, South Sulawesi. Seaweed farms are allocated to some regions within sub-districts: a) Awangpone, b) Cenrana, c) Tellu Siattingnge, d) Tanete Riattang Timur, e) Barebbo, f) Sibulue, g) Mare, h) Tonra, i) Salomeko, j) Kajuara. The seaweed farming programme was implemented between 2012 and 2016 and should be in its evaluation phase from 2017–2031. |

Note: Ministerial abbreviations used in the table: Ministry of Marine Affairs and Fisheries (MMAF), Ministry of Finance (MoF), Ministry of Industry (MoI), Ministry of Cooperative, Small–Medium Enterprises (MoCSME), Ministry of Maritime and Investment Affairs (MoMIA), and Ministry of Trade (MoT).

and interviews with key informants. The policies cover regulations related to the seaweed supply chain, including production, distribution, consumption, and research and development. The policy documents listed in Table 2.1 can be obtained through the website of the Information and Law Documentation Network (Jaringan Dokumentasi dan Informasi Hukum (JDIH)) of the respective ministry or local government. For example, MMAF Regulation No. 17/2019 can be found on the Ministry of Marine Affairs and Fisheries (MMAF) JDIH website (https://jdih.kkp. go.id).

### References

Indrati, M. F. 2021. *Konstitusi dan Konstitusionalisme*. Jakarta: Mahkamah Konstitusi Republik Indonesia.

Permani, Risti, Yanti Nuraeni Muflikh, Nunung Nuryartono, Scott Waldron, Alexandra Langford, Syamsul H. Pasaribu, and Fikri Sjahruddin. 2023. *The Policy Landscape and Supply Chain Governance of the Indonesian Seaweed Industry: A Focus on South Sulawesi.* Melbourne, Australia: Australia-Indonesia Centre. https://pair.australiaindonesiacentre. org/wp-content/uploads/2023/06/FINAL-REPORT_ENG_TWP-3_The-policy-landscape-and-supply-chain-governance-of-the-Indonesian-seaweed-industry_-A-focus-on-South-Sulawesi-2.pdf

Rumiarta, I. N. P. B. 2015. Kedudukan Peraturan Menteri Pada Konstitusi. *Kerta Dyatmika,* 12 (2): 15.

Supryadi, A. and Amalia, F. 2021. "Kedudukan Peraturan Menteri Ditinjau Dari Hierarki Peraturan Perundang Undangan Di Indonesia." *Unizar Law Review (ULR)* 4 (2): 00.

# Appendix 3

## Institutions in the Indonesian seaweed industry

*Fikri Firmansyah Sjahruddin, Yanti N. Muflikh, Scott Waldron, and Risti Permani*

A wide range of organisations are active in the Indonesian seaweed industry through policy, regulation, research and development, education and training, and industry support services. The list of organisations compiled in Table A3.1 is based on supply chain studies by Permani et al. (2023), Hogervorst and Kerver (2019), and Waters et al. (2019).

*Table A3.1* List of organisations involved in Indonesian seaweed industry

| | |
|---|---|
| Ministries | Ministry of Marine Affairs and Fisheries (*Kementerian Kelautan dan Perikanan*) |
| | Ministry of Industry (*Kementerian Perindustrian*) |
| | Ministry of Trade (*Kementerian Perdagangan*) |
| | Ministry of Cooperatives and Small–Medium Enterprises (*Kementerian Koperasi dan Usaha Kecil – Menengah*) |
| | Ministry of Finance (*Kementerian Keuangan*) |
| | Coordinating Ministry for the Economy (*Kementerian Koordinator Bidang Perekonomian*) |
| | Coordinating Ministry of Maritime and Investment Affairs (*Kementerian Koordinator Bidang Maritim and Investasi*) |
| Government departments and agencies | Agency for Development Planning, Research, and Development of South Sulawesi Province (*Badan Perencanaan Pembangunan Daerah, Penelitian, dan Pengembangan Provinsi Sulawesi Selatan*) |
| | Department of Marine Affairs and Fisheries of South Sulawesi Province (*Dinas Kelautan dan Perikanan Provinsi Sulawesi Selatan*) |
| | Department of Marine and Fisheries of Takalar Regency (*Dinas Kelautan dan Perikanan Kabupaten Takalar*) |
| | Fisheries Department of Luwu Regency (*Dinas Kelautan dan Perikanan Kabupaten Luwu*) |
| | Department of Fisheries and Marine Affairs of Bantaeng Regency (*Dinas Kelautan dan Perikanan Kabupaten Bantaeng*) |
| Investment and trading | Indonesia Investment Coordinating Board (*Badan Koordinasi Penanaman Modal*) |
| | Commodity Futures Trading Regulatory Agency, Ministry of Trade (*Badan Pengawas Perdagangan Berjangka Komoditi*) |

(*Continued*)

*Table A3.1* (Continued)

| | |
|---|---|
| Research, development, and extension agencies | Centre of Excellence for Development and Utilisation of Seaweed, University of Hasanuddin (*Pusat Unggulan Ilmu Pengetahuan dan Teknologi – Pengembangan dan Pemanfaatan rumput Laut*) |
| | Centre for Coastal and Marine Resources Studies, IPB University (*Pusat Kajian Sumberdaya Pesisir dan Laut, Institut Pertanian Bogor*) |
| | Southeast Asian Regional Centre for Tropical Biology |
| | National Research and Innovation Agency (*Badan Riset dan Inovasi Nasional*) |
| | Seaweed Cultivation Research Centre Gorontalo (*Loka Riset Budidaya Rumput Laut Gorontalo*) |
| | Mariculture Research and Extension Centre for Fisheries – Gondol – Bali (*Balai Besar Riset Budidaya Laut dan Penyuluhan Perikanan – Gondol – Bali*) |
| | Fisheries Recovery Resources Agency (*Balai Riset Pemulihan Sumberdaya Perikanan*) |
| | Mariculture Centre – Lombok (*Balai Perikanan Budidaya Laut – Lombok*) |
| | Mariculture Centre – Batam (*Balai Perikanan Budidaya Laut – Batam*) |
| | Brackish Water Aquaculture Centre – Jepara (*Balai Besar Perikanan Budidaya Air Payau – Jepara*) |
| | Brackish Water Aquaculture Centre – Situbondo (*Balai Besar Perikanan Budidaya Air Payau – Situbondo*) |
| | Brackish Water Aquaculture Centre – Takalar (*Balai Besar Perikanan Budidaya Air Payau – Takalar*) |
| | Brackish Water Aquaculture Centre – Ujung Batee, Aceh (*Balai Besar Perikanan Budidaya Air Payau – Ujung Batee, Aceh*) |
| | Brackish Water Aquaculture Research and Extension Centre – Maros (*Balai Riset Perikanan Budidaya Air Payau dan Penyuluhan Perikanan – Maros*) |
| | Training and Extension Centre for Fisheries – Banyuwangi (*Balai Pelatihan Dan Penyuluhan Perikanan – Banyuwangi*) |
| | Training and Extension Centre for Fisheries – Ambon (*Balai Pelatihan Dan Penyuluhan Perikanan – Ambon*) |
| | Training and Extension Centre for Fisheries – Medan (*Balai Pelatihan Dan Penyuluhan Perikanan – Medan*) |
| | Training and Extension Centre for Fisheries – Bitung (*Balai Pelatihan Dan Penyuluhan Perikanan – Bitung*) |
| Universities | University of Pattimura (*Universitas Pattimura*) |
| | University of Udayana (*Universitas Udayana*) |
| | University of Sam Ratulangi (*Universitas Sam Ratulangi*) |
| | University of Hasanuddin (*Universitas Hasanuddin*) |
| | IPB University (*Institut Pertanian Bogor*) |
| | University of Brawijaya (*Universitas Brawijaya*) |
| | University of Hang Tuah (*Universitas Hang Tuah*) |
| | University of Airlangga (*Universitas Airlangga*) |
| | University of Gadjah Mada (*Universitas Gadjah Mada*) |
| Industry organisations | Indonesian Seaweed Association (*Asosiasi Rumput Laut Indonesia*) |
| | Indonesia Seaweed Industry Association (*Asosiasi Industri Rumput Laut Indonesia*) |
| | Indonesia Seaweed Farmer Association (*Asosiasi Petani Rumput Laut*) |
| | Indonesian Independent Workers Union Cooperative (*Koperasi Serikat Pekerja Merdeka Indonesia*) |
| | Tropical Seaweed Innovation Network (TSIN) |
| | Resources Network (*Jaringan Sumberdaya* (Jasuda)) |

| Standardisation and certification | National Standardisation Agency (*Badan Standarisasi Nasional*) |
| | Standardisation and Agro-Industry Services Centre – Bogor (*Badan Standarisasi dan Pelayanan Jasa Agro-industri – Bogor*) |
| | Fish Quarantine, Quality Control, and Safety of Fishery Products Agency (*Balai Karantina Ikan, Pengendalian Mutu dan Keamanan Hasil Perikanan*) |
| | Ecocert |
| | Control Union |
| | Agriculture Certification Thailand |
| | BIOCert Indonesia |
| | Kiwa |
| | Certification of Environmental Standards (CERES) |
| Not-for-profit organisations and seaweed programme donors | Swiss Import Promotion Programme |
| | Yayasan Konservasi Alam Nusantara (YKAN) |
| | World-Wide Fund for Nature (Yayasan WWF Indonesia) |
| | United Nations Industrial Development Organisation |
| | Germany's Import Promotion Desk |
| | Rikolto |
| | Kalimajari Foundation |
| | Netherlands Embassy |
| | Centre for the Promotion of Imports from developing countries (CBI) |
| | Partnership for Australia-Indonesia Research (PAIR) |
| | Australia-Indonesia Partnership for Promoting Rural Incomes through Support for Markets in Agriculture (PRISMA) |
| | Australia-Indonesia Centre (AIC) |
| | Australian Centre for International Agricultural Research (ACIAR) |

## References

Hogervorst, R. and Kerver, K. 2019. *Seaweed Extracts – Indonesia*. The Hague, Netherlands: Centre for the Promotion of Imports from Developing Countries.

Permani, Risti, Yanti Nuraeni Muflikh, Nunung Nuryartono, Scott Waldron, Alexandra Langford, Syamsul H. Pasaribu, and Fikri Sjahruddin. 2023. *The Policy Landscape and Supply Chain Governance of the Indonesian Seaweed Industry: A Focus on South Sulawesi.* Melbourne, Australia: Australia-Indonesia Centre. https://pair. australiaindonesiacentre.org/wp-content/uploads/2023/06/FINAL-REPORT_ ENG_TWP-3_The-policy-landscape-and-supply-chain-governance-of-the-Indonesian-seaweed-industry_-A-focus-on-South-Sulawesi-2.pdf

Waters, Tiffany, Hilda Lionata, Tommi Prasetyo Wibowo, Robert Jones, Seth Theuerkauf, Subhan Usman, Imran Amin, and Ilman Muhammad. 2019. *Coastal Conservation and Sustainable Livelihoods through Seaweed Aquaculture in Indonesia: A Guide for Buyers, Conservation Practitioners, and Farmers (Version 1).* Arlington, VA and Jakarta: The Nature Conservancy.

# Appendix 4

## Companies in the Indonesian carrageen sector

*Irsyadi Siradjuddin and Boedi Julianto*

This appendix summarises information on the Indonesian carrageenan processing sector, which is designed to accompany the analysis presented in Chapters 1 and 2. Data was collected and cross-verified through a number of sources. Data on twelve companies that are members of the Indonesian Seaweed Industry Association (ASTRULI) was collected from the association. Details on another twenty non-members were collected via desktop research, field visits, and interviews. This included interviews with thirteen carrageenan processors, various trading companies, and discussions at the Strategic Seaweed Conference in Surabaya, September 2022. Data was collected between July and November 2022. It is important to note that the sector is in a constant state of change, with companies entering, exiting, and varying production.

*Table A4.1* Companies in the Indonesian carrageen sector

| Company (ASTRULI members marked *) | Location | Product | Capacity (tons/year) | Approximate utilisation | Actual production (tons year) | Investment |
|---|---|---|---|---|---|---|
| Agar Kembang | Kupang | ATC | 600 | 60% | 360 | Domestic |
| Algae Sumba Timur Lestari | Sumba | ATC | 1,200 | 50% | 600 | Domestic |
| Algalindo Perdana* | Pasuruan | RC | 1,000 | 50% | 500 | Domestic |
| Amarta Carragenan Indonesia * | Jatim | SRC | 1,500 | 50% | 750 | Domestic |
| Anugerah Mapan Jaya (operates PT Indoamkmur Hydrocolloid) | Maros | SRC | 1,200 | 30% | 360 | Domestic |
| Asia Mina Sejahtera (acquired Giwang Citra Laut October 2022) | Makassar | ATC, SRC | 1,000 | 0% | 0 | Domestic |
| Bantimurung Indah* | Makassar Banten | SRC, ATC | 1,200 | 50% | 600 | Domestic |
| Batulicin Algae Perdana | Tanah Bumbu | SRC | 600 | 30% | 180 | Domestic |

*Table A4.1* (continued)

| Company (ASTRULI members marked *) | Location | Product | Capacity (tons/ year) | Approximate utilisation | Actual production (tons year) | Investment |
|---|---|---|---|---|---|---|
| Biota Ganggang Laut | Pinrang | SRC, RC | 8,000 | 60% | 4,800 | Foreign-invested (China) |
| Buanatama Fajar Abadi* | Karawang | ATC, SRC | 600 | 30% | 180 | Domestic |
| Cahaya Cemerlang* | Makassar | ATC, SRC, RC | 1,200 | 30% | 360 | Domestic |
| Centram | Pasuruan | RC | 1,000 | 70% | 700 | Domestic |
| Fuyuan Biologi Teknologi | Situbondo | RC | 3,000 | 30% | 900 | Foreign-invested (China) |
| Galic Arta Bahari* | Bekasi | SRC | 1,000 | 40% | 400 | Domestic |
| Galic Bina Mada* | Cikarang | RC | 1,000 | 60% | 600 | Domestic |
| GreenOne Biotechnology | Situbondo | ATC | 4,000 | 60% | 2,400 | Foreign-invested (China) |
| Gumindo Perkasa Industri* | Cilegon | SRC | 1,500 | 50% | 750 | Domestic |
| Hongxin Algae International | Situbondo | ATC, SRC | 3,000 | 60% | 1,800 | Foreign-invested (China) |
| HW Marine | Situbondo | ATC | 1,000 | 60% | 600 | Foreign-invested (US) |
| Hydrocolloid Indonesia* | Cibinong | SRC | 1,200 | 30% | 360 | Domestic |
| Indoflora Cipta Mandiri* | Malang | RC | 1,200 | 50% | 600 | Domestic |
| Indonusa Algaemas Prima* | Malang | ATC, ATS | 1,200 | 50% | 600 | Domestic |
| Kappa Carrageenan Nusantara* | Pasuruan | ATC, RC | 600 | 50% | 300 | Domestic |
| Karaginan Indo Mandiri | Blitar | SRC | 600 | 30% | 180 | Domestic |
| Karaginan Indonesia | Semarang | SRC | 1,000 | 50% | 500 | Domestic |
| Ocean Carrageenan Indonesia | Mojokerto | ATC, SRC | 600 | 60% | 360 | Domestic |
| Phoenix Mas | Lombok | SRC | 180 | 20% | 36 | Domestic |
| Rote Karaginan Nusantara | Rote | ATC, SRC | 1,200 | 10% | 120 | Domestic |
| Sansiwita | East Java | SRC | 500 | 0% | 0 | Domestic |
| Sea6 Energy Indonesia | Bluleng | SRC, emerging products | 600 | 60% | 360 | Foreign-invested (India) |
| Segoro Algae | Bandung | RC | 600 | 30% | 180 | Domestic |
| Wahyu Putra Bimasakti | Makassar | ATC, SRC | 1,000 | 0% | 0 | Domestic |

Note: Refined Carrageenan (RC); Semi-Refined Carrageenan (SRC); Alkali Treated Cottonii; (ATC); Alkali Treated Spinosum (ATS)

# Index

Note: **Bold** page numbers refer to tables and *italic* page numbers refer to figures.

For Product Safety Concerns and Information please contact our EU
representative  GPSR@taylorandfrancis.com
Taylor & Francis Verlag GmbH, Kaufingerstraße 24, 80331 München, Germany

www.ingramcontent.com/pod-product-compliance
Lightning Source LLC
Chambersburg PA
CBHW052120230326
41598CB00080B/3913

*9 7 8 1 0 3 2 0 2 5 4 9 0*